# Advances in Management Research

## Mathematical Engineering, Manufacturing, and Management Sciences

*Series Editor: Mangey Ram, Professor, Assistant Dean (International Affairs), Department of Mathematics, Graphic Era University, Dehradun, India*

The aim of this new book series is to publish the research studies and articles that bring up the latest development and research applied to mathematics and its applications in the manufacturing and management sciences areas. Mathematical tools and techniques are the strength of engineering sciences. They form the common foundation of all novel disciplines as engineering evolves and develops. The series includes a comprehensive range of applied mathematics and its application in engineering areas, such as optimization techniques, mathematical modeling and simulation, stochastic processes and systems engineering, safety-critical system performance, system safety, system security, high assurance software architecture and design, mathematical modeling in environmental safety sciences, finite element methods, differential equations, reliability engineering, and similar topics.

**Sustainable Procurement in Supply Chain Operations**
*Edited by Sachin Mangla, Sunil Luthra, Suresh Jakar, Anil Kumar, and Nirpendra Rana*

**Mathematics Applied to Engineering and Management**
*Edited by Mangey Ram and S.B. Singh*

**Mathematics in Engineering Sciences**
Novel Theories, Technologies, and Applications
*Edited by Mangey Ram*

For more information about this series, please visit: www.crcpress.com/Mathematical-Engineering-Manufacturing-and-Management-Sciences/book-series/CRCMEMMS

# Advances in Management Research

## Innovation and Technology

Edited by
Avinash K. Shrivastava, Sudhir Rana, Amiya Kumar Mohapatra, and Mangey Ram

CRC Press is an imprint of the
Taylor & Francis Group, an informa business

CRC Press
Taylor & Francis Group
6000 Broken Sound Parkway NW, Suite 300
Boca Raton, FL 33487–2742



Printed on acid-free paper

International Standard Book Number-13 978-0-367-22688-6 (Hardback)



**Library of Congress Cataloging-in-Publication Data**

Names: Shrivastava, Avinash K., editor. | Rana, Sudhir, editor. | Mohapatra, Amiya K., editor. | Ram, Mangey, editor.
Title: Advances in management research: innovation and technology / edited by Avinash K. Shrivastava, Sudhir Rana, Amiya Kumar Mohapatra and Mangey Ram.
Description: Boca Raton, FL: CRC Press/Taylor & Francis Group, 2020. | Series: Mathematical engineering, manufacturing, and management sciences | Includes bibliographical references and index. | Summary: "This book covers advancements across business domains in knowledge and informational management. It provides research trends in the fields of management, innovation, and technology, and is comprised of research papers that show applications of IT, analytics, business operations in industry, and in educational Institutions"—Provided by publisher.
Identifiers: LCCN 2019032662 (print) | LCCN 2019032663 (ebook) | ISBN 9780367226886 (hardback; acid-free paper) | ISBN 9780429280818 (ebook)
Subjects: LCSH: Management. | Operations research. | Commercial statistics. | Economics. | Investments.
Classification: LCC HD31.2.A38 2020 (print) | LCC HD31.2 (ebook) | DDC 658.4/038—dc23
LC record available at https://lccn.loc.gov/2019032662
LC ebook record available at https://lccn.loc.gov/2019032663

**Visit the Taylor & Francis Web site at**
**www.taylorandfrancis.com**

**and the CRC Press Web site at**
**www.crcpress.com**

# Contents

# Preface

By breaking down the walls in R&D, firms have radically changed the nature and structure of products and services. Competitive forces have resulted in advancements in the methods by which organizations generate and deploy new products/services and processes. One of the major areas of emphasis in recent years has been interdisciplinary research. Cross-disciplinary collaboration has become increasingly prominent in not only transforming business models, but also paving the way for organizational strategic decision makers to act in a timely manner. In today's connected world all stand to gain from continuing research that fosters industrial practice, global outreach, and inclusive growth. Despite the growing recognition of the importance of multidisciplinary research across industries and sectors and the growing body of knowledge, there is limited synthesis of literature across the management, social sciences, and research and technology disciplines. Technology continues to change quickly, and companies are continuously investing in research to present new technologies. Thus, the focus on research innovation by academicians as well as practitioners. Therefore, it would be rational to explore advances that have taken place in management research and technology that can be of direct use to business organizations, practitioners, government policymakers, academicians, and society at large.

This book is an endeavor to bring together diverse themes related to management practices, innovation, and technological advancement to advance the existing body of knowledge in both theoretical and managerial domains. The book act as acts as a platform to better understand the interactions between innovation and research and technology and the economy and society. The volume contains chapters on information and communication technology, digital platforms, economic empowerment, ASEAN, work–life balance, exchange rate movements, capital adequacy, time and frequency analysis, brand equity, transportation problems, and responsible global citizens from the scholars represented more than ten countries.

We are grateful to all the authors for strengthening this volume with their research contributions and thankful for the reviewers for taking time from their busy schedules and providing valuable feedback. We really hope this volume will be useful to both academicians and practitioners and will act as a platform in providing information on management research, innovation, and technology under one umbrella.

# Editor Biographies

**Dr. Avinash K. Shrivastava** completed his B.Sc (H) in mathematics and received his Master's, M.Phil, and Ph.D. degrees in operational research from the Department of Operational Research, University of Delhi. Dr. Shrivastava is currently Assistant Professor at the International Management Institute, Kolkata, West Bengal. He teaches classes on various technical and management subjects, including business statistics, business mathematics, advanced operations research, quantitative applications in management, operations management, business research methodology, and multicriteria decision-making to undergraduate and postgraduate students. He has presented papers at conferences of international repute and has also won accolades for best paper presentation in the same. He has published extensively in high-indexed international journals. He is the managing editor of the *International Journal of System Assurance Engineering and Management* (IJSAEM) and guest editor of two ABDC category journals. He is also a life member of the Society for Reliability, Engineering, Quality and Operations Management (SREQOM). He is on the editorial board and serves as reviewer of various international journals.

**Dr. Sudhir Rana** believes in driving and motivating academics and research in such a way that it can be utilized in an enthusiastic and dynamic environment to foster versatile personalities. He is a marketing area faculty. His teaching and research interests include international marketing, business development sales and negotiation, internationalization, and customer relationship management. He holds an MBA and Ph.D. with MHRD Scholarship from the Government of India. His research profile is followed by publications into ABDC/ABS/Scopus/Web of Sciences ranked and indexed journals published by Emerald, Inderscience, Sage, and Neilson Journal Publishing, etc. Dr. Rana is a versatile academician. He performs various roles as guest faculty, scholar, editor, editorial board member, guest editor, author, guest speaker/panel speaker/moderator, conference speaker/session chair/committee member, convener, and workshop expert at various platforms in India and overseas. His association with journals such as the *Journal of International Business Education*, *International Journal of Business and Globalisation*, *International Journal of Indian Culture and Business Management*, along with editorship in *FIIB Business Review* always motivate him to ensure that quality research gets their best home.

**Dr. Amiya Kumar Mohapatra** is Associate Professor and Chairperson-IQAC at Fortune Institute of International Business (FIIB), New Delhi. He holds a Ph.D. and four Master's degrees and has completed an FDP from IIM Indore. Dr. Mohapatra has also qualified UGC-NET in three subjects (economics, management, and public administration). He has been engaged in teaching as well as research in the areas of economics, finance, and public policy. His main research focus in recent years has been to explore economics from the policy perspective. He has co-authored 5 reference books and published in 21 edited volumes. He has published more than 75 research papers/articles in various indexed journals/magazines/books and has also presented his research work at IIM Ahmedabad, IIM Bangalore, IIM Indore, IIT Delhi, Jawaharlal Nehru University, and Delhi University, to name a few. He is also guest editor and editorial board member of many reputed national and international journals. He has organized and led more than 20 national and international conferences and seminars.

**Dr. Mangey Ram** received his Ph.D. with a major in mathematics and a minor in computer science from G. B. Pant University of Agriculture and Technology, Pantnagar, India. He is currently a professor at Graphic Era Deemed to be University, Dehradun, India. He is Editor-in-Chief of the *International Journal of Mathematical, Engineering, and Management Sciences* and the guest editor and member of the editorial board of various journals. He is a regular reviewer for international journals, including IEEE, Elsevier, Springer, Emerald, John Wiley, Taylor & Francis Group, and many other publishers. He has published 131 research publications in IEEE, Taylor & Francis, Springer, Elsevier, Emerald, World Scientific, and many other national and international journals of repute and has also presented his works at national and international conferences. His fields of research include reliability theory and applied mathematics. Dr. Ram is a senior member of the IEEE; a life member of the Operational Research Society of India, the Society for Reliability Engineering, Quality and Operations Management in India, and the Indian Society of Industrial and Applied Mathematics; and a member of the International Association of Engineers in Hong Kong and Emerald Literati Network in the UK. He has been a member of the organizing committee of a number of international and national conferences, seminars, and workshops. He has been conferred with the "Young Scientist Award" by the Uttarakhand State Council for Science and Technology, Dehradun, in 2009. He received the "Best Faculty Award" in 2011 and recently the "Research Excellence Award" in 2015 for his significant contributions in academics and research at Graphic Era.

# 1 Impact of Credit Access on Economic Empowerment of Married Women in Ethiopia

*Manoj Kumar Mishra*

## CONTENTS

## 1.1 OVERVIEW

### 1.1.1 Background

Ethiopia is the second most highly populated country in Africa after Nigeria. Although it has the fastest-growing economy in the region, the annual per capita income of Ethiopia is low, 691 USD, and 23.4% of the population was under the poverty line in 2014/15(GTP 2 plan, 2016). To alleviate poverty through employment creation, the government has encouraged domestic saving and private investment. Moreover, increasing access to credit is important to overcoming poverty through job creation and income generation among those who lack of access to credit and other financial services (Derbew, 2015).

Monetary financial institutions (MFIs) give priority to rural poor women because women with access to credit are more likely to increase spending to improve household welfare than men are (Narain, 2009). Women also are more credible in loan repayment than men (Gobezie, 2010). Moreover, labor markets favor of men and most of the productive resources are controlled by men, especially in developing countries. The first mission of the Amhara Credit and Saving Institution (ACSI) is to increase access to credit by poor people in order to improve their productivity and income. ACSI gives special emphasis to women, as women are mostly marginalized and have limited access to financial services (ACSI, 2017).

Women are important contributors to agricultural development and the rural economy in Ethiopia. Women constitute 40% of the world labor force and 43% of the agricultural labor force (Box, 2015). Women are involved in all farm activities in Ethiopia, from clearing of land to harvesting, processing, and marketing of products. Women plant different crops, rear animals, and keep poultry near their homes. In other words, women undertake multiple responsibilities in Ethiopia, working outside the home on farmlands and on non-farm activities as equal as their male counterparts. Women are also involved in activates that men are hardly involved in in Ethiopia, such preparing and providing food to maintain family welfare and, of course, giving birth to children (Fabiyi, Danladi, Akande, & Mahmood, 2007). Just as a bird cannot fly with a single wing, countries that do not include women in their economic, social, and political agendas will be unable to achieve sustainable development.

### 1.1.2 Problem Statement

Women are the backbone of rural agriculture and the national economy in Ethiopia. They constitute over 50% of the total population in Ethiopia (Tegegne, 2012). To reduce poverty and achieve sustainable development, empowering women is essential. Women's economic empowerment (WEE) is one of the most important factors contributing to equality between women and men. A specific focus on women is necessary given that women are a majority among economically disadvantaged groups. Therefore, WEE is a priority in promoting gender equality and so as to bring sustainable development (Bayeh, 2016).

In particular, married women have less control over household resources and poor access to productive resources than single women, which hinders their movement towards economic empowerment (Tekaye & Yousuf, 2014).

In rural areas, many women do not have their own land or access to credit and usually engage in day labor. Due to a lack of working capital, women and men do not have equal opportunities to participate in income-generating activities. Women who lack access to credit and capital are more impoverished and have lower quality of life than women who have sufficient access to capital and own land (Tadesse, 2014).

Previous studies have indicated that MFIs have a greater impact on WEE (Eshetu, 2011; Tekaye & Yousuf, 2014) through their effect on women's income, saving, and decision-making. Researchers have also assessed the association between access to microfinance and WEE. However, studies evaluating the impact of credit specifically on married WEE in West Gojjam are limited. These studies could not separately show the contribution of credit access to women's income and income earned by other family members. Earlier studies considered household income as an indicator of WEE. Therefore, this study was initiated to fill this research gap by examining the impact ACSI credit access on the economic empowerment of married women, as decisions of married women are more influenced by men than other women and girls. This study also measured additional married women decision-making indicators. This study applied the propensity score matching (PSM) technique to estimate the effect of microcredit access on married WEE in Jabi Tehinan Woreda located in West Gojjam Zone of Amhara Region, Ethiopia.

### 1.1.3 Objectives of the Study

The main objective of this study was to assess the impact of microcredit access on married WEE. The specific objectives of the study were:

1. To estimate the impact of credit access on the saving of married women
2. To estimate the impact of credit access on increasing the income of married women
3. To identify determinants of married women's participation in the decision-making process

### 1.1.4 Scope and Limitation of the Study

Women's empowerment has many dimensions (i.e. political, sociocultural, economic, etc.), but this research addressed only economic empowerment of married women. ACSI provides financial services in both rural and urban areas, but this study focused on the impact of credit access on rural married women. The study was conducted on married women who are clients and non-clients to ACSI in Amhara regional state, specifically in Jabi Tehinan Woreda.

### 1.1.5 Significance of the Study

The pace of growth and development increases with the equal participation and mobilization of men and women and by deploying their human talents, knowledge, and skills effectively. However, getting women to participate in social, economic, and political processes is problematic in the case of Ethiopia. Empowering women in all development aspects paves the way towards sustainable development. Thus, increasing women's access to microcredit is among the policy tools in Ethiopia that can be used to empower women economically. The findings of this study will inform policymakers, politicians, and the government as a whole as to what the next task will be to empower rural women and suggest possible interventions to addresses limitations of empowering women through increasing their access to microcredit in Ethiopia, particularly in Amhara regional state.

### 1.1.6 Organization of the Chapter

This chapter has been organized into five sections. The first section presents an overview that presents the background of the study, statement of the problem, objective of the study, scope and limitation, and significance of the study. The second section reviews the theoretical and empirical literatures that are related to the impact of credit on WEE. The third section presents the methodology of the study, with emphasis on description of the study area, sampling design, and data analysis techniques. The fourth section presents the results and discussion. The last section presents and summarizes the main findings of the study and offers policy recommendations.

## 1.2 LITERATURE REVIEW

The history of microfinance is related with Grameen Bank, which was located in Bangladesh. The founder of this bank was Professor Mohammed Yunus. In 1976 he began to lend small amounts of money to poor households, who were organized into small peer-monitoring groups, and many of them poor women's groups, in a few nearby villages. These groups were effective and the demand for credit increased. Then in 1984 Grameen Bank become a government-regulated bank. Innovation in microfinancial services based on the bank's credit peer-monitoring model targeted at poor women has been a focus in the developing world. The 1990s was a period of rapid expansion of access to microcredit (Hulme & Moore, 2007). In 2006, Yunus and Grameen Bank were honored with the Nobel Prize, in recognition of Yunus's revolutionary social work and long-term vision of eliminating world poverty (Atikus, 2014).

Hulme and Moore (2007) found that most MFIs provide financial services to poor women, rather than poor men, because women have better repayment and success rates. Credit can empower women economically; enhance their ability to invest in productive human and physical capital; smooth consumption; improve the health, nutrition, and educational status of their families and in particular their children. Therefore, women's participation in microfinance has been considered instrumental for women's socioeconomic wellbeing and their families.

### 1.2.1 Microfinance and Women's Empowerment Globally

A study conducted by Syeda (2014) in Pakistan on the impact of microfinance on women's economic, familial, political, and social empowerment revealed that increasing financial services for women enhances their status of living; improves their self-confidence; and increases their control over resources; they are able to earn a livelihood for themselves and for their families and live lives of dignity and respect. Phyu (2012) used the breakdown position of women in the Nash bargaining solution as a framework for investigating the effects of microfinance on WEE in Myanmar. He analyzed the result by using PSM to solve the problem of selection bias and logistic regression method of analysis. He determined that the respondent's age, partner's education, partner's age, and family size do not affect women's decision-making. However, increasing household income has a negative effect on the women's decision-making in the household; in this study, 10% of respondents were dependent and they had to take the loan for the spouse's business to increase household income. Finally, the result confirmed that microfinance had a positive and significant effect in explaining women's independence in household decision-making, which was expressed in child-related decisions, household item decisions, and non-food item expenditures.

Swin and Wallentin (2009) also investigated the relationship between WEE and microfinance services. They used a general structural model and robust maximum likelihood to analyze the impact. They used empowerment as the latent variable. The researchers noted that WEE will only be achieved when women challenge existing deep-rooted factors. Finally, they concluded that microfinance services had a positive impact for the change of these factors and WEE.

Ringkvist (2013) investigated the empowerment of women through microfinance using decision-making power as an indicator of empowerment. Ringkvist used a multivariate regression technique and index-based approach to measure empowerment. The results confirmed that member women were more empowered than non-members. Factors such as being a client of a microfinance programme, food security, and being head of household were found to have a significant and positive effect on empowerment of women. On the other hand, age, total loan amount, length of education, and length of years as a microfinance member had no significant effect on empowerment of women.

### 1.2.2 Microfinance and WEE in Africa

The expansion of credit programmes in Africa dates back to the mid-1980s, but it increased dramatically in the late 1990s with the establishment of saving clubs and credit unions for women in Africa (Mayoux, 1999). Researchers have assessed the impact of microfinance on WEE in different African countries.

Jebili and Bauwin (2015) studied WEE through microfinance in Tunisia by examining the business development process of men and women. They found that even when women start their own businesses they still have to shoulder the same domestic responsibilities, which could imply for them an even greater workload and perhaps domestic conflict. Participating in income-generating activities could even result in an increase of women's dependency towards men. Women have very little control over resources.

A study by Van et al. (2012) in sub-Saharan Africa found some evidence of microcredit empowering women. Findings from Zimbabwe found no indication that participation in microfinance led to greater control over the earnings from the business; for both married men and women, there was more consultation and joint decision-making with the partner. On the other hand, rural microcredit programmes had a greater impact on financial management skill, owning a bank account, taking pride in contributing to household income, and gaining ownership of some selected household assets. When we see the overall financial inclusion of Africa, South Africa and Rwanda lead first and second respectively (Atikus, 2014).

### 1.2.3 Microfinance and WEE in Ethiopia

Most of the studies in Ethiopia have confirmed that MFIs have a positive impact on the economic empowerment of women; that is, the level of economic empowerment of clients is higher than for non-clients. According to Tekaye and Yousuf (2014), education level, experience in income-generating activities, and skill exposure have a positive and significant relationship with WEE. In contrast, the dependency ratio, time spent on household work, marital status, and diversification of loan use have a significant negative relationship with WEE.

Eshetu (2011) studied the impact of microfinance on WEE using cross-sectional data and a binomial logit method of analysis. She used economic decision-making power on large sales as an indicator of economic empowerment. The result of this study revealed that mature clients have improved access and control over assets, improved income, better asset possession, and that 50% of mature clients had personal cash savings. But the coefficient of income, savings, and asset possession were insignificant when used to determine decision-making on large sales. This is because unless the women themselves controlled and made decisions on the resource they generated, improvement on the above three variables by itself was not a sign of WEE. She also noted that married women clients had less decision-making influence on household large sales than widowed and divorced women, and she concluded that mature clients did not make decisions on large sales.

However, a study conducted using similar empowerment indicators and methods of analysis by Taye (2014) found that the relative high age of respondents, education, being a microfinance client, household income, and personal cash savings were positively related with the empowerment of women and that these variables were significant. In contrast to Eshetu's conclusion, he argued that mature microfinance clients were able to make decisions by themselves on large sales such as oxen, sheep, goats, cows, etc. He also explained that microfinance has had a great role in expansion of business and income of women, respect for women's rights, reduction in HIV infection, positive change on gender roles, investment of extra income on children, and providing a route to sustainable development. Based on the evidence, he concluded that microfinance has a positive impact on WEE.

Researchers have employed different dimensions to measure women's empowerment. Ahmed (2013) used economic empowerment, personal empowerment, household (family) empowerment, and political empowerment as measurements of empowerment. He used binary logit model and PSM to analyze the results. Explanatory variables such as the respondent's age, education status, marital status, duration as microfinance client, age at marriage, and non-formal education positively and significantly affect cumulative women's empowerment. He confirmed that microfinance had a positive effect on all of the empowerment indicators, as evidenced by the fact that microfinance clients were 33% more empowered than non-clients. Variables that negatively affected the empowerment of women included religion, access to media, and marriage. Because of their religious rules, Muslim women cannot participate in microfinance programmes. Married women also have less power in household decision-making and are less mobile.

In areas that are prone to war, drought, and other natural disasters like the Tigray region, MFIs have great role in alleviating poverty. Women are more severely affected by such difficulties, because more of the burden and responsibility is on their shoulders. Therefore, empowering women economically is a top priority, and financial services are one way of reducing poverty from the family to the society level. Meron and Samson (2015) found that MFIs are performing the lion's share in poverty reduction and WEE. They found that factors such as family size, being a participant of a household, and engagement in livestock production are positively related and significant to the sale of agricultural byproduct and positively related to economic empowerment.

## 1.3 METHODOLOGY, DATA, AND MODEL

### 1.3.1 Description of the Study Area

Amhara regional state is located in northwestern Ethiopia and is one of nine regional states of Ethiopia. It is sub-divided into 11 zones, 140 woredas, and about 3,429 kebeles, which is the smallest administrative unit. Based on a census conducted by the central statistics agency in 2007, the total population was 17,221,976. From the total population, 12.27% are urban inhabitants. The ethnic group breakdown is 91.2% Amhara, 3.46% Agaw/Awi, 2.62% Oromo, 1.39% Agaw/Kamyr, and 0.41% Argoba (CSA, 2013).

### 1.3.2 Background of ACSI

ACSI has been increasing its reach and improving its financial services and activities. Based on a 2017 institutional report, it has around 1,375,413 clients (63.4% female) in the region and 445 branches and 1,259 satellite offices. It has also outstanding loan birr 13,645,553,507; total net savers 8,173,202 (46.5% are women), total net saving birr 12,295,727,459.39 (26.56% women's saving).

Financial services include saving, insurance, credit, cash transfer, payment, and other services. It offers services for urban and rural peoples, with special attention given to rural agricultural communities. It supports rural agricultural productivity by providing loans for new technologies and inputs of agriculture and seeks to reduce urban youth unemployment by offering loans for starting capital.

### 1.3.3 West Gojjam Zone

West Gojjam zone is located in the Amhara region of Ethiopia. It is bordered on the south by the Abay River, which separates it from the Oromia region and Benishangul-Gumuz; on the west by Agewawi; on the northwest by North Gondar; on the north by Lake Tana and the Abay River, which separates it from the South Gondar; and on the east by East Gojjam. Finote Selam is the capital of this zone.

### 1.3.4 Jabi Tehinan Woreda/District

Jabi Tehinan woreda is bounded by Burieworeda in the west, Dembecha woreda in the east, Sekela and Quarite woreda in the north, and Dembecha and Burie woreda in the south. The average temperature is 23° Celsius.

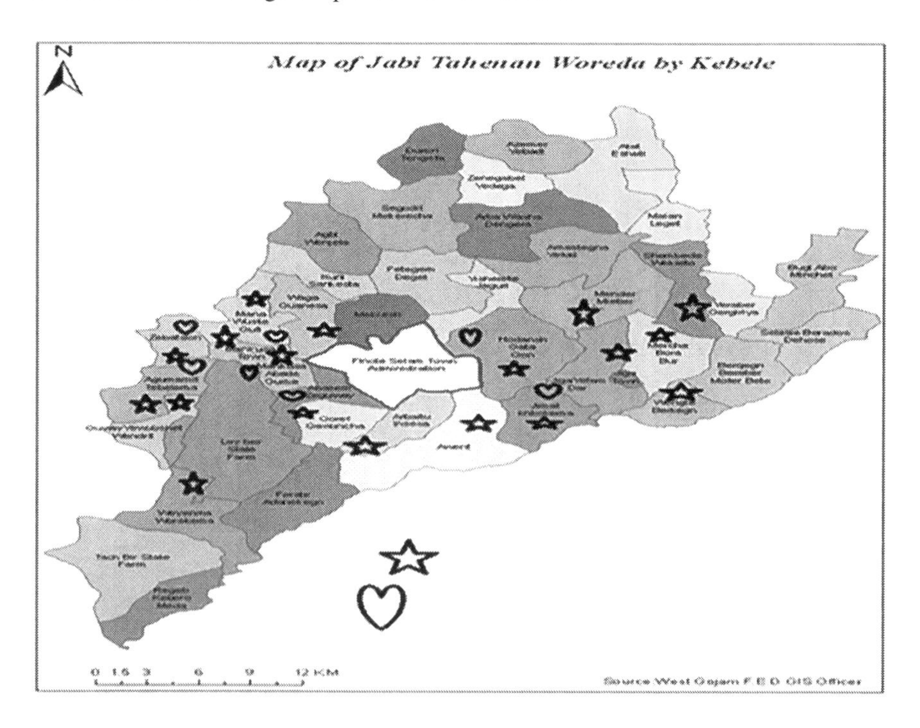


**FIGURE 1.1** Map of Jabi Tehinan Woreda.

*Source*: Jabi Tehinan Woreda Finance and Economic Office, 2018.

Based on Jabi Tehinan woreda administration information from 2016, the population of the area is as follows: male = 107,511, female = 110,935, total = 218,446 (50.78% female). From the total population, 16.4% live in urban areas and 83.5% live in rural areas. Of the population living in urban and rural areas, 56.4% and 49.02% are female, respectively.

Agriculture is the main source of income of Jabi Tehinan woreda, with around 84% of the population engaged in agriculture. The woreda is known for different types of crops, including teff, maize, sorghum, bean, and others. It also known for cultivation of fruits and vegetables such as mango, papaya, avocado, carrot, redroot, sugarcane, etc.

#### 1.3.4.1 Sampling Techniques and Sample Size

To select sample respondents from the study area, a two-stage random sampling procedure was conducted. First, out of a total 38 rural kebeles (i.e. the lowest administrative unit in Ethiopia), 5 kebeles (Abasem Zegwi, Hodansh Gategon, Arebayitu Ensisa, and Awunt Month) were randomly selected. The total target household size from these five kebeles was 2,172 (223 married women are clients and 1,949 are non-clients). Second, the sample sizes of rural households were selected based on the total household size of each kebele. The list of clients and non-clients was obtained from the same selected rural kebeles.

To decide on the sample size, we had to determine the desired level of precision, or sample error; confidence interval; and degree of variability. The formula developed by Kothari (2004) to determine the sample size for a finite population was used for this study:

$$n = \frac{z^2 * p * q * N}{e^2 \left(N - 1\right) + z^2 * p * q}$$

where $n$ = sample size; $z$ = standard normal variable at the required confidence level, 1.96 (from table); $p$ = estimated characteristics or proportion of the target population, 0.8; $q = 1 - p$, or 0.2; $N$ = population size, 2,172 (more homogeneous or less variable population; i.e., all women's were married, living in rural areas, and in the similar agroecology); e = margin of error, 0.05 for 95% level of confidence. The sample size was calculated to be:

$$n = \frac{2172 * 0.8 * 0.2 * 1.96^2}{\left(0.05\right)^2 \left(2172 - 1\right) + 1.96^2 2 * 0.2 * 0.8} = 233$$

Therefore, the total sample size was 233 (133 non-clients and 100 clients), which was selected from each kebele by a simple random sampling method based on household size. For focus group discussion, one group was selected randomly from each kebele to collect information about the general ACSI service delivery system and overall

WEE information. One group had seven married women, four non-clients and three clients. In addition, interviews were conduct with husbands of ACSI clients by selecting two men from each kebele and two ACSI officers to take information about the loan use and management in the household and the institution's efforts towards WEE. Total sample size (i.e., calculated sample size = 233; focus group = 35; and key informant interview with husbands of clients = 10) was 278.

### 1.3.5 Method of Data Analysis

#### 1.3.5.1 Descriptive and Inferential Statistics

The descriptive analysis focused on description of both dependent and independent variables analyzed by percentage, mean, standard deviation, diagram, and frequency distribution. The statistical significance of the categorical and continuous explanatory variables was tested by chi-square test and *t*-test respectively. The statistical significance of the outcome variables was also checked by *t*-test and chi-square test.

#### 1.3.5.2 Econometric Model

Different econometric tests were applied to examine model fitness, multicollinearity, and heteroscedasticity. The link test suggested by Pregibon (1980) was applied for the specification of independent variables. The omission of variables may also introduce an error. Therefore, the Ramsey RESET test or *ovtest* command was used to look for the symptoms of omitted variables, such as unusually large or small coefficients that have an incorrect sign (Wooldridge, 2015). Multicollinearity refers to the existence of a perfect or exact linear relationship among explanatory variables in regression models. Two measures were applied to test the presence of multicollinearity.

#### 1.3.5.3 Propensity Score Matching (PSM)

In this research, married women who are clients of ACSI or those who are taking credit from ACSI were the treatment group and non-clients were the control group. To investigate the impact of ACSI credit on client married WEE, it is essential to be able to compare the two groups. The treatment effect for an individual $i$ can be written as:

$$\tau i = Yi\ (1) - Yi(0) \tag{1.1}$$

where:

$Yi(1)$ = outcome of treatment (economic empowerment of $i$th woman when she is client)

$Yi(0)$ = outcome of untreated individual (economic empowerment of $i$th woman when she is non-client

$\tau i$ = change in outcome because of treatment

The fundamental evaluation seeks to estimate the mean impact of the programme, obtained by averaging the impact across all the individuals in the population. This parameter is known as the average treatment effect (ATE), which is the effect of treatment on WEE:

$$\text{ATE} = E(\tau i) = E(Y1 - Y0) \tag{1.2}$$

where $E(.)$ represents the average or expected value.

The ATE of an individual $i$ can be written as:

$$\text{ATE} = E(Yi(1)/D = 1) - E(Yi(o)/D = 0) \tag{1.3}$$

where:

E ($Yi(1)/D = 1$): Average outcome for treated individual woman who is client of ACSI

E ($Yi(o)/D = 0$): Average outcome for untreated individual woman who is non-client

Another quantity of interest is the average treatment effect on those treated, which measures the impact of the programme on those individuals who participated:

$$\begin{aligned} \text{ATT} &= E(Y1 - Y0/D = 1) \\ &= E(Y1/D = 1) - E\ (Y0/D = 1) \end{aligned} \tag{1.4}$$

The second term, E (Y0/D = 1), is the average outcome that the treated individuals would have obtained in the absence of treatment, which is not observed, but we can observe the term $E\ (Y0/D = 0)$.

$$\begin{aligned} \Delta &= E\ (Y1/D = 1) - E(Y0/D = 0) \\ &= E\ (Y1/D = 1) - E(Y0/D = 1) + E(Y0/D = 1) - E(Y0/D = 0) \\ &= \text{ATT} + E\ (Y0/D = 1) - E(Y0/D = 0) \\ &= \text{ATT} + \text{SB} \end{aligned} \tag{1.5}$$

Where $SB$ = selection bias, the difference between the counterfactual for treated individuals and the observed outcome for the untreated individuals. If $SB = 0$, then ATT can be estimated by the difference between the mean observed outcomes for treated and untreated.

$$\text{ATE} = E(Y/D = 1) - E(Y/D = 0)$$

#### 1.3.5.4 Description of Variables

**TABLE 1.1**
**Summary of Variables and Expected Signs**

| No. | Name of Variable | Symbol | Dependent/ Independent | Variable Type | Measurement | Expected sign |
|---|---|---|---|---|---|---|
| 1 | Credit participation | treat | Dependent/ independent | Dummy | 1 if woman participated; 0 otherwise | + |
| 2 | Amount of saving | tasav | Outcome Dependent | Continuous | ETB | |
| 3 | Income | totaloi | Outcome Dependent | Continuous | ETB | |
| 4 | Decision making | | Outcome Dependent | Dummy | 1 if woman can participate; 0 otherwise | |
| 5 | Age of woman | age | Independent | Continuous | Year | + |
| 6 | Education level of woman | educs | Independent | Dummy | 1 if literate; 0 if illiterate | + |
| 7 | Family size | fs | Independent | Continuous | Number | + |
| 8. | Dependency ratio | dr | Independent | Continuous | Number | - |
| 9 | Animal fattening and rearing activity | afra | Independent | Dummy | 1 if woman participated; 0 otherwise | + |
| 10 | Infrastructure | infra | Independent | Dummy | 1 if woman had access to infrastructure; 0 otherwise | + |
| 11 | Land size | lands | Independent | Continuous | Land size measured in hectare | + |
| 12 | Distance to urban center | dturban | Independent | Dummy | 1 if walking distance is less than 2 hrs; 0 otherwise | + |
| 13 | Husband's education | educD | Independent | Dummy | 1 if literate; 0 otherwise | + |

*Source*: Own expectation with review literature.

*A. Binary Logit Model*

**TABLE 1.2**
**Decision-Making Power Indicators**

| No. | Decision-Making Indicator | Symbol | Type of Variable | Measurement |
|---|---|---|---|---|
| 1 | Control over large household assets | CoaD | Dependent (Dummy) | 1 if she participates in decision; 0 otherwise |
| 2 | Decision on self-free movement | DSFM | Dependent (Dummy) | 1 if she participates in decision; 0 otherwise |
| 3 | Decision on use of family planning | DUFP | Dependent (Dummy) | 1 if she participates in decision; 0 otherwise |
| 4 | Decision to visit and support relatives | DTVSR | Dependent (Dummy) | 1 if she participates in decision; 0 otherwise |
| 5 | Decision to the type of crops to be planted | DTBSF | Dependent (Dummy) | 1 if she participates in decision; 0 otherwise |
| 6 | Decision to buy quality seed and fertilizer | DTBSF | Dependent (Dummy) | 1 if she participates in decision; 0 otherwise |
| 7 | Decision to participate in business activity | DTINCG | Dependent (Dummy) | 1 if she participates in decision; 0 otherwise |

## 1.4 RESULTS AND DISCUSSION

This section presents the findings of the study in line with the specific objectives; that is, the impact of credit access on married women's income and saving and their decision-making power.

As presented in Table 1.3 the total household size of randomly selected kebeles from Jabi Tehinan woreda was 2,172. From these, 100 married women who were ACSI clients or credit users and 133 married non-client women were randomly selected and interviewed. The clients and non-clients selected from each selected kebele are proportional to the size of households in each kebele. Additionally, supportive data was collected from two ACSI officers and 10 husbands of clients through key informant interviews.

Household decisions and discussions are important for managing and improving the family's living standard. The head of the household or the person who takes responsibility for the household may lead this activity. The results in Table 1.4 show that client decision-making power is greater than that of non-clients except for married women control over large household assets. A significant difference was found on decision-making power between clients and non-clients at the 1% significance level. However, the decision-making power of clients on large household assets is not significantly different from that of non-clients.

**TABLE 1.3**
**Total Population and Sample Respondents by Selected Kebeles**

| Kebele Code | Kebele | Total Household Size | Sample Respondents | |
|---|---|---|---|---|
| | | | Client | Non-client |
| 1 | Awunt Month | 456 | 21 | 28 |
| 2 | Hodansh Gategone | 478 | 22 | 29 |
| 3 | Jiga Yelmidar | 478 | 22 | 29 |
| 4 | Arbayitu Ensisa | 391 | 18 | 24 |
| 5 | Abasem Zegaye | 369 | 17 | 23 |
| | Total | 2,172 | 100 | 133 |

*Source*: Finote Selam branch of ACSI office.

**TABLE 1.4**
**Decision-Making Power of Married Women**

| Activities | Who Makes Decision? | | | | | | Chi-square |
|---|---|---|---|---|---|---|---|
| | Respondent | | Jointly | | Husband | | |
| | Non-client | Client | Non-client | Client | Non-client | Clients | |
| Who decides when you want to engage in income-generating activity? | 7 | 16 | 49 | 49 | 77 | 25 | *** |
| Who decides when you want to freely move outside the house? | 6 | 24 | 74 | 51 | 53 | 25 | *** |
| Who decides on use of family planning services? | 14 | 52 | 95 | 45 | 24 | 3 | *** |
| Who decides to visit and support relatives and parents? | 13 | 21 | 76 | 67 | 44 | 12 | *** |
| Who decides the type of crops to be planted on farm fields? | 3 | 5 | 30 | 42 | 99 | 53 | *** |
| Who decides to buy quality seed and fertilizer? | 3 | 6 | 18 | 37 | 112 | 57 | *** |
| Who controls large household assets? | 0 | 1 | 90 | 70 | 43 | 29 | 0.41 |

*Source*: Survey data.
*** 1% significance level

### 1.4.1 Econometric Analysis

#### A. Determining Common Support Region

The common support graph on Figure 1.2 shows the estimated propensity scores for clients and non-clients of ACSI. The upper half of the graph (i.e., red) shows treated on support or the propensity score distribution for credit users fall in common

**TABLE 1.5**
**Logistic Regression Results**

| ACSI Participation (Treat) | Coef. | Std. Err. | Z |
|---|---|---|---|
| Age | 0.057** | 0.0261 | 2.19 |
| Infrastructure | −0.110 | 0.137 | −0.8 |
| Distance to urban | −0.625** | 0.264 | −2.36 |
| Animal rearing and fattening activity | 0.937*** | 0.319 | 2.94 |
| Education status of women | 1.302*** | 0.362 | 3.59 |
| Land size | −0.289 | 0.273 | −1.06 |
| Dependency ratio | 1.714 | 1.108 | 1.55 |
| Family size | 0.375*** | 0.113 | 3.3 |
| Education status of husband | 0.186 | 0.188 | 0.1 |
| _cons | −4.325 | 1.403 | −3.08 |

Number of observations = 233
Prob. > Chi-squared = 0.000
Pseudo R2 = 0.2007

*Source*: Survey data.
*** 1% and ** 5% level of significance.

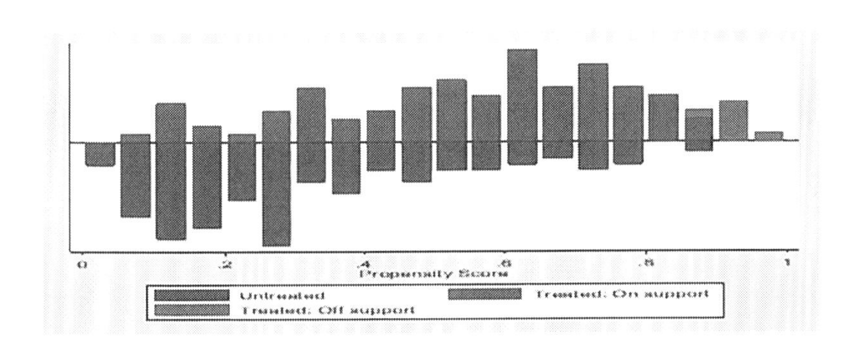


**FIGURE 1.2** Common support graph.

*Source*: Survey data.

support region while the lower half of the graph (i.e., blue) shows that the propensity score distribution for non-clients of ACSI fall within the common support region. The green-colored region represents the propensity score distribution of clients who did not match; that is, the participant groups that do not have appropriate comparison.

### 1.4.2 Factors Affecting Married Women's Decision-Making

The result of the logit model shows that access to credit was likely to increase women's decision-making measured as decision to self-free movement outside their

**TABLE 1.6**
**Factors Affecting the Decision-Making Power of Married Women**

| | Decision-Making Power Indicators | | | | | | |
|---|---|---|---|---|---|---|---|
| **Variables** | **Coah** | **DSFM** | **DUFP** | **DTVSRP** | **DFTCP** | **DTBSF** | **DINCG** |
| age | 0.205<br>(0.04) *** | 0.97 | 0.006 | −0.022 | −0.007 | 0.013 | 0.032 |
| treat | −0.21 | 3.1<br>(0.23) *** | 2.61<br>(0.14)*** | 1.56<br>(0.24)*** | 0.96<br>(0.2)*** | 1.37<br>(0.23)*** | 1.36<br>(0.32)*** |
| infra | −0.385 | 1.37 | −0.359 | −0.163 | −0.35 | −0.499 | 0.76<br>(0.18)* |
| dturban | −0.296 | 0.553<br>(−0.12)* | −1.92<br>(0.13)*** | −0.81<br>(−0.13)** | 0.13 | 0.087 | 0.377 |
| afra | −0.073 | 1.34 | −0.358 | −0.14 | 0.059 | −0.130 | 0.068 |
| educs | 0.73<br>(0.14)** | 0.678 | 0.288 | 0.064 | −0.04 | −0.148 | −1.14 |
| lands | −0.467 | 1.643<br>(0.12)* | 0.163 | −0.09 | −0.22 | −0.031 | 0.26 |
| dr | 1.22 | −0.662 | 0.065 | −0.25 | −0.97 | −0.902 | 1.65 |
| fs | −0.31<br>(−0.06)** | −0.012 | 0.964 | 0.023 | −0.1 | 0.044 | −0.09 |
| Constant | −4.79 | 3.1 | 0.84 | 2.067 | 0.16 | −1.49 | 0.11 |
| No. of obs. | 233 | 233 | 233 | 233 | 233 | 233 | 233 |
| LR chi2(10) | 60.94 | 20.8 | 33.87 | 22.3 | 14.07 | 24.59 | 35.64 |
| Prob > chi2 | 0.00 | 0.02 | 0.002 | 0.013 | 0.04 | 0.006 | 0.00 |
| Pseudo R2 | 0.204 | 0.07 | 0.208 | 0.088 | 0.05 | 0.091 | 0.112 |

*Source*: Survey data.
Values in the brackets are marginal effects and others are coefficients *** 1%, ** 5%, and * 10% significance level.

home, use of family planning, visit and support of relatives and parents, type of crop to be planted, seed and fertilizer, and participate in business activities. It is significant at the 1% significance level and has a positive relationship with decision-making indicators. Access to credit did not have significant effect on married women's control over large household assets. This result was consistent with findings of Eshetu (2011) in Ethiopia.

As shown on Table 1.6, the result of marginal effect revealed that married women who were clients had a 23% chance of making the decision for self-free movement outside their home; 32% chance to make the decision to participate in their own business activities; 24% chance of making the decision to visit and support relatives and parents; 20% chance of making decision on type of crop to be planted; and 23%

chance of making the decision to buy quality seed and fertilizer, keeping other variable constant.

## 1.5 CONCLUSION AND SUGGESTIONS

### 1.5.1 Conclusion

The results from the descriptive analysis showed that a woman takes ACSI credit, her saving habit and income also increase. The results from the econometric analysis revealed that the woman's age, family size, distance from urban center, animal rearing and fattening activity, and the woman's education status were significant factors in determining participation in ACSI credit. Infrastructure, land size, dependency ratio, and husband's education status were not significant. From determining factors, distance to urban center were negatively related with participation. Age, education, family size, and animal rearing and fattening activities were positively related with participation.

The result from ATT estimation also indicated that ACSI has a positive as well as significant impact on compulsory and voluntary saving of married women. Accordingly, clients of ACSI save annually 1587.32 ETB birr more than non-clients. But from these result clients of ACSI voluntary saving of ATT estimation was 560.71 ETB more than non-clients. Moreover, ACSI credit has a positive and significant impact on participating in one's own business activities. The ATT result shows that clients of ACSI can gain annual income of 1659.8 ETB more than non-clients.

Age, education status, and family size are the determinants of women's decision-making power on large household assets at the 1% and 5% levels of significance. Generally, ACSI credit is not the determinant factor for married women's decision-making on large household assets, but it is a major determinant factor for women's decisions on income-generating activity, free movement outside the house, visit and support by families and relatives, family planning services, type of crops to be planted, and buying quality seed and fertilizer at the 1% significance level.

### 1.5.2 Recommendation

Women require frequent advice about how to generate income, which type of business is appropriate for women, which type of business can operate in rural areas, and how they can operate different types of income-generating activities. Therefore, ACSI and other concerned bodies (i.e., governmental and non-governmental organizations, female and children affairs offices, technical and vocational enterprise development offices, agriculture and rural development offices) need to cooperate and pool resources to provide skill development and business startup capital to women. The institution should improve the information-processing system for pacing married women's economic empowerment.

## BIBLIOGRAPHY

ACSI. (2017). *Institutional profile*. Retrieved from www.acsi.org.et

Adhikari, D. B., & Shrestha, J. (2015). Economic impact of microfinance in Nepal: A case study of the Manamaiju village development committee, Kathmandu. *Economic Journal of Development Issues*, *15*(1–2), 36–49.

Ahmed, M. (2013). *Microfinance and rural women's empowerment: A cross-sectional level analysis: Evidence from rural woreda's of Harari Region* (Doctoral dissertation), Mekelle University.

Amha, W. (2012). *The development of deposit-taking microfinance institutions (MFIs) in Ethiopia: Performance, growth challenges and the way forward.*

Atikus Insurance. (2014). *Micro finance report*. Retrieved from www.atikusinsurance.come

Ayele, G. T. (2015). Microfinance institutions in Ethiopia, Kenya and Uganda: Loan outreach to the poor and the quest for financial viability. *African Development Review*, *27*(2), 117–129.

Batliwala, S. (1994). The meaning of women's empowerment: New concepts from action.

Bayeh, E. (2016). The role of empowering women and achieving gender equality to the sustainable development of Ethiopia. *Pacific Science Review B: Humanities and Social Sciences*, *2*(1), 37–42.

Caliendo, M., & Kopeinig, S. (2008). Some practical guidance for the implementation of propensity score matching. *Journal of Economic Surveys*, *22*(1), 31–72.

Derbew, K. (2015). Challenges facing Officials of Microfinance: The case of Amhara Credit and Saving Institution (ACSI). *Ethiopia, Research Journal of Finance and Accounting*, *6*(15), 42–49.

Eshetu, H. (2011). Impact of micro-finance on Women's Economic Empowerment: A Case-study of Amhara Credit and Saving Institution (ACSI), Kobo Woreda, Amhara National Regional State, Ethiopia. VDM-Verlag Dr. Müller.

Fabiyi, E. F., Danladi, B. B., Akande, K. E., & Mahmood, Y. (2007). Role of women in agricultural development and their constraints: A case study of Biliri local government Area, Gombe State, Nigeria. *Pakistan Journal of Nutrition*, *6*(6), 676–680.

Gebrehiwot, A. (2002). Microfinance institutions in Ethiopia: Issues of portfolio risk, institutional arrangement, and governance. In N. Zaid et al. (Eds.), *Micro finance theory policy and experience*. Proceedings of the International Workshop on The Dimension of Micro Finance Institutions in Sub Saharan Africa: Relevance of International Experience. Mekelle University.

Geleta, E. B. (2016). Microfinance and women's empowerment: An ethnographic inquiry. *Development in Practice*, *26*(1), 91–101.

Gobezie, G. (2010. Empowerment of women in rural Ethiopia: A review of two microfinance models. *Praxis: The Fletcher Journal of Human Security*, *25*, 23–38.

Goel, V. (2015). *Impact of microfinance services on economic empowerment of women: An empirical study.*

Growth and transformation plan 2(2015/16–1019/20). (2016, May). *Federal democratic of Ethiopia.*

Hulme, D., & Moore, K. (2007). Why has microfinance been a policy success in Bangladesh? In *Development success* (pp. 105–139). London: Palgrave Macmillan.

Hunt, A., & Binat, S. M. (2017). *Women's economic empowerment at international level: In depth analysis*. Directorate General for Internal Policies.

Jebili, B., & Bauwin, A. (2015). *Women's economic empowerment through microfinance services in Tunisia.*

Kabir, N. (2016). *Women's economic empowerment and inclusive growth: Labour markets and enterprise development*. School of Oriental and African Studies.

Khandker, S. R., Koolwal, G. B., & Samad, H. A. (2009). *Handbook on impact evaluation: Quantitative methods and practices*. World Bank Publications.

Kindane, A. (2007). *Outreach and sustainability of the Amhara credit and saving institutions (ACSI) Ethiopia* (Unpublished MA thesis), Norwegian University of Life Sciences.

Kothari, C. R. (2004). *Research methodology: Methods and techniques*. New Age International.

Lombardini, S., Bowman, K., & Garwood, R. (2017). *A 'how to' guide to measuring women's empowerment: Sharing experience from Oxfam's impact evaluations*.

Malaki, A. (2001). Sustainable banking with the poor: Microfinance handbook. An institutional and financial perspective. *Ibero Americana (Sweden), 31*(2), 125–131.

Mandal, K. C. (2013, April). Concept and types of women empowerment. In *International forum of teaching and studies* (Vol. 9, No. 2, p. 18). American Scholars Press, Inc.

Mayoux, L. (1999). Questioning virtuous spirals: Micro-finance and women's empowerment in Africa. *Journal of International Development, 11*(7), 957.

Mekonnen, A., & Asrese, K. (2014). Household decision making status of women in Dabat District, North West Ethiopia, 2009 Gc. *Science Journal of Public Health, 2*(2), 111–118.

Meron, Z., & Samson, A. (2015). Microfinance and women economic empowerment nexus: Eastern zone, Tigray Region, Ethiopia. *International Journal of Science and Research (IJSR)*, 414–417.

Modi, A., Patel, J., & Patel, K. (2014). Impact of microfinance services on rural women empowerment: An empirical study. *Journal of Business and Management, 16*(11), 68–75.

Mogues, T., Cohen, M. J., Lemma, M., Randriamamonjy, J., Tadesse, D., & Paulos, Z. (2009). *Agricultural extension in Ethiopia through a gender and governance lens*.

Mohapatra, S., & Sahoo, B. K. (2016). Determinants of participation in self-help-groups (SHG) and its impact on women empowerment. *Indian Growth and Development Review, 9*(1), 53–78.

Mokaddem, L. (2009). *Concept note on microfinance scaling up in Africa: Challenges Ahead and way forward*. Retrieved May 27, 2011.

Morduch, J. (2009). *Borrowing to save: Perspectives from portfolios of the poor*. Financial Access Initiative.

Morduch, J. (2017). Microfinance and economic development.

Narain, S. (2009). *Gender and access to finance*. The World Bank.

Narayan-Parker, D. (Ed.). (2005). *Measuring empowerment: Cross-disciplinary perspectives*. World Bank Publications.

Nwanesi, P. K. (2006). *Development, micro-credit and women's empowerment: A case study of market and rural women in Southern Nigeria*.

Paul, R. K. (2006). *Multicollinearity: Causes, effects and remedies*. New Delhi: IASRI.

Puskur, R. (2013). *Gender and governance in rural services: Insights from India, Ghana, and Ethiopia*.

Ringkvist, J. (2013). *Women's empowerment through microfinance: A case study on Burma*.

Rosenbaum, P. R., & Rubin, D. B. (1983). The central role of the propensity score in observational studies for causal effects. *Biometrika, 70*(1), 41–55.

Roy, A. D. (1951). Some thoughts on the distribution of earnings. *Oxford Economic Papers, 3*(2), 135–146.

Rubin, D. B. (1974). Estimating causal effects of treatments in randomized and nonrandomized studies. *Journal of Educational Psychology, 66*(5), 688.

Stern, N. (2005, April 27). *Making development happen: Growth and empowerment*. Presidential Fellows Lecture. Washington, DC: World Bank.

Syeda, N. (2014). *Impact of micro finance on women empowerment: A case study of select districts of Khyber Pakhtoonkhwa* (Doctoral dissertation), University of Peshawar, Peshawar.

Tadesse, B. (2014). *Income and asset building impact of micro credit provision to rural women: The case of menschen für menschen supported saving and credit cooperatives in Merhabete Woreda, Amhara National Regional State, Ethiopia* (Doctoral dissertation), Haramaya University.

Taye, C. (2014). *The Impact of Microfinance Financial service on the economic empowerment of women: The case study of Wisdom Micro Financing Institution Sc, At Woliso Woreda* (Doctoral dissertation), St. Mary's University.

Tegegne, M. (2012). *An assessment on the role of women in agriculture in Southern Nation Nationality People's Region: The case of Halaba Special Woreda, Ethiopia* (Doctoral dissertation), Indira Gandhi National Open University.

Tekaye, D., & Yousuf, J. (2014). *Women's economic empowerment through microcredit intervention: The case of Chinaksen Woreda, Oromia National Regional State* (Doctoral dissertation), Haramaya University.

Thaler, R. H., & Benartzi, S. (2004). Save more tomorrow: Using behavioral economics to increase employee saving. *Journal of Political Economy*, *112*(S1), S164–S187.

United Nations. (2010). *Achieving gender equality, women's empowerment and strengthening development cooperation: Dialogues at the economic and social council.* New York: United Nations.

Wooldridge, J. M. (2015). *Introductory econometrics: A modern approach.* Nelson Education.

Yehuwalashet, F. (2011). *Outreach services and sustainability: The case of Amhara credit and saving institution.*

# 2 Relationship between Work–Life Balance and Job Satisfaction of Reception and Concierge Employees in Five-Star Hotels in Colombo

*A. Chamaru De Alwis and Shavindri Lankeshni Samarasekera*

**CONTENTS**



## 2.1 BACKGROUND OF THE STUDY

The success of the hotel industry depends heavily on the quality of service provided to customers, and this quality is influenced by the people who are involved in day-to-day operations. Unlike other industries, the hotel trade is highly labour intensive and high customer contact. In order to successfully compete with industry rivals. Organizations should not only have a sufficient number of employees, but also employees who have the essential skills motivation. At present, the industry has recognized human capital as the most important and critical asset to gain competitive advantage (Bednarska, 2012).

However, the business world is increasingly competitive, regardless of the nature of the industry, and firms are constantly competing. This competition often requires employees to work like machines during their work shifts, and when employees fail to manage their work and personal life in an equal manner, it can lead to unpredictable consequences at unexpected situations. The hotel industry is no exception to this situation, and sometimes it is worse when compared with other industries. Even though hotels need their employees' maximum performance, it can be difficult to get the most out of employees because they are human beings with emotions who are susceptible to the high stress generated from working long hours (Peshave & Gujarathi, 2014). Working conditions in the hotel industry can have a negative impact on employee work–life balance, impacting the employees' personal lives as well as the organization itself (Deery, Jago, & Jago, 2015).

Work–life balance is a popular topic today as employees in the hotel trade face enormous pressure due to the nature of their jobs as well the competitive business environment. Even though other industries face similar problems of employee work–life balance, the hotel industry is unique due to the provision of round-the-clock service (Mohanty & Mohanty, 2014). Many job roles in the hotel industry tend to negatively affect, directly and indirectly, on the quality of family life and the psychological and physical wellbeing of employees (Bednarska, 2012).

According to many researchers, poor work–life balance situations occur mainly due to inflexible job characteristics, which is a common aspect in the hotel industry, ultimately impacting employee job satisfaction, commitment, and performance (Hughes, Bozionelos, & Hughes, 2007). Likewise, many researchers have proved that work–life imbalance is not a rare situation in the hotel industry and that it is a common phenomenon across the whole hotel trade. Work–life imbalance creates a more stressed and unsatisfied workforce within the organization, which negatively affects the day-to-day operations of hotels as well as customer satisfaction and organizational success (Kidd & Eller, 2012).

The prime purpose of any hotel is to ensure customer satisfaction in order to achieve the organization's ultimate goals and targets. Because human resources is an important asset towards achieving those goals, it is important for organizations to have a satisfied and loyal pool of employees (Jayawardhana, Silva, & Athauda, 2013).

To create an effective and productive environment, hotel management needs to ensure work–life balance for its employees. As almost all job roles in the hotel industry are pressure driven, the possibilities for work–life imbalance situations are rather high. But because management expects employees' maximum productivity with loyalty and commitment to their operation's processes, many hotels have started to give their attention to work–life balance and job satisfaction of employees (Mohanty & Mohanty, 2014).

Once an organization is able to achieve work–life balance for its employees, employee turnover will be minimized. Conversely, if work–life balance in organizations is poorly managed, employees will experience increased mental dissatisfaction; job stress; and negative behaviour, attitudes, and performance at both work and home. These conflicts between work, home, and life are linked with job dissatisfaction and turnover (Deery et al., 2015).

At present, with the increased call for work–life balance in the hotel industry, some organizations have started to implement strategies towards effectively managing work–life balance. However, in practice, these measures are ineffective due to the failures in implementing and sustaining them. Nevertheless, employment practices pertaining to areas of work–life balance are rated as the second most important on their ability to enhance employee productivity in hotels (Peshave & Gujarathi, 2014).

At the inception of the research work, the researcher visited selected five-star hotels in Sri Lanka and observed receptionists and concierge employees' behavior and commitment to work and the quality of customer service while on duty. During these observations the researcher noticed that although employees start their day energetically, as the time passes the way they talk to customers and the way they answer the telephone are not as enthusiastic as in the morning but rather lethargic in attitude. Also, when the hotel becomes congested with guest arrivals and departures, the way they interact with customers differs from that on non-busy days. It was observed that employees' patience and politeness drops when the hotel gets busy, and as a result they get blamed by the unhappy customers.

Through the researcher's observations it was concluded that even if these employees in reception and concierge are trained to act positively in stressful situations, in reality they fail to do so. Employees were willing to perform only their assigned duties, but nothing beyond that. Researchers have also stated that when there is dissatisfaction towards their job, employees stick to perform their specified tasks only (Deery et al., 2015).

The hotel industry is constantly changing due to the nature of the competitive environment; managers strive to get the maximum output from fewer employees in order to maximize the profit. As a result, employees are suffering from work–life imbalance problems and conflicts in managing work and family life. Even though tourism and the hotel industry have become significant industries towards Sri Lanka's economic development, this has a number of drawbacks, and imbalance in work–life relationship is one of them (Jayawardhana et al., 2013). It is a well-known fact that hotel and tourism industry employees suffer severely due to work and family issues (Ranasinghe, Lanka, Deyshappriya, & Lanka, 2011).

Though they do not maintain records relating to employee work–life balance, the management of most hotels considers work–life balance to be one of five fundamental elements to be considered within the organization. As per front office managers' opinion, high turnover, frequent absenteeism and low performance are interlinked with work–life imbalance.

Long, isolated working hours; low pay, and often low status; and heavy workloads in hotel-sector jobs has led to low employee work–life balance, resulting in four major problems: high turnover, low performance, increased grievances and chronic absenteeism. Moreover, researchers say that poor work–life balance in the hotel industry has led to higher-than-average skill shortages, low productivity, high labour turnover, and hard-to-fill vacancies (Deery et al., 2015).

Thus, in light of these issues, the researcher developed the research problem of the current study: What factor(s) is/are most influential on receptionists and concierge employees in hotels experiencing poor work–life balance?

A number of researchers (Fayyazi & Aslani, 2015; Saeed & Farooqi, 2014; Haar, Russo, Suñe, & Ollier-Malaterre, 2014; Mukhtar, 2012) have examined work–life balance and job satisfaction and have found that there is a relationship between these two variables. However, the researcher did not come across any studies that tested the relationship between selected dimensions of work–life balance and job satisfaction. Accordingly, in this research, the researcher aimed to discover whether there was any relationship between involvement balance, satisfaction balance and time balance and how imbalance in these variables affects the job satisfaction of hotel employees.

## 2.2 SIGNIFICANCE OF THE STUDY

Due to inherent characteristics of the hotel industry, most hotels face the issue of achieving success through developing and maintaining a productive workforce. Employees of this industry are also struggling to achieve a better life by fulfilling their career expectations. Therefore, this chapter is significant in several ways to the hotel industry in the local context.

Hotels in the hospitality industry suffer from problems such as employee job dissatisfaction, low performance, high absenteeism, high turnover, grievances, and many more that have a huge influence on the success of an organization. These are also some of the foremost indicators of an imbalanced work–life relationship. Hence, this study, which focused on factors affecting poor work–life balance in five-star hotels in Sri Lanka, will be advantageous to the entire hotel industry in improving the work–life balance of their workforce and thus reducing the common problems of high absenteeism, high turnover, low performance, dissatisfaction, grievances and so on. Moreover, from the employees' point of view, it benefits them to construct a healthy work–life relationship and have a balanced and greater life by way of supporting management to eliminate the factors that influence poor work–life balance. Employees will have good knowledge about how they should manage their work life and family life, and thus it will help to mitigate a number of social matters, such as family issues due to stress, depression, and anger, which have the potential to lead to divorces and suicides in the long run.

Improvement of work–life balance will improve employee job satisfaction and thereby lead employees to meet individual performance targets as well as organizational goals and objectives. It will form a pool of loyal employees, which, in turn, benefits the organization, because retaining dedicated and loyal employees is a strength.

## 2.3 WORK–LIFE BALANCE AND THE HOTEL INDUSTRY

In considering work–life balance in the hotel industry, Cooper (1998), suggests that the concept of presenteeism is extremely important. Cooper argues that *presenteeism* is an overwhelming need to put in more hours or appear to be working for very long hours. According to Cullen and McLaughlin (2006), the culture of almost all hotels promotes this phenomenon, and it ultimately results in poor work–life balance among employees. Dohrty (2004) examined work–life balance for women involved in the hotel industry and determined that the very long hours and lack of flexibility had a heavy negative influence on their work–life balance. She argues that the culture of long hour is counterproductive for the whole hotel industry and that it is one of the root causes of poor employee work–life balance (Deery et al., 2015).

As work–life balance has rapidly emerged as a popular topic for all industries, regardless of size and type, a number of solutions and strategies have been developed to address the situation. Hartel et al. (2007) argues that organizations can successfully implement various work–life balance initiatives that may help employees to have a better balance between their work and personal life and simultaneously gain improvements in wellbeing and provide many different organizational benefits as well. Organizations are now introducing a variety of family-friendly policies and principles that include flexible working hours, working from home, job sharing, part-time work, compressed work week, parental leaves, telecommunication, on-site child care facilities, family apartments etc.

Many employers also provide a range of benefits related to employee's health and wellbeing, including extended health insurance for employees and departments, personal days, and access to programmes or services to encourage fitness and physical and mental health of employees. Further, other practices may support children's education and health and employee participation in volunteer work, or facilitate phased retirement (Lazăr, Osoian, & Raţiu, 2010). However, although many organizations have taken steps to improve work–life balance, most of the time they are just written policies and ideas and fail in implementation. Because of that, the problem remains (Faisal, 2015).

## 2.4 JOB SATISFACTION AND THE TOURISM INDUSTRY

The tourism sector is highly labour intensive, with the performance of its human resources being a significant and determining factor in its sustainability. Therefore, this dependence on human resources for its proper functioning and growth demands both effective and efficient workforce practices in order for employees to perform at optimum levels, thereby enabling the sector to remain viable in a rapidly changing and fiercely competitive global environment. As organizations in the tourism sector have become increasingly aware that mismanagement of resources can lead to their demise, they have focused on cost minimization while simultaneously maintaining quality. Thus informed, these organizations are encouraged by the literature to consider employee performance as a means to gain competitive advantage.

Subsequently, researchers have investigated workplace issues that are likely to impact on employee attitudes in order to better understand, and therefore to develop strategies to improve, both individual and organizational performance. One of the issues includes the tacit acceptance that turnover behaviour is quite appropriate and an accepted element of life within the industry. The profound impact that turnover behaviour has on an organization's costs has increased the need for continual management awareness in order to mitigate any likely adverse effects to the enterprise.

Theorists have proven that low levels of job satisfaction, which is impacted by routinization, role conflict, and lack of promotional opportunity, can have harmful effects on the organization. In the same way, it can impact not only an employee's decision to quit an organization but also the organization's bottom line. Extrinsic job satisfaction is related to turnover intention, implying that managerial style and reward strategies present both challenges and opportunities to industry stakeholders. Hotel organizations must maintain constant surveillance regarding employee satisfaction towards the job by optimizing the manner in which employees are managed and the manner in which they are rewarded. The failure of management is likely to aggravate turnover and the subsequent costs to operations (Zopiatis, Constanti, & Theocharous, 2014).

## 2.5 IMPACT OF WORK–LIFE BALANCE ON JOB SATISFACTION OF HOTEL EMPLOYEES

The quality of the interpersonal interaction between customers and service employees is critical in satisfying customers, ultimately influencing the bottom line of the company. Positive attitudes and emotions in service employees during service encounters can create a favourable impression on customers. They are then more likely to purchase a product, do return business with the company and speak well of the company. Because of this, most companies in today's highly competitive business environment have begun to focus heavily on managing their employees' emotional behaviour, prescribing implicit and explicit display rules for the appropriate emotional expressions that their employees should use during customer encounters.

A major challenge of working in the hospitality industry is coping with work timing and shifts. This is particularly highlighted as the core business is more during holidays and festivals. For hospitality service professionals, these times are the busiest, and it becomes impossible for them to avail themselves of any leave or time off during this period. This naturally creates work–life imbalance and family conflict as the expectations of their families are ignored (Biswakarma et al., 2015).

In the hospitality industry, employees may have to fake their true emotions according to the situation in many customer-contact situations. Employees who repeatedly suppress their true emotions suffer a continuing discrepancy between their inner feelings and outward expressions. This emotional discrepancy leads to emotional discomfort and job stress that, in turn, causes job dissatisfaction. In hotels where face-to-face and voice-to-voice interactions between service providers and customers continually occur, front office employees are particularly vulnerable to job dissatisfaction due to the work stress. However, although current hotel human resources managers are aware of this situation, few hotel organizations effectively implement strategies to control work–life balance and prevent turnover (Lee & Ok, 2012). Thus, a deeper and clearer understanding of the factor(s) that is/are most influential on front office employees in hotels experiencing poor work–life balance is critical in attempting to create strategies for controlling work–life balance and its outcomes.

## 2.6 CONCEPTUAL FRAMEWORK

This study explored the relationship between work–life balance and job satisfaction of receptionists and concierge staff of hotels. Three dimensions of work–life balance were considered as independent variables to determine what factor(s) had an influence on the dependent variable of job satisfaction.

Greenhaus, Collins, and Shaw (2003) argue that there are three dimensions of work–life balance: (1) time balance, which is allocating equal time to work and family; (2) involvement balance, defined as mental involvement with work and family issues; and (3) satisfaction balance, which is equal satisfaction with family and work. Greenhaus et al. (2003) tested the relationship between these three work–life balance dimensions and quality of life and determined that individuals with high work–life balance do not necessarily experience a higher quality life than those with work–life conflict (Fayyazi & Aslani, 2015).

A number of studies (Fayyazi & Aslani, 2015; Saeed & Farooqi, 2014; Haar et al., 2014; Mukhtar, 2012) have examined the relationship between work–life balance and job satisfaction with or without involvement of different mediators, but the researcher did not come across any studies that tested the relationship between Greenhaus's work–life balance dimensions and job satisfaction. Because Greenhaus's study was also focused on examining the relationship between work–life balance and quality of life, in the current study, the researcher's objective was to find the relationship of Greenhaus's work–life balance dimensions with job satisfaction and to identify the factors that were most and least influential on employees' job satisfaction. Even though the independent variables are similar to Greenhaus's study, the dependent variable is dissimilar. Thus the researcher developed an original model for the current study:

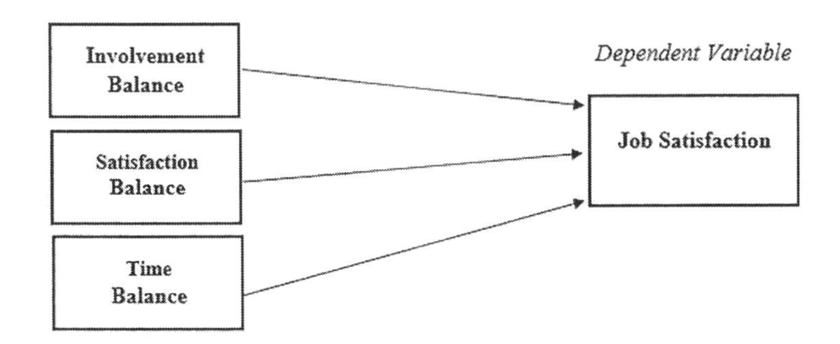


**FIGURE 2.1** Conceptual framework.

*Source*: Developed by the researcher based on literature.

### Hypothesis

Based on the conceptual framework the following hypotheses were developed by the researcher:

- **$H_{1A}$**: There is a relationship between involvement balance and total involvement in work and family roles in predicting employee job satisfaction.
- **$H_{1B}$**: There is a relationship between satisfaction balance and total satisfaction with work and family roles in predicting employee job satisfaction.
- **$H_{1C}$**: There is a relationship between time balance and total time devoted to work and family roles in predicting employee job satisfaction.

## 2.7 VARIABLES AND DIMENSIONS

### 2.7.1 Job Satisfaction

Despite its wide usage in scientific research, as well as in everyday life, there is no consensus on what job satisfaction represents. Different authors have different approaches towards defining job satisfaction. The researcher used Schleicher, Watt, and Greguras's (2004) definition of job satisfaction as the working definition for the current study: "Job Satisfaction can be defined as the degree to which individual's beliefs or thoughts are concerning various components of their jobs".

As cited in Bruck, Allen, and Spector (2002), Locke (1969) stated that job satisfaction is a positive emotional state coming from an individual's subjective experience with his or her job. It reflects the degree to which a person's wants, needs, or expectations are met at work. Furthermore, job satisfaction is defined as the extent to which the expectations that an individual hold for a job match what one actually receives from the job. Job satisfaction can be characterized as an attitude concerning the extent to which people like or dislike their jobs (Bruck et al., 2002).

Aziri (2011) cited Hoppock (1935), who defined job satisfaction as "any combination of psychological, physiological and environmental circumstances that cause a person truthfully to say I am satisfied with my job". According to this approach, although job satisfaction is under the influence of many external factors, it remains something internal that has to do with the way the employee feels; that is, job satisfaction presents a set of factors that cause a feeling of satisfaction.

Moreover, Aziri (2011) also cited Vroom's (1964, p. 15) definition of job satisfaction, which focuses on the role of the employee in the workplace. Vroom defined job satisfaction as "effective orientations on the part of individuals toward work roles which they are presently occupying".

One of the most often cited definitions on job satisfaction is the one given by Spector (1997), according to whom job satisfaction has to do with how people feel about their job and its various aspects. It has to do with the extent to which people like (satisfaction) or dislike (dissatisfaction) their job. That's why job satisfaction and job dissatisfaction can appear in any given work situation (Aziri, 2011). Later on, Spector (2000) simply defined job satisfaction as "the extent to which people like their jobs" (Lee & Ok, 2012).

In general, job satisfaction refers to an individual's positive emotional reactions to a particular job. It is an affective reaction to a job that results from a person's comparison of actual outcomes with those that are desired, anticipated, or deserved (Mukhtar, 2012). According to Robbins (2003, p. 72), job satisfaction refers "to an individual's general attitude toward his or her job". A person with a high level of job satisfaction holds a positive attitude about the job, while a person who is dissatisfied with his or her job holds negative attitudes about the job (Buitendach & Rothmann, 2009).

Greenberg and Baron (2008), for instance, view job satisfaction as a feeling that can produce a positive or negative effect towards one's roles and responsibilities at work, and they added that it is important to understand that there is no single way to satisfy all workers in the workplace (Diala & Nemani, 2011).

Schleicher et al. (2004, p. 167) in their research, "Re-examining the Job Satisfaction–Performance Relationship: The Complexity of Attitudes", stated that "job satisfaction is relevant to the consistency between the affective and cognitive components of one's attitude" and used two measurements to assess both these components. The Overall Job Satisfaction Scale (OJS) was used to assess the affective component, to understand how participants felt about their jobs, and the Minnesota Satisfaction Questionnaire (MSQ) was used to assess the cognitive component, which asked participants what they believed about various components of their jobs.

Because the present study is using MSQ as the job satisfaction measurement, the working definition for job satisfaction can be stated as "the degree to which individual's beliefs or thoughts are concerning various components of their jobs" (Schleicher et al., 2004).

### 2.7.2 Work–Life Balance

The researcher used Greenhaus et al.'s (2003) work–life balance dimensions as the working definition of the study. This definition of work–life balance is a combination of three dimensions: involvement balance, satisfaction balance, and time balance. These dimensions were used as the independent variables of the current study. Greenhaus et al. (2003) identified these three dimensions as major elements of work–life balance and defined each dimension individually:

Involvement Balance can be defined as having an equal level of psychological involvement in work and family roles.

Satisfaction Balance is defined as having an equal level of satisfaction with work and family roles.

Time Balance is when someone is having as equal amount of time devoted to work and family roles.

Work–life balance emphasizes proper prioritizing between work (career life and ambition) and lifestyle (pleasure, leisure, health, family and spiritual development). It heavily impacts employee performance and mainly decides by the dimensions of job content, psychological/personal factors, and family factors (Deery et al., 2015).

Work–life balance can be defined as an individual's ability to meet both their work and family commitments, as well as other non-work responsibilities and activities. Work–life balance is related to reduced stress and greater life satisfaction, with some indication that the relationship is strengthened over time.

Traditionally, work–life balance has been seen as an issue for individual employees, with organizational efforts at improving work–life balance focusing on programmes aimed to help employees better manage their home life (e.g. childcare or counselling). Proponents argue that work–life balance contributes to employee engagement (job satisfaction and organization commitment), which, in turn, contributes to higher productivity and lower organizational turnover. As cited in Parkes and Langford (2008), De Cieri et al. (2005, p. 92) argue that any organization aiming to increase competitive advantage must "develop the capability to attract, motivate and retain a highly skilled, flexible and adaptive workforce" by "an approach to HR and work–life balance strategies that cater for the diverse needs of the workforce".

Given the high level of interest in work–life balance among researchers, practitioners, and commentators, the researcher aimed to test whether there is an impact of work–life balance on employee job satisfaction and what factor(s) (dimensions) of work–life balance might moderate or mediate the relationship between work–life balance and employee job satisfaction. Furthermore, the researcher studied work–life balance dimensions introduced by Greenhaus et al. (2003) with job satisfaction. Greenhaus et al. (2003) identified three dimensions of work–life balance: (1) time balance, (2) involvement balance, and (3) satisfaction balance. Furthermore, work–life balance can be viewed as a continuum where at one end the imbalance is in favour of family and at the other end the imbalance is in favour of work. Work–life balance that refers to equal commitment and time allocation to work and personal life issues is considered to be in the middle of the continuum (Fayyazi & Aslani, 2015).

Greenhaus et al. (2003, p. 513) defined work–life balance as "The extent to which an individual is equally engaged in—and equally satisfied with—his or her work role and family role". Furthermore, Greenhaus et al. (2003) defined three components of work–life balance, and the researcher used those as working definitions for the present study.

**TABLE 2.1**
**Working Definitions**

| Variable | Definition |
|---|---|
| **Independent Variables** | |
| Involvement balance | An equal level of psychological involvement in work and family roles |
| Satisfaction balance | An equal level of satisfaction with work and family roles |
| Time balance | An equal amount of time devoted to work and family roles |
| **Dependent Variable** | |
| Job satisfaction | The degree to which individual's beliefs or thoughts are concerning various components of their jobs. |

*Sources*: Greenhaus et al. (2003), Schleicher et al. (2004).

### 2.7.3 Population and Sample

The population of the research consisted of receptionist and concierge staff in five-star hotels in the Colombo area. According to the Annual Statistical Report issued by the Sri Lanka Tourism Development Authority in 2016, the highest occupancy rate by region is in in the Colombo Region, which is a combination of Colombo City and Greater Colombo. Therefore, the researcher limited the population to within the Colombo Region.

As per the statistics included in the Annual Report under "Capacity and Nights in all Accommodation Establishments by Category", the most "Total Guest Nights", "Foreign Guest Nights", and "Local Guest Nights" have been occupied by five-star hotels. Therefore, it is evident that the busiest employees in the hotel trade in Sri Lanka are in five-star hotels in the Colombo Region, and thus were selected as the most suitable population for the study.

As mentioned on the Sri Lanka Tourism website, which is registered and approved under the Sri Lanka Tourism Promotion Bureau, five-star hotels offer "superior standard and an extensive range of first-class guest services". Moreover, it indicates that there are five five-star hotels in Colombo District: the Cinnamon Grand Hotel, the Cinnamon Lakeside Hotel, the Hilton Colombo, the Taj Samudra, and The Kingsbury (as at July, 2018).

The total number of receptionist staff (80) and concierge staff (30) of these five hotels was 110. Thus, the population was 110 ($N = 110$). A total of 110 receptionist and concierge staff were selected as the sample of research.

### 2.7.4 Data Collection Method

As this was a quantitative research study based on primary data, data collection was mainly done via a constructed questionnaire survey. Data were collected via a standard questionnaire based on the existing measurement scales in the literature that have already been tested and validated widely by previous research in the domain of work–life balance and job satisfaction. A structured, self-administered questionnaire was designed to collect data from employees in reception and concierge; the design of the questionnaire was guided by the objectives of the study. Work–life balance was assessed using the 24-item scale suggested by Thomas and Ganster (1995). The scale examines three dimensions of work–life balance: time balance, involvement balance (input balance), and satisfaction balance (output balance). Coefficient alpha for the original measurement scale is 0.92.

A job satisfaction measure known as the Minnesota Satisfaction Questionnaire short form was used to measure job satisfaction. It consists of 20 items on a 5-point Likert scale, ranging from "1 = very dissatisfied" to "5 = very satisfied", that asked participants what they believed about various components of their jobs. As noted by Schleicher et al. (2004), coefficient alpha for this scale is 0.88.

In this research, the first section of the questionnaire, questions 1 to 24, were included to determine the level of work–life balance of receptionist and concierge employees. Respondents were required to mention their level of agreement on a 5-point Likert scale that ranged from "strongly agree" to "strongly disagree".

Questions 25 to 44 on the questionnaire attempted to determine the job satisfaction of receptionist and concierge employees of hotels. It consisted of 20 items on a 5-point Likert scale, ranging from "1 = very dissatisfied" to "5 = very satisfied" that asked participants about their satisfaction with various components of their jobs.

The last section of the questionnaire included six demographic questions (questions 45 to 50). The respondents had to choose only one option from the given options.

A primary survey was carried out covering all five hotels. Three respondents were randomly selected from each hotel; thus, a total of 15 questionnaires were distributed. The researcher timed all the respondents while they were engaged in filling out the questionnaire, and per respondent it took between 10 to 15 minutes to complete it. Standard questionnaires were used for the current study and did not involve any language translations or change of words of the items of the original measurement scales. The questionnaire distributed was in simple English language, and the respondents of the primary survey declared that they did not experience difficulty in understanding the questions and completing the questionnaires. After completion, all distributed questionnaires were returned by all receptionist and concierge staff and were effectively used to test the reliability.

Validity and the adequacy of the data were ensured through exploratory factor analysis (EFA). From the total sample of 110, 80 responses were considered to check the validity of the study. The Kaiser-Meyer-Olkin (KMO) test and Bartlett's test were used to check the adequacy of the sample. The KMO measure of sampling adequacy should be above 0.7 (KMO > 0.7) and the significant ($p$-value) should be less than 0.05 for the sample to be considered as adequate.

The survey response rate was 55.2% for receptionist staff and 100% for concierge staff. Each questionnaire was scrutinized by the researcher to examine the response pattern and identify abnormalities in the completion of questionnaires.

The collected primary data were analyzed with the support of IBM SPSS (Statistical Package for Social Science) version 20. Frequency tables and descriptive statistics were used to elaborate on the sample composition and the individual behaviour towards variables.

A normality test was used to determine the distribution of the data. Normally distributed data are considered as good and decent to use in research. The normality of the distributed data was assessed by the Kolmogorov-Smirnov normality test using SPSS.

A linearity test can be used to determine whether the relationship between the independent variable and the dependent variable is linear. A linearity test is a requirement in correlation and linear regression analysis. The purpose is to check whether there is a linear relationship between the free variable and the dependent variable.

A parametric statistical procedure was used to identify the relationship between normally distributed variables. The Pearson product-moment correlation coefficient and Spearman correlation coefficient were used to understand the nature and the relationship between the dimensions of the independent variables and the dependent variable.

## 2.8 RESULTS AND DISCUSSION

### 2.8.1 Relationship between Involvement Balance and Job Satisfaction

The Pearson product-moment correlation coefficient was applied to measure the relationship between involvement balance and employee job satisfaction. Involvement balance ($M$ = 2.97, *Std. D* = 0.63) was found to be significantly correlated with employee job satisfaction of receptionists ($M$ = 3.58, *Std. D* = 0.57).

There was a moderate positive correlation between the two variables: $r = 0.397$, $n = 99$, $p = 0.000$. $H_{1B}$ was accepted. In other words, there is a moderate positive correlation between involvement balance and job satisfaction. That moderate correlation is statistically significant, as the significance value is less than 0.01.

### 2.8.2 Relationship between Satisfaction Balance and Job Satisfaction

The Pearson product-moment correlation coefficient was generated to identify the relationship between satisfaction balance and job satisfaction of all respondents. Satisfaction balance of employees ($M$ = 3.18, *Std. D* = 0.56) was found to be positively correlated with job satisfaction ($M$ = 3.58, *Std. D* = 0.57). There was a moderate positive correlation between the two variables: $r = 0.479$, $n = 99$, $p = 0.000$. $H_{1C}$ was accepted. In other words, there is a moderate positive correlation between satisfaction balance and job satisfaction. That moderate correlation is statistically significant, as the significance value is less than 0.01.

### 2.8.3 Relationship between Time Balance and Job Satisfaction

Hypothesis 1 ($H_{1A}$): There is a relationship between time balance and total time devoted to work and family roles in predicting employee job satisfaction.

**TABLE 2.2**
**Correlation between Involvement Balance and Job Satisfaction**

| Correlations | | JSMean | Involvement Mean |
|---|---|---|---|
| *JSMean* | Pearson Correlation | 1 | .397** |
| | Sig. (two-tailed) | | .000 |
| | *N* | 99 | 99 |
| *Involvement Mean* | Pearson Correlation | .397** | 1 |
| | Sig. (two-tailed) | .000 | |
| | *N* | 99 | 99 |

** Correlation is significant at the 0.01 level (two-tailed).

**TABLE 2.3**
**Correlation between Satisfaction Balance and Job Satisfaction**

| Correlations | | JSMean | Satisfaction Mean |
|---|---|---|---|
| *JSMean* | Pearson Correlation | 1 | .479** |
| | Sig. (two-tailed) | | .000 |
| | *N* | 99 | 99 |
| *Satisfaction Mean* | Pearson Correlation | .479** | 1 |
| | Sig. (two-tailed) | .000 | |
| | *N* | 99 | 99 |

** Correlation is significant at the 0.01 level (two-tailed).

**TABLE 2.4**
**Correlation between Time Balance and Job Satisfaction**

| Correlations | | | JSMean | Time Mean |
|---|---|---|---|---|
| *Spearman's rho* | *JSMean* | Correlation Coefficient | 1.000 | .467** |
| | | Sig. (two-tailed) | | .000 |
| | | *N* | 99 | 99 |
| | *Satisfaction Mean* | Pearson Correlation | .467** | 1.000 |
| | | Sig. (two-tailed) | .000 | |
| | | N | 99 | 99 |

** Correlation is significant at the 0.01 level (two-tailed).

**TABLE 2.5**
**Correlation between All Respondents' Job Satisfaction and Selected Variables**

| Hypothesis | Relationship | Correlation | M | SD | N | Sig |
|---|---|---|---|---|---|---|
| $H_{1A}$ | with involvement balance | 0.397 | 2.97 | 0.63 | 99 | 0.000* |
| $H_{1B}$ | with satisfaction balance | 0.479 | 3.18 | 0.56 | 99 | 0.000* |
| $H_{1C}$ | with time balance | 0.467 | 2.85 | 0.57 | 99 | 0.000* |

* Correlation is significant at the 0.01 level.

Based on the linearity test executed earlier, the Spearman correlation coefficient was performed to identify the relationship between time balance and employee job satisfaction. Time balance ($M$ = 2.85, *Std. D* = 0.57) was found to be significantly correlated with employee job satisfaction of receptionists ($M$ = 3.58, *Std. D* = 0.57).

There was a moderate positive correlation between the two variables: $r = 0.467$, $n = 99$, $p = 0.000$. $H_{1C}$ was accepted. In other words, there is a moderate positive correlation between satisfaction balance and job satisfaction. That moderate correlation is statistically significant, as the significance value is less than 0.01.

## 2.9 CONCLUSION

There is a positive relationship between involvement balance, satisfaction balance, and time balance and employee job satisfaction. Thus, as expected, all three variables had positive relationships with job satisfaction, and it was evident that there is a work–life balance problem among receptionists and concierges in all five-star hotels in the Colombo district. Consequently, the three hypotheses were accepted, proving that "there is a moderate relationship between involvement balance and job satisfaction", "there is a moderate relationship between satisfaction balance and job satisfaction", and "there is a moderate relationship between time balance and job satisfaction". Thus, in conclusion, the statistical analysis of the study proved that all three hypotheses were accepted. In concluding the present study, a number of recommendations are presented to help enhance employee work–life balance and thereby increase employee job satisfaction for better guest service.

## 2.10 RECOMMENDATIONS

According to data, most of the employees are struggling and unhappy with their current working schedules. The majority of the employees are not happy with the working shifts. They also emphasized that they have very limited time to spend with their family members because of the tied working shifts and schedules. Thus, the researcher suggests that hotel management should reconsider their current scheduling arrangements. Currently, the human resources departments in all selected hotels have recognized that work–life balance is one of most affected areas in the operational process and that their working shifts and schedules make the situation worse. Hence, a special programme, "job sharing", has been introduced in some hotels, by which employees are able to change their shifts in a flexible manner by discussing it with other staff members. However, the researcher came to find that this particular programme is not functioning well at all hotels and that some managers are not willing to execute the job-sharing concept within their hotels. Therefore, the researcher suggests that it is important to establish a job-sharing system at all the hotels and to allow employees to enjoy the benefits of such a programme.

Moreover, except those who have joined the organization as management trainees, employees at most of the hotels specialize in one department, and some will perform the same job role during their entire career. Oftentimes they lack good knowledge of other job roles in the department. Hence, they perform the same task again and again during the workday, and by the end of the day they are full of stress, which can ultimately lead to career disappointment as well. Thus, the researcher suggests that management train employees on all the functions of each department and develop a programme to rotate them among different units of the department after a particular time period. For example, receptionists must be given the chance to work in concierge, bell service, mail, and information and also as cashiers and night auditors. The workload is a little bit different from one job role to another in the department, and they will be able to experience high-pressure work shifts as well as some flexible working schedules under different job roles. At the same time, some hotels have introduced a programme called "work at home", where employees are allowed to do their work at home without coming to the hotel when they are facing an urgent situation. However, this programme could be enjoyed by only few departments because all the hotel functions cannot be completed from home. Especially because front office responsibilities cannot be performed from home, if the employees can be rotated among other departments, such as rotating the front office with the back office, all the employees will be able to enjoy such an arrangement. Such an arrangement could positively impact work and family roles and, in doing so, increase employee satisfaction, especially for employees who have stated that they are worried as to whether they should work less and spend more time with their children.

A front office is a group of employees that work together as a team to coordinate service for guests. In hotels or resorts, a concierge assists guests by performing various tasks such as making restaurant reservations, booking hotels, arranging for spa services, recommending night life hotspots, booking transportation (like taxi, limousines, airplanes, boats, etc.), coordinating porter service (luggage assistance request), procuring tickets to special events, and assisting with various travel arrangements and tours of local attractions. Concierges also assist with sending and receiving parcels. Thus, they are set to play an integral role in the hotel 24 hours a day. But in some hotels, only three to five people are employed in concierge, and every day they have to rush to complete their duties. Because the guest capacity in a five-star hotel is much higher, concierge employees complain that three people in concierge is not sufficient to perform these job duties. While collecting the data, the researcher learned that, apart from their regular duties, managers frequently ask them to perform duties that are not under their job role, and thus at the end of the day they are physically and emotionally drained. The data set actually proved their point because it was identified that concierge employees have more problems balancing their work and family roles than do receptionists. As a solution, hotels can reassign the roles played by the concierges only for their required job role. Concierges deployed for other services can be replaced with other contract employees. This will improve their involvement as well as satisfaction balance by reducing stress at the office.

For the questions regarding the family, the majority of the married employees were unhappy with the amount of family time they had and unhappy about taking care of their children and think that their job makes it difficult to be the kind of spouse or parent that they would like to be. Aligning with these outcomes, the researcher assumes that many of the employees' families do not have a good awareness or understanding about the role of an employee in the hotel industry, especially in a five-star hotel. Thus, the researcher suggests that management involve employees' family members in different activities such as get togethers, parties, dinners, or trips to make employees' spouses aware about the daily routine work roles and how important it is to for the employees to be available for guests at night. Thereby, the families could give their support to make their husband's or wife's job successful by understanding the working conditions and reducing family conflicts. Management should also develop facilities for families who have infants, because some employees stated that they are not comfortable with the arrangements for their children and that it involves lot of effort to make arrangements for their children. These employees are not able to set their mind at ease while at work and thus are not able to give their maximum output towards the organization. Establishing daycare facilities, especially for breastfeeding mothers, will foster loyalty of employees and their family members towards the organization and improve work–life balance.

## REFERENCES

Aziri, B. (2011). Job satisfaction: A literature review. *Management Research and Practice, 3*(4), 77–86. https://doi.org/

Bednarska, M. (2012). Features of the labor market in tourism as development barriers tourism economy. In M. Bednarska & G. Gołembski (Eds.), *Contemporary challenges for the tourism economy*, EU Scientific Journal in Poznań, No. 225 (pp. 47–60). Poznan: Ed. EU in Poznan.

Biswakarma, S. K., Sandilyan, P. R., & Mukherjee, M. (2015). Work life balance for hospitality employees – A comparative case study of two five star hotels in Kolkata. *ELK Asia Pacific Journal of Human Resource & Organisational Behaviour, 1*(2), 26–41, ISSN 2394-0409; DOI: 10.16962/EAPJHRMOB/issn.2394-0409.

Bruck, C. S., Allen, T. D., & Spector, P. E. (2002). The relation between work-family conflict and job satisfaction: A finer-grained analysis. *Journal of Vocational Behavior, 60*(3), 336–353. https://doi.org/10.1006/jvbe.2001.1836

Buitendach, J. H., & Rothmann, S. (2009). The validation of the Minnesota job satisfaction questionnaire in selected organisations in South Africa. *SA Journal of Human Resource Management, 7*(1), 1–8. https://doi.org/10.4102/sajhrm.v7i1.183, https://doi.org/10.1177/1096348007031003080l\r10.1177/1096348007299919

Cooper, C. L. (1998). The changing nature of work. *Community Work and Family, 1*(3), 313–317.

Cullen, J., & McLaughlin, A. (2006). What drives the persistence of presenteeism as a managerial value in hotels?: Observations noted during an Irish work–life balance research project. *International Journal of Hospitality Management, 25*, 510–516.

De Cieri, H., Holmes, B., Abbott, J., & Pettit, T. (2005). Achievements and challenges for work/life balance strategies in Australian organisations. *The International Journal of Human Resource Management, 16*(1), 90–103.

Deery, M., Jago, L., & Jago, L. (2015, November). Work-life balance in the tourism industry : Case study. In *Where the "bloody hell" are we?* (pp. 1–13). Queensland: Griffith University.

Diala, I., & Nemani, R. (2011). Job satisfaction: Key factors influencing information technology (IT) professionals in Washington, DC. *International Journal of Computing Technology Applications, 2*(4), 827–838.

Faisal, F. (2015). Relationship between work-life balance & job performance amo. https://doi.org/2013490002

Fayyazi, M., & Aslani, F. (2015). The impact of work-life balance on employees' job satisfaction and turnover intention: The moderating role of continuance commitment. *International Letters of Social and Humanistic Sciences, 51*, 33–41. https://doi.org/10.18052/www.scipress.com/ILSHS.51.33

Greenberg, R. U., & Baron, I. F. (2008). Pay enough or don't pay at all. *The Quarterly Journal of Economics, 115*, 791–810.

Greenhaus, J. H., Collins, K. M., & Shaw, J. D. (2003). The relation between work-family balance and quality of life. *Journal of Vocational Behavior, 63*(3), 510–531. https://doi.org/10.1016/S0001-8791(02)00042-8

Haar, J. M., Russo, M., Suñe, A., & Ollier-Malaterre, A. (2014). Outcomes of work – Life balance on job satisfaction, life satisfaction and mental health: A study across seven cultures. *Journal of Vocational Behavior, 85*(3), 361–373. https://doi.org/10.1016/j.jvb.2014.08.010

Hartel, C., Fujimoto, Y., Strybosch, V., & Fitzpatrick, K. (2007). *Human resource management: Transferring theory into innovative practice*. NSW, Australia: Pearson Education Australia.

Hoppock, R. (1935). *Job satisfaction*. New York: Harper and Brothers, p. 47.

Hughes, J., Bozionelos, N., & Hughes, J. (2007). Work-life balance as source of job dissatisfaction and withdrawal An exploratory study on the views of male. https://doi.org/10.1108/00483480710716768

Jayawardhana, A. A. K. K., Silva, S. De & Athauda, A. M. T. P. (2013). Business strategy, market orientation and sales growth in hotel industry of ancient cities in Sri Lanka. *Tropical Agricultural Research, 24*(3), 228–237.

Kidd, J., & Eller, C. (2012). Work Life Balance and Work-Family Conflict in the Hospitality Industry. *Relationship Between Work-Life Balance & Job Performance among Employee at Kulliyyah of Dentistry, IIUM Kuantan, 6*(1), 2010.

Lazăr, I., Osoian, C., & Raţiu, P. (2010). The role of work-life balance practices in order to improve organizational performance. *European Research Studies Journal, 13*(1), 201–213.

Lee, J. H. J., & Ok, C. (2012). Reducing burnout and enhancing job satisfaction: Critical role of hotel employees' emotional intelligence and emotional labor. *International Journal of Hospitality Management, 31*(4), 1101–1112. https://doi.org/10.1016/j.ijhm.2012.01.007

Locke, E. A. (1969). What is job satisfaction?, *Organizational Behavior and Human Performance, 4*(4), 309–336.

Mohanty, D. K., & Mohanty, S. (2014). An empirical study on the employee perception on work-life balance in hotel industry with special reference to Odisha. *Journal of Tourism and Hospitality Management, 2*(2), 65–81. https://doi.org/10.15640/jthm.v2n2a5

Mukhtar, F. (2012). *Work life balance and job satisfaction among faculty at Iowa State University* (Graduate Theses and Dissertations), Iowa State University.

Parkes, L. P., & Langford, P. H. (2008). Work – Life balance or work – Life alignment ? *Journal of Management and Organisation, 14*(3), 267–284. https://doi.org/10.1017/S1833367200003278

Peshave, M. A., & Gujarathi, R. (2014). An analysis of work-life balance (WLB) Situation of employees and its impact on employee productivity with special reference to the Indian hotel industry abstract. *Asian Journal Management, 5*(1), 69–74.

Ranasinghe, R., Lanka, S., Deyshappriya, R., & Lanka, S. (2011). Analyzing the significance of tourism on Sri Lankan economy. An Econometric Analysis, 1–19.

Robbins, S. P. (2003). *Organizational behaviour concepts, controversies, application* (8th ed). New Jersey: Prentice-Hall International.

Saeed, K., & Farooqi, Y. A. (2014). Examining the relationship between work life balance, job stress and job satisfaction among university teachers. *International Journal of Multidisciplinary Sciences and Engineering, 5*(6), 9–15. https://doi.org/10.6007/IJARPED/v3-i4/965

Schleicher, D. J., Watt, J. D., & Greguras, G. J. (2004). Reexamining the job satisfaction-performance relationship: The complexity of attitudes. *Journal of Applied Psychology, 89*(1), 165–177. https://doi.org/10.1037/0021-9010.89.1.165

Spector, P. E. (1997). *Job satisfaction: Application, assessment, causes, and consequences*. Thousand Oaks, CA: Sage Publication Inc.

Spector, P. E. (2000). *Industrial & organizational psychology*. New York: John Wiley & Sons.

Sri Lanka Tourism Development Authority. (2016). *General statistical report, 74*. Retrieved from www.sltda.lk/sites/default/files/annual-statical-report-2016.pdf

Thomas, L. T., & Ganster, D. C. (1995). Impact of family supportive work variables on work family conflict and strain: A control perspective. *Applied Psychology*, *80*(1), 6–15.

Vroom, V. H. (1964). *Work and motivation*. New York: Wiley

Zopiatis, A., Constanti, P., & Theocharous, A. L. (2014). Job involvement, commitment, satisfaction and turnover: Evidence from hotel employees in Cyprus. *Tourism Management*, *41*, 129–140. https://doi.org/10.1016/j.tourman.2013.09.013

# 3 The Evolution of Digital Platforms

*Aneesh Zutshi, Tahereh Nodehi, Antonio Grilo, and Belma Rizvanović*

## CONTENTS



## 3.1 INDUSTRIES DISRUPTED BY PLATFORMS

Most large businesses of the 20th century were built around provisioning of products and services to their clients (Alstyne, Van Parker, & Choudary, 2016). Multinational firms focused their attention on the innovation of better or more economic products, services, or commodities and reaching economies of scale. With the advent of the Internet and its ubiquitous usage throughout 1990s, early experiments for a completely new form of business model started to be explored (Still et al., 2017; Carter, Vonno, & Singh, 2017; Wang & Yin, 2017). In 1994, eBay was founded by the French-born Iranian-American computer programmer Pierre Omidyar as a personal hobby for art and antique collectors to buy and sell products online (Joint Research Centre Technical Reports, 2019). Around the same time Amazon was founded as an online bookstore that allowed other booksellers to sell through their site. What started as modest experiments redefined the way traditional businesses work. Both Amazon and eBay did not produce any product, did not buy inventory, and did not brand the products that they sold (Constantiou, Marton, & Tuunainen, 2017). They were merely platforms that connected buyers and sellers, and facilitated the interaction between them. They mimicked a form of brokerage service, but through the use of web technology, at a scale without any human intervention. Despite the burst of the dot com bubble in 2002, these companies continued to grow into multibillion-dollar enterprises and are thriving even today after more than two decades (Luca, 2017). Since then the platform business model has seen numerous variations and use cases and has managed to disrupt industry after industry (John Elliott & Tanguturi, 2018). As an example, Booking.com, founded in 1996 in the Netherlands, allows hotels around the world to sell rooms online. Before the popularity of online hotel booking sites, travellers had little information about the quality of experience to expect, especially when visiting unfamiliar cities (Gyódi, 2017). The famous names in the hotel industry leveraged on this to build international brands, such as the Hilton for premium customers or the Ibis for budget travellers. The introduction of the platform business in the hotel industry led to abundant choice for consumers and the possibility for smaller hotels to provide their offers to travellers globally. Customers could not only see pictures of their hotels before booking, but could also read reviews and ratings from other clients before making their choice. Quality service could be valued and valorized in ways that was not possible before. Booking.com disrupted the business of traditional travel agencies while helping both travellers as well as hotel owners.

More than a decade later came another major disruption in the accommodation and travel industry, with the emergence of Airbnb.com in 2008 (Gyódi, 2017). It enabled homeowners to rent rooms, apartments, or houses for short-term stays, thus competing with the hotel industry. This unlocked rental spaces in cities, often significantly, contributing to the growth of the tourism industry in many cities. It has also led to increases in rental and real estate prices and has affected the construction industry.

Similar disruptive trends have been seen in multiple industries, where startups with platform business models have disrupted significant sectors of the economy and brought about shifts in user behaviour (Parker, Alstyne, & Choudary, 2016). The emergence of social media is a prime example of a platform where social interaction and content is generated by the user, while platforms like Facebook and Instagram provide a user-friendly interface for social communication while monetizing on the ad revenue.

Platform businesses have evolved utilizing advances in digital technologies, such as the advent of smartphones and mobile devices (Parker, Alstyne, & Jiang, 2017). Both, iOS and Android, the two major smartphone OS (owned by Apple and Google respectively), utilize a platform model where third-party developers are able to create mobile apps that can be sold through the app stores and Apple or Google receive a share of revenue of every application sold.

With the advent of new location-based services, GPS, and mobile devices, startups like Uber and Taxify have become multibillion-dollar enterprises and have disrupted the entire taxi industry worldwide. Similarly, Waze, an Israeli startup, now owned by Google, uses user location-based data to provide a free GPS service that is enhanced through user-provided data on real-time traffic conditions and disruptions.

Figure 3.1 provides a set of examples of digital platforms across multiple industry segments. These include travel, agriculture, media, gaming, and education, amongst others. As new technologies continue to emerge, newer platform models continue to evolve, leading to newer forms of disruption and newer business opportunities. The integration of the internet of things (IoT), Big Data, biometrics, augmented and virtual realities, and 3D printing is opening vistas for newer forms of platform models.

Digital platforms are transforming market competition in all industries around the world, and platform-based companies are gaining market share rapidly. Many traditional companies and financial institutions are trying to embrace the platform model in order to stay ahead of the game (Pollari, 2018). For instance, Germany's Fidor Bank, recently acquired by French banking group BPCE, solicits the participation of community users and partners to enable both traditional lending and lending via peer-to-peer capabilities (Morvan, Hintermann, & Vazirani, 2016).

**FIGURE 3.1** Examples of digital platforms across industries.

With the emergence of digital platforms, a new form of gig economy has emerged where talent is able to provide services remotely on platforms like Upwork or Freelancer. New economic models are emerging around these forms of remote working, which is being used as an alternative form of employment by digital nomads and digital workers.

## 3.2 CHARACTERIZATION OF A DIGITAL PLATFORM

A digital platform allows for the interaction and exchange of information, product, or finance amongst its users, in order to provide value to its users. Platforms that have only one type of user interacting with each other are known as single-sided platforms; examples include Facebook and YouTube. However, a large number of platforms connect two types of users, such as buyers and sellers in the case of e-marketplaces, or producers and consumers in the case of electricity marketplaces.

One of the defining characteristics of digital platforms is the existence of *network effects* (Economides & Tåg, 2012; Alstyne & Parker, 2017). The larger the number of producers in the platform, the more attractive the platform becomes for consumers,

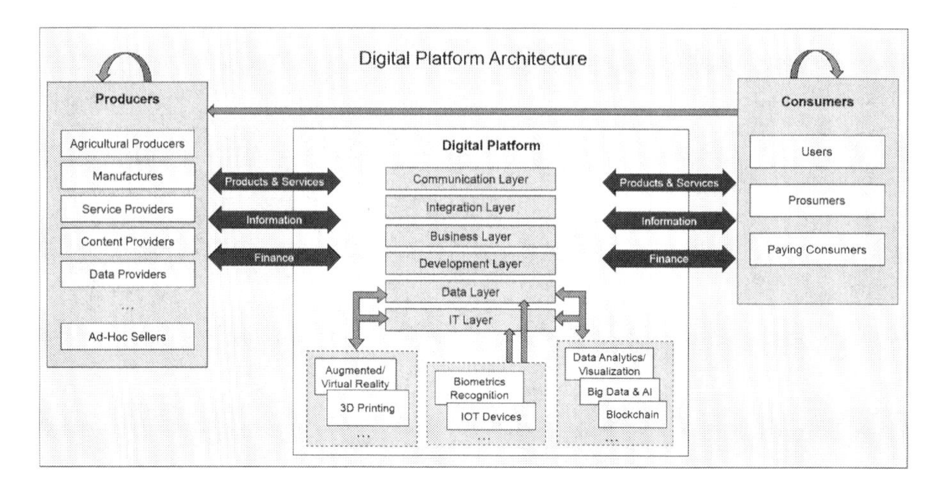


**FIGURE 3.2** Conceptual representation of a digital platform.

whereas the larger the number of consumers, the more attractive the platform is for producers (Farrell & Klemperer, 2007). Thus, as a successful platform gains a critical mass, the network effects drive more and more users to the platform, making it grow exponentially (Ismail et al., 2014). This is one of the key reasons for several businesses to explore the platform business model because of the tremendous opportunities in reaching economies of scale (Currier, 2018).

In Figure 3.2 we present a conceptual representation of a digital platform architecture. On one hand, we have producers of products, services, or finance which include manufacturers, service providers, sellers, or data providers. On the other hand, we have consumers that could either be free users or paying consumers (prosumers) who play an active role in the creation of the content/product that they consume. In a single-sided platform, the same user at different times can play the role of a producer as well as that of a user, for example, such as on YouTube or Facebook.

Digital platforms are getting more sophisticated as they integrate newer business models and interface with newer forms of technologies. A typical digital platform architecture includes the following components:

1. **Communication layer:** This involves the various interfaces and process workflows as producers and consumers engage with the platform. For instance, the users of Waze GPS are usually drivers who are providing inputs while driving. Thus, the interface must be minimalistic in ways such that the driver can provide inputs with minimum distraction. A carefully designed communication layer is vital for ensuring quick adoption. This layer includes the design of a user-centric user experience.
2. **Integration layer:** Digital platforms today do not function as standalone pieces of software. They interface with various other units such as data processing units, external analytical systems, data sources, and advertisement servers. This involves integration with external APIs and providing

inputs to other external service providers. For example, travel websites use APIs to integrate with flight booking databases such as Amadeus. These integrations are vital for the smooth functioning of a platform and seamless integration with external sources. At the design stage, it is vital to carefully define the type of integration required depending in the platform use cases.

3. **Business layer:** This layer is where the business models are defined. Monetizing a digital platform is a careful balancing act. Many platforms aim to grow exponentially, making full use of the network effects that they leverage to stay ahead of the competition. If the startups are well funded, immediate revenue maximization may not be as important as maximum growth in users. Thus, the business objectives must be carefully determined before a business model is defined. The next question of critical importance is whom to charge. Many platforms scale so fast that advertising revenue is sufficient for long-term sustainability. However, this is not the case with many mid-sized platforms. Key decisions may include whether to charge on a per transaction basis or a subscription basis. If a freemium model is proposed, the definition of the free component is a key issue that needs to be addressed by the business layer.
4. **Development layer:** Many platforms allow for third parties to develop applications, products, or services. These developer tools are vital components of platforms. For platforms like Facebook, Android, and iOS, the developer layer is fundamental for the growth of the applications and hence the value of the platform.
5. **Data layer:** This layer involves multitiered data discovery environments with applied intelligence to extract business value and meaning. The data services should also include advanced analytics, visualization, and AI/machine learning to allow data scientists to create new algorithms and models to deliver entirely new insights.
6. **IT layer:** This layer defines the architecture for storage of data, database design, and integration with cloud infrastructure. The advent of the blockchain technology can be an essential interface for secure data storage and validation, as well as secure financial transactions for platforms of the future. This layer forms the physical or virtual interface with the communication or integration layers which can include the input and output with other interfaces and newer technologies such as biometric-based identification devices, IoT data or output interfaces such as augmented or virtual reality.

## 3.3 KEY STRATEGIES FOR IMPLEMENTING A PLATFORM BUSINESS MODEL

Based on the huge success of several platform-based businesses, numerous startups as well as several large businesses have tried to rush into developing a platform-based business model. Around 15% of Fortune 100 companies have already built some sort of platform-based business model (Morvan et al., 2016). However, the very network effects that benefit large established platforms make the market entry of newer platforms even more difficult. Businesses must evaluate whether a platform model is feasible for their business. It is not always prudent to rush into being a platform operator, especially in a crowded space. Plenty of businesses are finding success in becoming a digital partner to other platforms, such as app developers, complementors, or affiliate providers, and reaping rewards from the platform economy without actually being platform owners. Producers and suppliers can leverage the power of not one but multiple platforms simultaneously to sell their products or services to newer business or end user customers.

The following are some of the key success strategies for new platform initiators, based on the literature reviewed:

1. **Focus on attaining critical mass:** Success in a platform business is a question of attaining economies of scale. Hence, the focus must be on rapid customer adoption and increase in daily users rather than on revenue generation. Platform businesses need deep pockets to run, and usage costs often act as barriers to reaching critical mass.
2. **Help producers/sellers provide value to buyers:** Although in a platform business the actual transaction or exchange is done by buyers and sellers, it is vital for the platform to actively support the sellers to provide a greater value to buyers. For example, Amazon does this by handling logistics and warehousing for sellers through the Amazon Fulfilled service, thus providing the buyers with a much improved and lower-cost logistics solution, thus facilitating the buying process.
3. **Personalized services:** Tools such as Big Data, artificial intelligence, and data analytics can provide a wealth of information about individual users that can offer a personalized experience to users. The ability to understand customer intent and translate that into meaningful outcomes can be a big advantage for the platform. Recommendation systems and automated workflows play an integral part in creating this user experience.
4. **Engagement through dynamic pricing:** Pricing strategies can differentiate platforms by presenting opportunities for greater flexibility and reward. A freemium approach means users have easy, generally free, access to a platform before deciding whether they want to be buyers. Alternatively, pay-as-you-go pricing can be combined with fixed subscription fees. Surge pricing is increasingly used to manage peak demand, in contrast with discount pricing in periods of low demand. For example, Airbnb has rolled out a smart pricing system for all hosts on its platform that adjusts room prices based on changes in demand in real time. Uber uses surge pricing to attract new drivers when there is a shortage.
5. **Developing trust:** Trust is at the heart of a building a vibrant platform. This has several components. The primary element is a secure authentication of users. User authentication can ensure that users that do not respect community rules or behave offensively can be blocked. Cyber security and prevention of the platform from being prone to hacking or malicious attacks is another key aspect. The third aspect is fraud prevention and compensation. Both Amazon and eBay compensate the buyers in case the product is

not delivered or is not as per seller description. With the advent of financial-based platforms and crypto currencies, security and trust issues assume even greater relevance.

## 3.4 BARRIERS TO THE ADOPTION OF A DIGITAL PLATFORM

Although digital platforms have disrupted numerous industries, their disruption potential varies across industries.

Figure 3.3 identifies the key characteristics for industry sectors with a high disruptive potential:

1. **High information intensity:** Because platforms allow for seamless exchange of information, thereby optimizing competition, it is highly relevant in all industries which are rich in information density.
2. **High gatekeeper costs:** Platforms open up sectors where a gatekeeper limits the supply of products. This is especially true if either the gatekeeper costs are too high or the gatekeeper is inefficient. In case of traditional taxis, the municipalities limited the number of licenses, creating an artificial scarcity. Amazon Kindle removed the publisher-selection procedure, thus allowing anyone to publish, and customer reviews and ratings to decide on the quality. Today, the academic publishing industry is at a crossroads where the peer-review process has become slow and inefficient, and thus a potential exists for a platform-based business model that can ensure fair-quality metrics.
3. **High fragmentation:** Industries with a large number of small fragmented players are ideal use cases for platform-based business models. The real

Platform Disruption Potential by Industry

| | Services | Education | Finance | Health | Energy | E-Govt | Best Scenario For Platform |
|---|---|---|---|---|---|---|---|
| Information Intensity | High | High | High | High | High | High | High |
| Gatekeeper Cost | High | High | High | High | High | High | High |
| Fragmentation | High | High | High | High | Medium | Low | High |
| Information Asymmetry | High | High | High | High | Medium | Medium | High |
| Freedom from Regulation | Medium | Medium | Low | Low | Low | Low | High |
| Failure Costs | Medium | Medium | High | High | High | High | Low |
| Asset Intensity | Low | Medium | Medium | High | High | Medium | Low |
| Disruption Potential | High | High | Medium | Medium | Low | Low | |

**FIGURE 3.3** Platform disruption potential.

*Source*: Adapted from Parker et al. (2016).

estate sector is one such example where a large number of small players are involved in the sale and rent of homes. This was one of the sectors to be first disrupted with the emergence of real estate platforms like Olx, and platforms like Uniplaces catering to niche segments like students. Real estate deals do require a physical support in activities such as negotiation and preparation of contracts. Hence, real estate dealers like Remax are still relevant and operate globally through franchise models, but have developed their own digital platforms to manage buyers and sellers with an additional physical interface.

One of the reasons for the low disruption potentials of digital platform in the e-governance space is because, unlike other sectors, here the services are provided by a single public institution or very few regulated organizations. Also, the sector is heavily influenced by regulations, which makes introduction of new platform businesses difficult. As more and more sectors of the government become deregulated or decentralized, new forms of platform innovation can come to fill this gap. Public healthcare is one such space where some governments are allowing patients more choice in terms of selection of healthcare services and claim reimbursements through voucher (Itala & Tohonen, 2017). The moment such reforms are initiated, digital platforms assume relevance.

4. **High information asymmetry:** Sectors where information is highly concentrated or not easily available are prime sectors for disruption. By connecting to service providers in real time, information related to service availability can mean quick capitalization on business potential. Uber relays information related to vacant drivers, connecting them with available customers in real time, thus bridging the information gap.
5. **High freedom from regulation:** Highly regulated industries are often accompanied by significant barriers to market entry, thus making platforms more difficult to operate. An example is online pharmacies, which cannot sell prescription drugs online without doctor's prescriptions. This makes online pharmacies a difficult business to be in. Finance, health, and energy sectors are industries with significant regulations compared to other industries, which explains why platforms have not disrupted these industries in the scale and speed with which disruptive innovation was seen in the entertainment and tourism industries.
6. **Low failure costs:** Industries such as health care, where failures can mean life or death, do not leave much scope for experimentation. Thus, clients are less prone to try other options even if they are available at lower costs. The same is true for large B2B contracts in sectors such as energy, where downtime and failures can be costly and companies are careful before changing business partners.
7. **Low asset intensity:** Asset-intensive industries have bigger risks and bigger vendor lock in, and hence have a less dynamic business environment. Implementation of platforms where business relationships are dynamic and temporary is easier. This is not the case in large infrastructure projects or the healthcare industry.

In one form or another, digital platforms are creeping into almost every industry segment, from logistics to health care to energy. The future of business is built around more choice, more dynamism, and more transparency, something that platforms promise. Even sectors where there exist certain barriers, as detailed in this section, regulation reforms, forms of interoperability, and transparency are creating conditions for newer forms of digital platforms to evolve.

## 3.5 THE FUTURE OF DIGITAL PLATFORMS

Platform-based business models have significantly improved user experience, choice, competition, and have innovated business models that create new jobs, new development ecosystems, and efficient business processes. The fact that the app ecosystems have led to hundreds of thousands of startups around the world is a testimony to the economic potentials of platforms as engines of economic growth.

There are also some concerns with the rise of digital platforms, which in effect are becoming large monopolies dictating economic rules of engagement. For instance, with the rise of Uber, municipalities no longer act as gatekeepers, deciding on taxi fares. This power now rests with Uber, which is often criticized for certain business practices such as excessive surge pricing or taking as high as 25% of the taxi fare as a commission. With great powers comes greater responsibility. As large platforms assume social functions, they still remain private companies with a large venture-backed private investment whose ultimate purpose is eventual profit maximization. Practices such as manipulation of consumers and markets, not maintaining strict privacy of user data, and predatory or unfair pricing are issues that have emerged from time to time.

As platforms grow in scale and influence, they could benefit from governance reforms where all stakeholders have a say in policy decisions and building fair platform regulations. Some experience from Open Source communities and decentralized ecosystems such as those being experimented over blockchain technology could be precursors to a new form of platform governance that could bring fairness and decentralized democracy as a form of sustainable platform growth. The platforms of the future could also explore forms of decentralized ownership that would make platform sustainability more important than mere profit motive.

## REFERENCES


Carter, P., Van Vonno, J., & Singh, S. (2017). *Introducing the platform enterprise digital enabling the real-time platform business*. Retrieved from https://www.sap.com/docs/download/cmp/2017/05/digital-platform-spotlight.pdf

Constantiou, I. D., Marton, A., & Tuunainen, V. (2017). Four models of sharing economy platforms. *MIS Quarterly Executive Journal, 16*.

Currier, J. (2018). The network effects manual: 13 different network effects (and counting). *NFX Venture Firm*. Retrieved from www.nfx.com/post/network-effects-manual

Economides, N., & Tåg, J. (2012). Network neutrality on the internet: A two-sided market analysis. *Information Economics and Policy, 24*, 91–104.

Farrell, J., & Klemperer, P. (2007). Coordination and lock-in: Competition with switching costs and network effects. *Handbook of Industrial Organization, 3*.

Gyódi, K. (2017). Airbnb and booking.com: Sharing economy competing against traditional firms? *DELab UW, 3*(3). Retrieved from delab.uw.edu.pl.

Ismail, S. et al. (2014). *Exponential organizations, Why new organizations are ten times better, faster, and cheaper than yours (and what to do about it).* ExO Partners LLC. Retrieved from https://www.amazon.com/Exponential-Organizations-organizations-better-cheaper-ebook/dp/B00OO8ZGC6

Itala, T., & Tohonen, H. (2017). *Difficult business models of digital business platforms for health data: A framework for evaluation of the ecosystem viability.* 19th IEEE Conference on Business Informatics, CBI 2017. pp. 63–69.

John Elliott, N. C., & Tanguturi, P. (2018). *Digital platforms will define the winners and losers in the new economy.* Accenture digital. Retrieved from https://www.amazon.com/Exponential-Organizations-organizations-better-cheaper-ebook/dp/B00OO8ZGC6

Joint Research Centre Technical Reports. (2019). *Digital platform innovation in European SMEs: An analysis of SME instrument business proposals and case studies.* Retrieved from https://publications.europa.eu/en/publication-detail/-/publication/785b88e3-408a-11e9-8d04-01aa75ed71a1/language-en/format-PDF/source-94966226

Luca, M. (2017). Designing online marketplaces: Trust and reputation mechanisms. *Innovation Policy and the Economy, 17,* 77–93.

Morvan, L., Hintermann, F., & Vazirani, M. (2016). *Five ways to win with digital platform* (pp. 1–33). Accenture. Retrieved from www.accenture.com/us-en/_acnmedia/PDF-29/Accenture-Five-Ways-To-Win-With-Digital-Platforms-Full-Report.PDF.

Parker, G., Van Alstyne, M. W., & Jiang, X. (2017). Platform ecosystems: How developers invert the firm. *MIS Quarterly, Special Issue: IT and Innovation*, 41, 255–266.

Parker, G. G., Van Alstyne, M. W., & Choudary, S. P. (2016). *Platform revolution, how networked markets are transforming the economy – And how to make them work for you.* Norton & Company. Retrieved from https://www.amazon.com/Platform-Revolution-Networked-Markets-Transforming/dp/0393249131

Pollari, I. (2018). *The rise of digital platforms in financial services.* KPMG International Cooperative. Retrieved from https://assets.kpmg/content/dam/kpmg/xx/pdf/2018/02/kpmg-rise-of-digital-platforms.pdf.

Still, K. et al. (2017). *Business model innovation of startups developing multisided digital platforms.* In Proceedings – 2017 IEEE 19th Conference on Business Informatics. pp. 70–75.

Van Alstyne, M., & Parker, G. (2017). Platform business: From resources to relationships. *GfK Marketing Intelligence Review*, 9, 24–30.

Van Alstyne, M., Parker, G., & Choudary, S. P. (2016). Pipelines, platforms and the new rules of strategy. *Harvard Business Review Magazine*, 94, 54. Retrieved from https://hbr.org/2016/04/pipelines-platforms-and-the-new-rules-of-strategy

Wang, X., & Yin, Z. (2017). *Internet platform business model renovate catering industries.* 4th International Conference on Industrial Economics System and Industrial Security Engineering (IEIS).

# 4 Impact of Exchange Rate Movements and World Output on Indian Exports

*Anisul M. Islam and Sudhir Rana*

## CONTENTS



## 4.1 INTRODUCTION

In relatively open economies such as India, foreign exchange rate policies and trends in the world economy are among the major influences in determining a nation's trade and macroeconomic performance. It perhaps stands true for the fact that they affect the investment and other economic decisions of the world. Regardless of the geographical boundaries between nations, the movement of goods and services, labor, capital, and technology throughout the world affects the economies of different countries. However, inter-country and inter-regional trade transactions normally require conversion of a currency with another currency (Genc and Artar, 2014) as reflected in the exchange rate. It is to be noted that governments in many countries tend to manage foreign exchange policies to influence different economic variables in order to generate favorable economic outcomes, such as in increasing exports, foreign investment, and economic growth and reducing trade deficits.

In order to maintain competitiveness, an exporter has to maintain its prices in accordance with the prices prevailing in the international market. As a result, while exchanging their revenues with an appreciating domestic currency, an exporter may suffer a decline in exports along with revenue losses as lesser amounts of domestic currency would be received by them in exchange for the export earnings denominated in the foreign currency. On the contrary, a depreciating currency is likely to have a positive impact on export sectors by promoting exports and thereby improve the trade balance (Abeysinghe and Yeok, 1998).

Given the importance of exchange rate movements on the economic performance of any nation, the purpose of this chapter is to investigate the long-run relationship between the trade-weighted real effective rate exchange (REER) and aggregate exports for India after controlling for changes in world income. The current chapter is amongst only a few studies that have endeavored towards an understanding of the long-run relationship between the REER and aggregate exports for India. Additionally, it is a chance for the authors to apply theoretical knowledge to a practical situation by utilizing the maximum likelihood–based multivariate co-integration analysis as described later. Further, diagnostic tests were done to ensure that the results are stable and reliable.

The plan of the chapter is as follows: The next section discusses some background information followed by a brief literature review. Next, the model is presented followed by the empirical methodology used in the chapter along with data and variables. Then empirical results are presented followed by concluding remarks.

## 4.2 BACKGROUND

Due to the emerging role of high technology in exports, increasing global customer supplier relationships and an increasing role of multinational corporation's network over the past several decades (Rana and Sharma, 2015), emerging global markets have gradually become methodologically important trading hubs as these countries have witnessed not just a rapid expansion in international trade but also a growing prominence in the global trade canvas. However, there are some recurrent concerns pertaining to the impact of exchange rate movements on trade and on a country's trade activities. For instance, the collapse of the Bretton Woods system in the early 1970s triggered a heavy wave of debates on whether exchange rate variability is a deterrent to global trade. Discussions on exchange rate effects on trade were rekindled after the 1997 Asian financial crisis and more recently the 2008 global financial crisis, as the latter had a heavy impact on global trade.

To look at the bigger picture, one must begin looking from the bottom of the pyramid such that the overall trade activity of a country is the result of the aggregated trade outcome arising out of decisions made by many individual firms at the micro level. Hence, in order to understand the effects of exchange rate changes on trade or trade balance, it is critical to analyze how the exchange rate and its movements affect the decisions of a wide range of individual firms. Such an analysis provides insights into heterogeneous responses across firms towards exchange rate movements and the related policy implications of the central bank's effort in managing and stabilizing foreign exchange variations.

In the Indian context, the Indian economy is an interesting case study to explore the issue of the impact of exchange rate fluctuations on exports. India, among many other economies, stood as one of the least open economies of the world during the 1960s and 1970s. Indeed, before the 1990s, India's exchange rate was more or less fixed. The year 1991 marked the beginning of an extensive regime shift. India launched its policy reform agenda and implemented a host of liberalization reforms, primarily targeting the foreign exchange market and its tradable sectors. By 1992–93, India shifted to a more market-oriented exchange rate system through devaluations and deregulations. Since then the exchange rate has mostly been under a managed floating regime with the Reserve Bank of India (RBI) intervening from time to time to stabilize the nominal exchange rate.

The goal of the policy makers, including the RBI, had been to keep the REER stable, but the REER in India in recent years has been appreciating slowly. Some anecdotal evidence suggests that until 1993–94 the relationship between REER and total exports had been exactly what the textbook prescribes—that is, exchange rate appreciation having a negative effect on exports—but in some periods, the relationship seems to have been reversed. In the conventional trade literature, exchange rate appreciation of the home currency tends to cause losses to the exporting sectors by degrading the competitiveness of the concerned industries in an international market. This is because when appreciation occurs it becomes costlier for foreign buyers to purchase the products from India as the foreign price of the domestic product becomes costlier with the appreciation of the domestic currency. The opposite is expected to happen when the home currency depreciates.

## 4.3 BRIEF REVIEW OF LITERATURE

In the realm of many multinational's competing globally, Indian economy and firms have emerged to have become more international than ever before (Rana, Saikia and Barai, 2018). Against the background just described, it would be interesting to study how the movements in the exchange rate have affected Indian firms' exporting decisions and to investigate whether the data show any indication of a weakening of the link between REER and exports for India. The trade effect of exchange rate movements has been an intensely debated issue since the breakdown of the Bretton-Woods system. There exists a large body of literature on examining the impact of exchange rate depreciation (or appreciation) on trade flows. These studies were conducted for different countries with different methodologies and data periods, and hence, not surprisingly, have different results. As such, existing empirical studies do not offer any firm conclusion on the effect of exchange rate on international trade and other flows (Côté, 1994; Cheung, 2005). Some studies provide results indicating a negative impact of currency appreciation on exports, whereas others provide the opposite (positive) impacts. For example, Cushman (1986, 1988), Akhtar and Hilton (1984), Kenen and Rodrik (1986), Thursby and Thursby (1987), Brada and Mendez (1988) Bahmani-Oskooee (1996), and Hongwei and Zhu (2001) found a negative impact of an appreciating exchange rate on trade, whereas Hooper and Kohlhagen (1978), Gotur (1985), Baily et al. (1986, 1987), Koray and Lastrapes (1989), McKenzie (1998), and Lee (1999) found no significant influence.

On the one hand, Veeramani (2008) found a positive relationship in that the Indian exports grew rapidly since 2000 despite the exchange rate appreciation. He nevertheless cautioned that this need not imply that the latter had no adverse impact on the former; that is, that the actual growth of exports could have been larger had the REER not appreciated. Ghosh (1990) has looked at the role of real exchange rates and relevant price elasticity of supply and demand factors and found a negative effect for the years 1973–74 to 1986–87, though not significant statistically. Additionally, a second model by the same author with a slightly varied specification resulted in a positive association between real exchange rate and export growth, although it too was statistically insignificant. Overall, the study concludes that the real exchange rate has played a marginal role in explaining the growth of exports. The study also suggests many non-price factors such as demand and supply variables which can help determine export performance. In his work, Nayyar (1976, 1988) too emphasized upon the importance of factors other than exchange rate such as world demand in the determining exports. On the contrary, Wadhva (1988), Virmani (1991), Krishnamurthy and Pandit (1995) found empirical evidence on the significant role of price on the export performance.

Following a non-structured but comprehensive model of India's exports during 1963–94, Srinivasan (1998) focused on negative elasticity of export performance with regard to the real exchange rate. This finding was also consistent with that of Joshi and Little (1994). Another study by Sinha and Roy (2007) indicated that India's export performance during the post-reform period was often led (caused) by the movements in the exchange rates. Additionally, they found that demand factors played a major role in explaining India's disaggregateed export performance during 1960–99.

In a different study, Dholakia and Saradhi (2000) found that export quantities respond to exchange rate depreciation. They reported that the export function has become more elastic after the reform periods. Mallik (2005) presented an overview of growth performance of India's exports since 1950–51, highlighting the performance of the post-1990 period. He focused on a number of non-policy factors such as deceleration in output growth and sluggish global demand. He reported that the Asian crisis and various restrictive trade practices adopted by the industrialized countries were responsible for the sluggish export performance of 1990s. He points out that Indian exports are still primarily supply-side driven, although demand-side factors like relative prices (including exchange rate movements) and world income are becoming increasingly important. He also argued that the strategic trade policy has a role in providing the enabling conditions for the export expansion.

Gervais, Schembri, and Suchanek (2016) examined the role of real exchange rate adjustment towards achieving a sustainable current account position. Raissi and Tulin (2018) conducted an empirical study on estimating the short-term and long-run price and income elasticity of Indian exports. Aydýn, Cýplac, and Yücel (2004) argued for the fact that real exchange rates play a significant role in determining imports and trade deficits but not exports. Sahinbeyoglu and Ulasan (1999) argued that the traditional export equations are insufficient in forecasting and that variables such as uncertainty indicators and investment have crucial roles in explaining exports.

Further, an inward-looking policy with a lack of strong domestic competition and the existence of high domestic demand results in a decrease in exports (Bhagwati and Srinivasan, 1975; Wolf, 1978). Some other studies have focused on the link between the exchange rate and exports via direct or indirect channels based on the fact that some exports rely on the intensity of imports, such as imported raw materials and other imported intermediate inputs. In such a situation, exports of commodities may depend upon the volume of imports as well as import prices, as import costs directly affect the export prices. Some empirical studies have found strong empirical evidence on the role of import prices on the export performance (Wadhva, 1988; Virmani, 1991; Krishnamurthy and Pandit, 1995). Further, a few studies, including Veeramani (2008) and Srinivasan and Wallack (2003) in the Indian context, have looked into the impact of exchange rate changes on overall exports. In a more recent study, Islam (2018) found strong comparative advantage of India's trade with Bangladesh resulting in a large trade surplus for India.

At a micro level, Rajaraman (1991) conducted an econometric exercise by studying the impact of movements in the real external value of the Indian rupee vis-à-vis currencies of competing exporters upon Indian exports of cut diamonds, carpets, and hand/machine tools during the period 1974–87. The study provides strong evidence that the exchange rate plays a significant role in explaining exports of those products.

## 4.4 MODELING AN EXPORT DEMAND FUNCTION FOR INDIA

Based on the literature review and given the necessity of reasonably long time-series data for co-integration analysis, this study focused on two critical explanatory factors in explaining Indian export performance. One such variable is the externally focused foreign market access factors (Redding and Venables, 2004) as captured by world real income. Another critical variable in the model that may reflect both internal and external factors is the real effective exchange rate (REER) variable. Although some may consider the REER variable as representing an external factor, it may also represent some internal factors such as domestic prices vis-à-vis foreign prices and the trade weights of India with its trading partners. In fact, this complex variable captures the relative competitiveness of home exports vis-à-vis India's foreign competitors.

As such, the global demand for aggregate Indian real exports (XR) is specified in this chapter on these two variables as shown by equation 4.1 below:

$$\mathrm{XR} = f\left(\mathrm{REER}, \mathrm{GDPRW}\right) \tag{4.1}$$

Note that although some other variables might be relevant for an export function, such as domestic production costs, infrastructure constraints, and institutional and policy factors, long time-series data for these variables are generally not available for India. As such, in this simple model, the number of variables is kept at a minimal level based on data availability for a long period in order to maximize the number of observations needed for a meaningful long-run co-integration analysis.

Using a log-linear functional form and using t as the time subscript t (t = 1, 2,. . ., t) and adding a random error term μ, α as an intercept term, $\beta$ as the exponent of REER, and $\delta$ as the exponent of GDPRW, equation 4.1 can be rewritten to obtain equation 4.2 as:

$$XRt = \alpha \; REER_t^{\beta} GDPRW_t^{\delta} e^{\mu t} \tag{4.2}$$

Taking logarithms on both sides and using L as prefix in front of the variable name to indicate the natural log transformation of the variables, the export function takes the following form:

$$LXR_t = \alpha + \beta \; LREER_t + \delta \; LGDPRW_t + \mu_t \tag{4.3}$$

The *a priori* sign expectations of the coefficients are as follows:

$$\alpha > ?; \beta < 0; \delta > 0 \tag{4.4}$$

Equation 4.3 is the empirically testable form of the long-run equilibrium real export demand function for India. The log-linear form was chosen because the coefficients of the variables can be interpreted directly as the elasticity coefficients. In addition, this form is helpful in reducing possible heteroskedasticity problems in the data (Maddala, 1992). Further, it is conventional now to use log-linear specification for most empirical co-integration studies (Johansen and Juselius, 1990).

The *a priori* sign expectations for each variable are given in equation 4.4. The intercept term α can be positive or negative and cannot be specified *a priori* because it will reflect the combined impact of omitted variables and the random factors. As elaborated in the Methodology section, the REER variable is measured as the weighted average of inflation-adjusted foreign currency price per unit of the home currency (Indian rupees) of the trading partners of India. Given this measurement of REER, an increase of this variable implies a real appreciation of the Indian currency and a decrease means a real depreciation. Conventional trade theory indicates that when the home currency appreciates against foreign currencies, the Indian products would become more expensive for foreigners to buy. Thus, Indian goods will be less competitive in the global markets, thus reducing demand for Indian export goods. This would imply an expected negative sign for the REER coefficient ($\beta < 0$). On the other hand, the coefficient of the world real GDP variable (GDPRW) is expected to have a positive sign, indicating that a rising world real income would increase world demand for foreign goods, including goods from India ($\delta > 0$). The underlying assumption, of course, is that the Indian export goods are considered by foreign consumers as normal goods as opposed to inferior goods.

## 4.5 ESTIMATION METHODOLOGY

Equation 4.3 can be estimated by the standard regression method if the variables XR, REER, and GDPRW are stationary in their levels and the residual term μ is uncorrelated and homoskedastic. However, if the variables are non-stationary in their levels, the standard regression methodology becomes inappropriate because the usual t and F tests become meaningless (Engle and Granger, 1987). More importantly, the estimated regression coefficients will be spurious (Granger and Newbold, 1974). In this situation, one needs to apply co-integration techniques to estimate the equation.

### 4.5.1 Time Series Properties of the Variables

A stationary series is generally characterized by a time-invariant mean and a time-invariant variance. Several alternative methods are available to test for non-stationarity of a time series. In this chapter, we will apply the following three well-known and widely used techniques for this purpose: (a) the augmented Dickey-Fuller test (ADF test); (b) the Phillips-Perron test (PP test); and (c) The KPSS test. An elaborate discussion for each can be found in Dickey and Fuller (1981), Phillips and Perron (1988), and Kwiatkowski, Peter, Phillips, and Shin (1992) respectively. If these tests find that the variables are non-stationary in their levels, then one needs to apply a co-integration test. In this study, the Johansen-Juselius (JJ) co-integration method will be applied.

### 4.5.2 Co-integration Tests: The Johansen-Juselius (JJ) Method

The multivariate co-integration techniques developed by Johansen (1988, 1991, 1992) and Johansen-Juselius (1990, 1992) using a maximum likelihood estimation procedure allows researchers to estimate simultaneously the system involving two or more variables to circumvent the problems associated with the traditional regression methods used in previous studies on this issue. Further, this method is independent of the choice of the endogenous variable because it treats all the variables in the model as endogenous within a vector autoregression (VAR) framework. More importantly, this method allows one to estimate and test for the presence of more than one co-integrating vector(s) in the multivariate system. In addition, it enables the researchers to test for various structural hypotheses involving restricted versions of the co-integrating vectors and speed of adjustment parameters using likelihood ratio tests.

Following Johansen (1988) and Johansen and Juselius (1990), a VAR representation of the $N$-dimensional data vector $X_t$ is specified as follows:

$$X_t = + \Pi_1 X_{t-1} + \ldots\ldots\ldots\ldots\ldots + \Pi_k X_{t-k} + \psi + e_t \qquad (t = 1,2,\ldots\ldots\ldots\ldots, T) \qquad (4.5)$$

where $e_1, \ldots, e_T$ are distributed as $N$-dimensional i.i.d. normal variables, $\Psi$ represents a vector of constants, and the $X_t$ is a vector of all the endogenous variables in the system. In this study, the vector $X_t$ is of dimension $N = 3$ because it contains three endogenous variables ($LXR_t$, $LREER_t$, and $LGDPRW_t$,).

Now using the notation $\Delta = 1 - L$, where L is the lag operator, the VAR system represented by equation 4.5 can be rewritten as the error correction model (ECM) as follows:

$$\Delta X_t = \Pi X_{t-1} + \varphi_1 \Delta X_{t-1} + \ldots\ldots\ldots\ldots \varphi_{k-1} \Delta X_{t-k+1} + \psi + e_t, \qquad (4.6)$$

where: $\varphi_I = -(I - \Pi_1 - \ldots - \Pi_i)$ $\quad (I = 1, \ldots, k-1)$,
and $\Pi = (I - \Pi_1 - \ldots - \Pi_k)$.

The main focus of the Johansen-Juselius technique is on the parameter matrix Π. The rank r of this matrix is r(Π), where ($0 < r < N$) will determine the number of co-integrating vectors in the VAR system. If the rank of this matrix is found to be r, then there are r linear combinations of the variables in the system that are stationary and that all other linear combinations are non-stationary.

The matrix Π can be rewritten as $\Pi = \alpha\beta$ (where α is the speed of adjustment vector (also called weights or loadings) and β is the co-integrating vector. The dimensions of α and β are ($N \times r$) and the system (equation 4.6) is subject to the condition that Π is less than full rank matrix, that is, $r < N$. The procedure boils down to testing for the value of r on the basis of the number of significant eigenvalues of Π. For this purpose, the maximum eigenvalue test ($\lambda_{max}$) and the trace test ($\lambda_{Trace}$) are applied. The $\lambda_{max}$ and $\lambda_{Trace}$ test statistics are given by the following equations:

$$\lambda_{max}\left(r, r+1\right) = -T \ln\left(1-\lambda_{r+1}\right) \tag{4.7}$$

$$\lambda_{Trace}\left(r\right) = -T\,\Sigma \ln\left(1-\lambda_{I}\right) \qquad \left(I = r+1,....,n\right) \tag{4.8}$$

where $\lambda_I$ is the estimated value of the *i*th characteristic root (eigenvalue) obtained from the estimated Π matrix, *N* is the number of variables, and T is the number of usable observations.

The $\lambda_{max}$ statistic tests the null hypothesis that the number of co-integrating vectors is r against the alternative of (r + 1) co-integrating vectors. If the estimated value of the characteristic root is close to zero, then the $\lambda_{max}$ will be small. The $\lambda_{trace}$ statistic, on the other hand, tests the null hypothesis that the number of distinct characteristic roots is less than or equal to r against a general alternative. In this statistic, $\lambda_{trace}$ will be small when the values of the characteristic roots are closer to zero and its value will be large, the further the values of the characteristic roots from zero. Note that the statistic has a sharper alternative hypothesis than the trace statistic and hence it may be preferred in selecting the number of co-integrating vectors. Also note that in actual test situation, the two statistics may give conflicting results.

Johansen and Juselius (1990) provide the critical values of the $\lambda_{max}$ and $\lambda_{trace}$ test statistics using simulation method for determining the rank of the Π matrix. The above mentioned test statistics are distributed as $\chi^2$ with appropriate degrees of freedom ($N - r$) where *N* is the number of variables and r is the value of the rank under the null hypothesis. In these likelihood ratio tests, the null hypotheses are accepted if the estimated values are less than the critical values at the appropriate significance level and the degrees of freedom.

The co-integration techniques are also useful in analyzing the short-run dynamics underlying the long-run equilibrium demand function given by equation 4.3. The estimated long-run equation can be used to estimate the corresponding vector error correction model that will give short-run coefficients as well as the speed of adjustment coefficient. More specifically, the sign and magnitude of the speed of adjustment coefficient will give important information about the short-run dynamics of the system; that is, its stability as well as the direction and speed of adjustment towards the long-run equilibrium path. If this coefficient is negative and less than unity, it will indicate that the system is stable. If the coefficient is statistically significant, this will provide additional evidence for the existence of the co-integration relationship involving the variables in the export function. The magnitude of the coefficient will indicate the average speed with which the system will go back to its long-run equilibrium path. A small coefficient will indicate slow adjustment and a large coefficient will indicate rapid adjustment.

## 4.6 VARIABLES, MEASUREMENTS, AND DATA

This section presents information about the data, measurement of the variables, data frequency, and time period covered.

### 4.6.1 Measurement of LXR, GDPRW, and the REER

The variable of primary interest is the real exports of India. This variable is constructed as follows. India's nominal exports X (in millions of US Dollars) is converted into inflation-adjusted real exports (XR) using India's GDP price deflator ($DEF_{India}$) as follows:

$$XR = \left(X_{India} / DEF_{India}\right) * 100 \tag{4.9}$$

In constructing the composite variable GDPRW, first, the nominal GDP of each country ($GDP_i$) is converted into a real GDP ($GDPR_i$) using that country's price deflator ($DEF_i$) as follows:

$$GDPRi = \left(GDPi / DEF_i\right) * 100 \tag{4.10}$$

Because there are many countries with which India has trading relations, the aggregate world real GDP (GDPRW) is constructed as the simple summation of $GDPR_i$ of all countries in the world less the GDPR for India, as follows:

$$GDPRW = \left(\Sigma GDPR_i - GDPR_{India}\right) \tag{4.11}$$

Note that India's real GDP is deducted because Indian income would not be affecting world demand for Indian exports. The GDPRW as calculated in equation 4.11 is now used to capture the world demand for Indian exports.

Measuring the REER variable is even more complex than measuring GDPRW. This is because various factors need to be incorporated in calculating this variable, such as the bilateral nominal exchange rate, home and foreign country prices, and home and foreign country trade shares. In order to construct the REER variable, assume that $E_i$ represents the bilateral nominal exchange rate measured as foreign currency price per unit of the home country currency (Indian rupees). Given this definition of the bilateral nominal exchange rate for the home country, an increase in the exchange rate will mean the appreciation of the home currency and a decrease in E would indicate a depreciation of the home country. Further, defining the foreign price index for country i as $P_i^*$ and home price index as P, the inflation-adjusted bilateral real exchange rate index for the home (Indian) currency against the currency of country i, $q_i$ can be defined (converted in an index form with a specified year as the base year) as follows:

$$q_i = E_i \left(P_i^*/P\right) \quad (4.12)$$

The index form is used in order to avoid dealing with different currencies measured in different units. Because there are many bilateral real exchange rates facing the home country with different foreign trading partner countries, a multilateral weighted average real exchange rate (REER) needs to be constructed using the bilateral real exchange rates of the home country's trading partners with appropriate weights assigned to the different bilateral real exchange rates. The resulting multilateral inflation-adjusted weighted average of the bilateral real exchange rates for the home country is known as the real effective exchange rate (REER). It is thus a weighted average of the inflation adjusted bilateral real exchange rates ($q_i$) between the home country currency vis-à-vis its trading partner currencies, weighted by the respective trade shares with its trading partners. Now defining the trade weight of the home country's trade with its trading partners is $w_i$, and using geometric averaging method, the trade-weighted inflation-adjusted multilateral real effective exchange rate REER can be defined, following Hinkle (2000), as follows:

$$REER_i = \Pi\left(q_i\right)^{wi} = \Pi\left[E_i\left(P_i^*/P\right)\right]^{wi} \quad (4.13)$$

with i = 1,2, . . . m trading countries, $0 < w_i < 1$, and $\Sigma w_i = 1$. Note that instead of the arithmetic average, the geometric averaging technique is normally preferred because a geometric average has some specific properties of symmetry and consistency that an arithmetic average does not have (Hinkle, 2000). These were not discussed here due to space limitations.

### 4.6.2 Data Sources and Data Frequency

The World Bank's World Development Indicators were used for most variables except the REER variable. The REER variable was taken from Darvas (2012). The data frequency was annual, and the time period covered was 1960 to 2017.

## 4.7 THE EMPIRICAL RESULTS

This section presents the empirical results based on the unit root and co-integrations tests along with various residual based diagnostic tests.

### 4.7.1 Time Series Properties of the Variables

Because the power of different unit root tests could differ, we conducted three separate unit root tests. Table 4.1 reports the results of the unit root tests using the ADF, Phillips-Perron, and KPSS tests. All three tests applied to all three variables clearly indicate one uniform result; that is, that all three variables are non-stationary in their level form and that these are stationary in their first differences. Hence, we

**TABLE 4.1**
**India Exports-Summary Results of Unit Root Tests**

***1. Augmented Dickey-Fuller Test: Null of Unit Root (Non-stationary)***

| Variables | Level | First Difference | Result (Level) |
|---|---|---|---|
| LXR | −2.23 (C,T) | −9.96*** (C) | Non-stationary |
| LREER | −1.22 (N) | −5.30*** (N) | Non-stationary |
| LGDPRW | −3.17 (C,T) | −5.42*** (C,T) | Non-stationary |

***2. Phillips-Perron Test: Null of Unit Root (Non-stationary)***

| Variables | Level | First Difference | Result (Level) |
|---|---|---|---|
| LXR | −2.23 (C,T) | −6.96*** (C) | Non-stationary |
| LREER | −1.26 (N) | −5.32*** (N) | Non-stationary |
| LGDPRW | −3.07 (C,T) | −5.38*** (C,T) | Non-stationary |

***3. KPSS (Kwiatkowski-Phillips-Schmidt-Shin) Test: Null of No Unit Root (Stationary)***

| Variables | Level | First Difference | Result (Level) |
|---|---|---|---|
| LXR | 0.21** (C,T) | 0.12 (C,T)) | Non-stationary |
| LREER | 0.20** (C,T) | 0.26 (C) | Non-stationary |
| LGDPRW | 0.22** (C,T) | 0.16 (C,T) | Non-stationary |

*Notes*: (1) The McKinnon critical values for ADF and PP tests (with both intercept and trend) are: (a) 1% = −4.07, (b) 5% = −3.46, and (c) 10% = −3.16 respectively; (with only intercept) are: (a) 1% = −3.48, (b) 5% = −2.88, and (c) 10% = −2.58; (without intercept and trend) are: (a) 1% = −2.59, (b) 5% = −1.94, and (c) 10% = −1.62 respectively. (2) Critical values for KPSS test (with both intercept and trend: (a) 1% = 0.22, (b) 5% = 0.15, and (c) 10% = 0.12 respectively; (with intercept only): (a) 1% = 0.74, (b) 5% = 0.46; and (c) 10% = 0.35 respectively; (without intercept and trend) (a) 1%=0.73; (b) 5% = 0.46; and (c) 10% = 0.35 respectively. (3) *** = significant at 1% level, ** = significant at 5% level, and * = significant at 10% level. (4) Letters in parentheses after the coefficients represent the following characteristic included during the unit root tests and in determining the critical values as appropriate: C = Intercept; T = Trend; and N = None (No Intercept; No Trend). (4) SIC was used to determine optimal lag length.

conclude that all three variables are non-stationary, that is, I(1). As a result, the standard regression model is not appropriate in examining the relationships. In this situation, one has to use the co-integration techniques to uncover the appropriate relationships.

### 4.7.2 Selection of Optimal Lag Length

Before conducting co-integration analysis, one needs to specify the appropriate lag length k of the VAR system so as to make the estimated model's residuals uncorrelated and homoskedastic. It is also well-known that co-integration tests are highly sensitive to the lag structure used in the estimation (Johansen and Juselius, 1990, 1992). Therefore, determining the appropriate lag length of the endogenous variables is a critical step before conducting any co-integration test. Using the

**TABLE 4.2**
**Test for Optimal Lag Structure from the VAR Model**

**VAR Lag Order Selection Criteria**

**Endogenous variables: LXR, LREER, LGDPRW**

**Exogenous variables: C**

**Sample: 1960–2020**

**Included observations: 53**

| Lags | LogL | LR | FPE | AIC | SC | HQ |
|---|---|---|---|---|---|---|
| 0 | –32.10919 | NA | 0.000755 | 1.324875 | 1.436401 | 1.367763 |
| 1 | 293.9205 | 602.8474* | 4.82e-09* | –10.63851* | –10.19241* | –10.46696* |
| 2 | 299.3018 | 9.341087 | 5.54e-09 | –10.50196 | –9.721274 | –10.20174 |
| 3 | 303.3912 | 6.635563 | 6.73e-09 | –10.31665 | –9.201388 | –9.887773 |
| 4 | 306.8473 | 5.216769 | 8.44e-09 | –10.10744 | –8.657607 | –9.549907 |

*Notes*: * indicates lag order selected by the criterion.
LR: Sequential modified LR test statistic (each test at 5% level)
FPE: Final prediction error
AIC: Akaike information criterion
SC: Schwarz information criterion
HQ: Hannan-Quinn information criterion

*Source*: Author calculations.

VAR system, the optional lag structure test was conducted and the results are shown in Table 4.2. Several different lag selection criterion were employed in determining the appropriate optimal lag length and the uniform result found in Table 4.2 is k* = 1.

### 4.7.3 Preliminary Selection of the Appropriate Deterministic Components Model

As discussed earlier, this study used the maximum likelihood–based Johansen-Juselius multivariate co-integration analysis. In the first step, we used the trace and maximum eigenvalue tests for a preliminary determination of rank (i = 0, 1, 2, 3) related to three endogenous variables and the selection of an appropriate deterministic components model (j = 1, 2, 3, 4, 5) corresponding to the five different deterministic components cases (models) of the co-integrating system as specified in Hansen and Jeselius (1995). In Table 4.3, out of the five possible deterministic component models, we selected Model 2, $r(\Pi) = 3$ (three possible co-integrating relations), and with no trends but intercepts in CE and no intercept in VAR, for conducting the co-integration test that follows. This preliminary analysis indicates that there could be a maximum of three co-integrating vectors (rank = 3) and that the system contained an intercept in CE, but no trend in CE or VAR. Further probe into the determination of the exact rank follows.

**TABLE 4.3**
**Preliminary Selection of the Rank and Model Selection Tests**

| | Model 1 | Model 2 | Model 3 | Model 4 | Model 5 |
|---|---|---|---|---|---|
| Trend specification | No trend | No trend | No linear trend | Linear trend | Quadratic trend |
| Intercept specification | No intercept in CE or VAR | Intercept in CE, no intercept in VAR | Intercept in CE, Intercept in VAR | Intercept in CE, No intercept in VAR | Intercept in CE, intercept in VAR |
| Trace | 2 | 3* | 2 | 1 | 0 |
| Maximum-eigenvalue | 2 | 3* | 0 | 0 | 0 |

*Notes*: (1) For five models with different trend and intercept specifications, see Hansen and Juselius (1995). (2) Further, the critical values for the selection of rank and deterministic components jointly for different models are given in Hansen and Juselius (1995); 90% critical values were used. (3) * indicates model and rank selected based on trace test. (4) ** indicates model and rank selected by maximum eigenvalue test. (5) Selections were based using 5% critical value.

*Source*: Author calculations.

### 4.7.4 Final Determination of the Rank of the Π Matrix

Based on the preliminary rank and model selection (rank 3, Model 2), additional tests were conducted to determine the exact rank of the Π matrix and the results are reported in Table 4.4. The trace test reported in Panel A of this table shows that possible null hypotheses of $r = 0$, $r = 1$, and $r = 2$ are rejected, thus indicating further that this system will have a maximum rank of three ($r = 3$) co-integrating equations. Similarly, the maximum eigenvalue test shows exactly the same rejections of all possible null hypotheses, thus reconfirming the previous results that the system has a maximum rank of three ($r = 3$).

### 4.7.5 Estimating the Final Long-run Co-integrating Export Function

We estimated the three co-integrating equations, and Table 4.5 reports the one that is most appropriate and relevant for this export function study, which is normalized on the LXR. We then rearranged terms to solve for LXR as the dependent variable, the variable of our research interest. This long-run co-integrating export function for India is reported in Table 4.5.

The results show that the REER variable has a positive sign and that the coefficient is statistically significant at the 10% level. The magnitude of the coefficient is greater than unity, thus indicating that the real effective exchange rate has a mildly elastic impact on Indian real exports. This result is contrary to our theoretical

## TABLE 4.4
## Final Determination of the Rank of r(Π) Based on $\lambda_{max}$ and $\lambda_{Trace}$ Test Statistics

**A. Trace Test**

| **Null Hyp: $H_0$: r** | **Eigenvalues ($\lambda_i$)** | **$\lambda_{Trace}$ statistic** | **0.05 critical value** $\lambda_{trace}$** | **Prob.** |
|---|---|---|---|---|
| 0 (None)* | 0.393152 | 55.80204 | 35.19275 | 0.0001 |
| 1 (At most 1)* | 0.275458 | 28.33076 | 20.26184 | 0.0031 |
| 2 (At most 2)* | 0.175426 | 10.60887 | 9.164546 | 0.0264 |

**B. Maximum Eigenvalue Test**

| **Null Hyp: $H_0$: r** | **Eigenvalues ($\lambda_i$)** | **$\lambda_{max}$ statistic** | **0.05 critical value** $\lambda_{max}$** | **Prob.** |
|---|---|---|---|---|
| 0 (None)* | 0.393152 | 27.47127 | 22.29962 | 0.0001 |
| 1 (At most 1)* | 0.275458 | 17.72189 | 15.89210 | 0.0031 |
| 2 (At most 2)* | 0.175426 | 10.60887 | 9.164546 | 0.0264 |

*Notes*:
* Denotes rejection of the hypothesis at the 0.05 level.
** MacKinnon-Haug-Michelis (1999) *p*-values.

*Source*: Author calculations.

## TABLE 4.5
## Maximum Likelihood Estimates of Co-integrating Vector (Normalized on LXR)

| **Variables** | **Co-integrating Equation** | **Standard Error (s.e.)** | ***t*-values** |
|---|---|---|---|
| *Dependent (LHS) Variable: LXR* | | | |
| LREER | 1.266947* | 0.70323 | 1.80160 |
| LGDPRW | 4.340255*** | 0.54045 | 8.03084 |
| CONSTANT | −118.8936*** | 19.8720 | 5.98297 |

*Notes*: *** Significant at 1% level; ** significant at 5% level; and * significant at 10% level.

expectation of a negative sign. The estimated result seems to indicate an opposite response pattern of Indian exports to movements in the REER. Given this unexpected result, the question arises about how to explain this observed phenomenon. We can speculate an explanation as follows. When the Indian currency appreciates (REER rises), exports of Indian goods might be negatively affected in the global markets as exports become less competitive in the world market. However, many of these Indian export goods might be highly dependent on imported raw materials, capital equipment, technology, and energy, among others. The appreciating exchange rate would have beneficial effects in reducing the costs of the imported inputs that go into the production of the export goods. The latter beneficial effect may over-compensate for the former adverse effect, thus resulting in an overall positive effect on Indian exports.

Consistent with the theoretical expectation, the world real income variable (GDPRW) has the expected positive sign and the coefficient is statistically highly significant at the 1% level. The magnitude of the coefficient is greater than unity, indicating that world income has a very strong and highly elastic effect on Indian exports. In other words, a 1% increase in world income would likely generate a 4.34% increase in Indian exports in real terms. The results further indicate that Indian export goods in the world market are considered not only as normal goods ($\delta > 0$), but also as luxury goods ($\delta > 1$). Thus, Indian exports seem to depend very highly and very sensitively to world income. On the flip side, if world income declines, Indian exports would likely experience a strong decline.

### 4.7.6 Estimated Short-run Dynamics

The vector error correction model (VECM) was specified and estimated based on the estimated long-run co-integrating equation and the optimal lag length of k = 1 reported earlier with the estimated error correction term ECT(−1). The estimated VECM is estimated by the OLS method and the results are reported in Table 4.6. The Durbin-Watson statistic shows that the residuals from this equation are free of serial correlation. However, the overall regression equation is not satisfactory as the $R^2$ and the adjusted $R^2$ values are quite low and the F-statistic is not statistically significant. In this table, the estimated coefficients of the one-period lagged differenced variables (Δ) represent the short-run elasticity coefficients. The short-run effect of the REER variable (ΔLREER(−1)) appears with a positive sign, but is not statistically significant. So, the short-effect seems to be non-significant from the real exchange rate movements. The short-run effect of the world real income variable (ΔLGDPRW(−1)) appears with an expected positive sign, but is not statistically significant. So, the short-run effect seems to be non-significant from the changes in world real income.

More importantly, the coefficient of the lagged error correction term, ECT(−1), is of special significance as the sign, magnitude, and statistical significance of this coefficient are of utmost interest. Looking at this coefficient, the coefficient of the ECT(−1) shows that the coefficient has the expected negative sign, indicating that the Indian export function is stable and returns back to equilibrium from any deviation from long-run equilibrium due to any external shock disturbing the system. The magnitude of the coefficient is −0.39, which indicates that the speed of adjustment towards long-run equilibrium from any disturbance would be corrected by 39% over a one-year period. It is to be noted, however, that the coefficient is not statistically significant and hence not that reliable.

**TABLE 4.6**
**Vector Error Correction Model with Short-run Dynamics**

| Explanatory Variables | Coefficient | Std. Error (s.e.) | *t*-Statistic | Prob. |
|---|---|---|---|---|
| C | 0.062139* | 0.033849 | 1.835799 | 0.0729 |
| ΔLRX(−1) | 0.291719 | 0.198291 | 1.471168 | 0.1481 |
| ΔLREER(−1) | 0.193392 | 0.182655 | 1.058788 | 0.2952 |
| ΔLGDPRW(−1) | 0.057989 | 0.739927 | 0.078371 | 0.9379 |
| ECT(−1) | −0.399863 | 0.280031 | −1.427921 | 0.1601 |
| Max Lags Used | 1 | | | |
| R-squared | 0.075620 | | | |
| Adjusted R-squared | −0.004760 | | | |
| F-statistic | 0.940777 | | | |
| Prob (F-statistic) | 0.448944 | | | |
| Akaike info criterion | −2.084039 | | | |
| Schwarz criterion | −1.894644 | | | |
| Hannan-Quinn criterion | −2.011665 | | | |
| Durbin-Watson stat | 1.984942 | | | |
| N | 51 | | | |

*Notes*: *** = significant at 1%, ** = significant at 5%, and * = significant at 10% level.

### 4.7.7 Analysis of the Residuals

It is critical to conduct diagnostic tests on the residuals from the VECM model. These test results are reported in Table 4.7. The first row gives the Jarque-Bera statistic (JB) to test the null hypothesis that the residuals are normally distributed. The reported test result supports that the null hypothesis could not be rejected; hence, it is concluded that the residuals are normally distributed as per the expected i.i.d. properties. The serial correlation Q-tests (row 2) with three alternative lags (k = 1, k = 3, and k = 5) shows that the null of no serial correlation is accepted. Thus, there is no serial correlation in the residuals. Further, the Breusch-Godfrey serial correlation LM test (row 3) also shows the absence of serial correlation in the VECM model. These results are consistent with the previously reported Durbin-Watson serial correlation test result. We also tested for any heteroskedasticity in the residuals. The Breusch-Pagan-Godfrey heteroskedasticity test (row 4) shows that the residuals are free of heteroskedasticity problem. Finally, the Ramsey reset test (row 5) shows that, although the export function model is simple, this simple model is specified well, as we could not reject the null hypothesis of "model is not misspecified". It is thus concluded that the residuals from the VECM model are well-behaved by any conventional econometric test.

**TABLE 4.7**
**Diagnostic Tests Based on the Residuals from the VECM**

| Test Name | Null Hypothesis | Test Type | Statistic | Prob. | Test Result |
|---|---|---|---|---|---|
| Jarque-Bera normality test | $H_0$: Errors are normally distributed | Jarque-Bera statistic | 0.427169 | 0.807643 | Accept $H_0$ |
| Serial correlation Q-test | $H_0$: Errors are not serially correlated | Q-stat (1 lag) | 0.0120 | 0.973 | Accept $H_0$ |
| | | Q-stat (3 lags) | 1.0485 | 0.790 | Accept $H_0$ |
| | | Q-stat (5 lags) | 1.3694 | 0.928 | Accept $H_0$ |
| Breusch-Godfrey serial correlation LM test | $H_0$: Errors are not serially correlated | F-statistic with D.F. (num., denom.) | 0.209642 (5; 41) | 0.9565 | Accept $H_0$ |
| Breusch-Pagan-Godfrey heteroskedasticity test | $H_0$: Errors are not heteroskedastic | F-statistic with D.F. (num., denom.) | 0.2865312.3099 (4; 46) | 0.8852 | Accept $H_0$ |
| Ramsey reset test for model specification with one fitted value | $H_0$: Model is not misspecified | F-statistic with D.F. (num., denom.) | 0.676149 (1; 45) | 0.4153 | Accept $H_0$ |

## 4.8 CONCLUSIONS

In this chapter, we estimated the export function for India using the co-integration technique. Co-integration results indicate that the Indian export demand function is strongly dominated by world income. The elasticity of exports with respect to world income is found to be significantly greater than unity. However, contrary to the expectations of conventional trade theory, the impact of appreciating REER was found to have an unexpected positive impact on Indian exports. The magnitude of the coefficient is greater than unity, thus indicating that the real effective exchange rate has a mildly elastic and positive impact on Indian real exports. This seemingly unexpected result could be due to possible high import intensity of Indian exports, as explained earlier. Further, analysis of the short-run dynamics shows that the export demand function is stable and adjusts to its long-run equilibrium path with a reasonable speed of adjustment. However, although the adjustment coefficient is of the expected negative sign to indicate stability, it was not found to be statistically significant.

The important finding of this study is that for a fast-growing country like India, the effect of real effective exchange rate on exports does not seem to follow the conventionally expected relationship, at least as far as aggregate exports are concerned. Hence, any discussion that focuses exclusively on exchange rate runs the risk of overlooking other factors that may hinder India's exports. For instance, world income was found to have stronger impact on Indian exports than REER. As such, trying to improve India's access to global markets should take priority over concerns about exchange rate appreciation. Thus, policy makers should not focus exclusively on an exchange rate policy to promote India's export activity. India's policy makers should focus more on a host of other critical factors, such as improving infrastructure, including highways and ports; liberalizing the labor market; removing or reducing various internal barriers to trade; and facilitating better integration of the Indian economy with the rest of the word, among others. Unfortunately, the role of these factors was not examined in this study due to the lack of sufficiently long time-series data for co-integration analysis. Perhaps future studies can focus on these other factors.

## REFERENCES


Abeysinghe, T., & Tan, L-Y. (1998). Exchange rate appreciation and export competitiveness: The case of Singapore. *Applied Economics*, *30*, 51–55.

Akhtar, M. A., & Hilton, R. S. (1984). *Exchange rate uncertainty and international trade: Some conceptual issues and new estimates for Germany and the United States.* New York: Federal Reserve Bank.

Aydýn Faruk, M., Çýplak, U., & Yücel, M. E. (2004). *Export supply and import demand models for the Turkish economy, the central bank of the republic of Turkey* (Research Department Working Paper No. 04/09).

Bahmani-Oskooee, M. (1996). Exchange rate uncertainty and trade flows of LDCs: Evidence from Johansen's cointegration analysis. *Journal of Economic Development, 21*, 23–35.

Bailey, M. J., Tavlas, G. S., & Ulan, M. (1986). Exchange-rate variability and trade performance: evidence for the big seven industrial countries. *Review of World Economics*, *122*(3), 466–477.

Bailey, M. J., Tavlas, G. S., & Ulan, M. (1987). The impact of exchange-rate volatility on export growth: some theoretical considerations and empirical results. *Journal of Policy Modeling*, *9*(1), 225–243.

Bhagwati, J., & Srinivasan, T. N. (1975). *Foreign trade regimes and economic development.* India and New York: Columbia University Press (for the NBER).

Brada, J. C., & Mendez, J. (1988). Exchange rate risk, exchange rate regime and the volume of international trade. *Kyklos*, *41*(2), 263–280.

Cheung, Y. W. (2005). An analysis of Hong Kong export performance. *Pacific Economic Review*, *10*(3), 323–340.

Côté, A. (1994). *Exchange rate volatility and trade* (Working Paper 94–95). Bank of Canada.

Cushman, D. O. (1986). Has exchange risk depressed international trade? The impact of third-country exchange risk. *Journal of International Money and Finance*, *5*(3), 361–379.

Cushman, D. O. (1988). US bilateral trade flows and exchange risk during the floating period. *Journal of International Economics*, *24*(3–4), 317–330.

Darvas, Z. (2012). *Real effective exchange rates for 178 countries: A new database* (No. MT-DP-2012/10). IEHAS Discussion Papers.

Dholakia, R. H., & Saradhi, R. V. (2000). Exchange rate pass-through and volatility: Impact on Indian foreign trade. *Economic and Political Weekly*, 4109–4116.

Dickey, D. A., & Fuller, W. A. (1981). Likelihood ratio statistics for autoregressive time series with a unit root. *Econometrica: Journal of the Econometric Society*, 1057–1072.

Engle, R. F., & Granger, C. W. J. (1987, March). (co-integration and error correction: representation, estimation, and testing. *Econometrica, 55*(2), 251–276.

Genc, E. G., & Artar, O. K. (2014). The effect of exchange rates on exports and imports of emerging countries. *European Scientific Journal, ESJ*, *10*(13).

Gervais, O., Schembri, L., & Suchanek, L. (2016). Current account dynamics, real exchange rate adjustment, and the exchange rate regime in emerging-market economies. *Journal of Development Economics*, *119*, 86–99.

Ghosh, J. (1990, March). Exchange rates and trade balance: Some aspects of recent Indian experience. *Economic and Political Weekly, 25*(9), 441–445.

Gotur, P. (1985). Effects of exchange rate volatility on trade: Some further evidence. *International Monetary Fund Staff Papers, 32*, 475–512.

Granger, C. W. J., & Newbold, P. (1974). Spurious regressions in econometrics. *Journal of Econometrics, 2*(1), 111–120.

Hansen, H., & Juselius, K. (1995). *CATS in RATS: Cointegration analysis of time series.* Estima.

Hinkle, L. E., Montiel, P. J., & Chinn, M. D. (2000). Book Reviews – Exchange rate misalignment: Concepts and measurement for developing countries. *Journal of Economic Literature, 38*(3), 651–651.

Hongwei, D. U., & Zen, Z. (2001). The effects of exchange rate risk on exports: Some additional empirical evidence. *Journal of Economics Studies, 28*(2), 106–121.

Hooper, P., & Kohlhagen, S. (1978). The effects of exchange rate uncertainty on the price and volume of international trade. *Journal of International Economics, 8*, 483–511.

Islam, Anisul M. (2018). Inter- and Intra-industry trade Relations between Bangladesh and India: Empirical Results. *FIIB Business Review*, 7, 4, 280–292.

Johansen, S. (1988). Statistical analysis of co-integration vectors. *Journal of Economic Dynamics and Control, 12*, 231–254.

Johansen, S. (1991, November). Estimation and hypothesis testing of co-integration vectors in gaussian vector autoregressive models. *Econometrica, 59*, 1551–1580.

Johansen, S. (1992). Determination of the cointegration rank in the presence of a linear trend. *Oxford Bulletin of Economics and Statistics, 54*, 383–397.

Johansen, S., & Juselius, K. (1990). Maximum likelihood estimation and inference on cointegration – With applications to the demand for money. *Oxford Bulletin of Economics and Statistics, 52*(2), 169–210.

Johansen, S., & Juselius, K. (1992, May). Testing structural hypothesis in a multivariate co-integration analysis of the PPP and the UIP for UK. *Journal of Econometrics, 53*, 211–244.

Joshi, V., & Little, I. M. D. (1994). *India: Macroeconomics and Political Economy, 1964-1991.* World Bank Publications.

Kenen, P. B., & Rodrik, D. (1986). Measuring and analyzing the effects of short-term volatility in real exchange rates. *The Review of Economics and Statistics*, 311–315.

Koray, F., & Lastrapes, W. D. (1989). Real exchange rate volatility and US bilateral trade – A VAR approach. *Review of Economics and Statistics, 71*, 708–712.

Krishnamurthy, K., & Pandit, V. (1995). *India's trade flows: Alternative policy scenarios: 1995–2000* (Working Paper No. 32). Delhi: Centre for Development Economics, Delhi School of Economics.

Kwiatkowski, D., Peter, C. B., Phillips, P. S., & Yong, C. S. (1992). Testing the null hypothesis of stationary against the alternative of a unit root. *Journal of Econometrics, 54*, 159–178.

Lee, J. (1999). The effect of exchange rate volatility on trade in durables. *Review of International Economics, 7*(2), 189–201.

MacKinnon, J., Haug, A., & Michaelis, L. (1999). Numerical distribution functions of likelihood ratio tests for co-integration. *Journal of Applied Economics, 14*(5), 563–577.

Maddala, G. S. (1992). *Introduction to econometrics* (2nd ed.). New York: Palgrave Macmillan.

Mallik, J. K. (2005). India's exports: Policy defeating exchange rate arithmetic. *Economic and Political Weekly, XL*(52), 5486–5496.

McKenzie, M. D. (1998). The impact of exchange rate volatility on Australian trade flows. *Journal of International Financial Markets, Institutions and Money, 8*, 21–38.

Nayyar, D. (1976). *India's exports and export policies in the 1960s.* Cambridge: Cambridge University Press.

Nayyar, D. (1988). India's export performance: Underlying factors and constraints. In R. E. B. Lucas & G. F. Papanek (Eds.), *The Indian economy: Recent development and future prospects.* New Delhi: Oxford University Press.

Phillips, P., & Perron, P. (1988, June). Testing for a unit root in time series regression. *Biometrika*, *75*, 335–346.

Raissi, M., & Tulin, V. (2018). Price and income elasticity of Indian exports – The role of supply-side bottlenecks. *The Quarterly Review of Economics and Finance*, *68*, 39–45.

Rajaraman, I. (1991, March). Impact of real exchange rate movements on selected Indian industrial exports. *Economic and Political Weekly*, 669–678.

Rana, S., Saikia, P. P., & Barai, M. K. (2018), Globalization and Indian manufacturing enterprises, *FIIB Business Review*, *7*(3), 167–175.

Rana, S., & Sharma, S. K. (2015). A literature review, classification, and simple meta-analysis on the conceptual domain of international marketing: 1990–2012. In S. Zou, H. Xu, & L. H. Shi (Eds.), *Entrepreneurship in International Marketing, Advances in International Marketing* (Vol. 25, pp. 189–222). Bingley: Emerald.

Redding, S., & Venables, A. (2004). *Geography and export performance external market access and internal supply capacity* (NBER Working Paper No. 9637). Washington, DC.

Sahinbeyoglu, G., & Ulasan, B. (1999). *An empirical examination of the structural stability of export function: The case of Turkey* (Research Department Discussion Paper No. 9907). Central Bank of the Republic of Turkey.

Sinha & Roy, S. (2007). *Demand and supply factors in the determination of India's disaggregated manufactured exports: A simultaneous error-correction approach* (Working Paper 383), CDS, Trivandrum.

Srinivasan, T. N. (1998). India's export performance: A comparative analysis. In I. J. Ahluwalia & I. M. D. Little (Eds.), *India's economic reforms and development: Essays for Manmohan Singh*. New Delhi: Oxford University Press.

Srinivasan, T. N., & Wallack, J. (2003). Export performance and the real effective, exchange rate. In A. O. Krueger & S. Z. Chinoy (Eds.), *Reforming India's external, financial, and fiscal policies*. Stanford: Stanford University Press.

Thursby, J. G., & Thursby, M. C. (1987). Bilateral trade flows, the Linder hypothesis, and exchange risk. *The Review of Economics and Statistics*, *69*, 488–495.

Virmani, A. (1991, February). Demand and supply factors in India's trade. *Economic and Political Weekly*, 309–314.

Wadhva, C. D. (1988). Some aspects of India's export policy and performance. In *The Indian Economy, Recent Developments and Future Prospects*. Delhi: Oxford University Press.

Wolf, T. A. (1978). Exchange-rate adjustments in small market and centrally planned economies. *Journal of Comparative Economics*, *2*(3), 226–245.

# 5 Information and Communication Technology as a Contingent Factor in India's Economic Growth–Remittances Nexus

*T. K. Jayaraman and Keshmeer Makun*

CONTENTS

## 5.1 INTRODUCTION

The stock of international migrants residing and working in countries away from their home countries was estimated at 258 million in 2017, having risen from 173 million in 2000 [United Nations Department of Economic and Social Affairs (UNDESA), 2017]. In 2017, the number of Indian-born migrants working overseas was 17 million, exceeding the corresponding number of Mexican-born persons (13 million). They are followed by migrants from Russia, China, Bangladesh, the Syrian Arab Republic, Pakistan, and Ukraine, ranging from 6 to 11 million each (UNDESA, 2017). Although their remittances (REM) to the countries of origin fell from US$444.3 billion in 2014 to US$429.3 billion in 2016 due to slow recovery from the Great Recession of 2008–12 in advanced countries, India continued to remain the largest REM recipient country: US$70.4 billion in 2014, US$72.2 billion in 2015, and US$62.7 billion in 2016, followed by the second-largest recipient, China with US$62.3 billion, US$63.9 billion, and US$61.0 billion during the corresponding years.

REM inflows from blue-collar Indian immigrants working in the Gulf Cooperation Council (GCC) nations and in other countries in the Middle East have dominated the annual REM inflows into India. Though in smaller amounts, unlike those from the white-collar Indian migrants in North America, they are sent on a regular basis, monthly or quarterly, to their families left behind. Aside from supplementing and enhancing household incomes, REM being in foreign currencies adds to India's real resources. In the absence of REM inflows, these annual additions to foreign exchange reserves would have to be earned through export of goods and services. In the context of India's weak export performance, REM provides a substantial support to India in building up its foreign exchange reserves.

The connection between REM and growth is through increasing aggregate demand by stepping up consumption by beneficiary families in rural India, which is dependent on monsoon-fed agriculture. Increments in expenditure on food, clothing, and medicines and education for children have been made easier by regular inflows of REM for reducing poverty to a great extent. However, any savings from steady annual inflows of REM without any opportunity for depositing the funds in banks tend to get frittered away on wasteful consumption. Depositing them in financial institutions, a process known as financialization of savings, increases bank reserves. Growth in credit is a logical consequence to financial deepening and financial sector development (FSD). It is well known that well-functioning financial markets lower transaction expenditures and would enable the savings from REM into productive investments.

Further, economic progress in recent years has also benefitted from the emergence of information and communication technology (ICT). Rapid spread of ICT since the late 1990s, particularly mobile voice and data networks, has brought notable gains in several spheres, including labour productivity, entrepreneurship, and innovations in business processes and service delivery. One of the most visible areas of improvement is the enhanced access to financial services and savings in transaction costs through new modes of online as well as mobile banking.

Despite the fact that India has been lagging behind (Kumar & Radcliffe, 2015) relative to similarly placed countries in the low- and middle-income group (LMIC),[1] as classified by the World Bank (2017), the impact of initiatives in recent years has been noticeable in Asia and Africa.[2] Although there were negative effects of the India's 2016 demonetization and the resultant cash crunch which were felt in some states in India, those states that have made notable progress in digitalization since the mid-2000s, the adverse effects were reported to be minimal.[3]

Earlier empirical studies (Jayaraman, Choong, & Kumar, 2012; Siddique et al., 2010) on the relationship between growth and REM in India did not pay attention on the influence of FSD and ICT. The present chapter attempts to fill the gap by focusing on these missing factors. The chapter is structured along the following lines: the next section provides a brief literature review of theoretical and empirical studies on the REM–growth nexus in the context of ongoing proactive measures in favour of FSD and ICT. Section 5.3 discusses growth trends in REM and FSD, and ICT spread. Section 5.4 deals with data, modeling, and econometric estimation methodology. Section 5.5 reports the results, and Section 5.6 provides conclusions and policy implications.

## 5.2 A BRIEF LITERATURE REVIEW

Among the three categories of capital transfers to the capital-starved LIMCs (Tables 5.1 and 5.2) REM has emerged to be the most reliable one (Mashayekhi, 2014). Further, REM in absolute amounts and in percentages of GDP of the region or countries are higher than respective figures of FDI and ODA inflows.[4] According to the latest estimates of the World Bank (2018), except for China, REM flows to LMICs (US$462 billion) were bigger than FDI streams in 2018 (US$344 billion). Quantitative studies either for a single country and panels of countries (Stahl & Habib, 1989; Leon-Ledesma & Piracha, 2001; Edwards & Ureta, 2003; Page & Adams, 2003; Hildebrandt & McKenzie, 2005; Yang, 2008; Giuliano & Ruiz-Arranz, 2009) have shown that inflows of REM have (1) helped the wellbeing of families that are left behind; (2) assisted the beneficiary families in upgrading their homes and improving their farming activities; (3) enabled households to pay children's education fees and bear the costs of old-age medical care; and (4) added to the recipient nation's foreign reserves. Increases in reserves also raise the credit worthiness of recipient countries, enabling them to borrow from international funding agencies for financing further growth-enhancing investments.

### 5.2.1 Role of Financial Sector Development (FSD)

It is estimated that only 50% of adults (15 years and older) in LMICs have access to financial institutions. Further, only 47% of women and 37% of youth have access to banking services, and only 34% of firms use bank loans, compared to 51% in developed nations. Almost 80% of micro, small to medium-sized businesses in rural areas do not have access to bank credit, compelling them to look for funds at much higher rates of interest from money lenders (Mashayekhi, 2014). Giuliano and Ruiz-Arranz (2009) observe that REM becomes a substitute in countries with weak financial sectors. In countries, with well-developed credit markets, REM might also be utilized for consumption spending. In the previous case of substitutability, in econometric analysis the sign of the coefficient of REM would be positive and significant and the interaction term's coefficient would be negative and significant. In the second and complementary case of the relationship, the coefficients of REM and interaction term would be both positive and statistically significant.

### 5.2.2 Financial Inclusion

The term *financial inclusion* became popular in the mid-2000s, following the big boost given by United Nation (UN). It also became a major component of poverty alleviation efforts. Recognizing the requirement for enlarging the availability of contacts to financial products and services, the UN included financial inclusion, characterizing it as "successful access to reasonable and sustainable financial services from formal suppliers", as one of its major goals in its post-2015 Development Programme (UN, 2015).

India has been in the forefront, well before the beginning of the new millennium (Chakrabarty, 2011) by adopting a proactive set of initiatives.[5] The purpose behind these initiatives was to cover the hitherto neglected segments of the population by mainstream commercial banks, such as peripheral farmers, landless laborers, oral lessees, the self-employed, urban slum residents, migrants, ethnic minority farmers, socially omitted groups, senior residents, women, and unorganized sector enterprises (Reserve Bank of India, 2008). Once a bank account is opened, it paves the way for access to all financial products. Further, having a bank account enables usual payments and acceptance of deposits as well as automatic receipt of REM transfers from overseas remittances at a small charge, besides making purchases on credit much easier and faster (Mohan & Ray, 2017).

### 5.2.3 Emergence of ICT

Developed countries that made notable investments in the ICT sector by the 1980s began to derive benefits by the mid-1990s through inventing cheaper devices for spreading ICT over different segments of economic activities. Understanding its significance in economic progress, the developing nations by late 1990s took serious interest to speed up development of the ICT sector, along with implementation of general economic reforms. The major focus was on moving away from government monopolies by inviting new competitors and promoting investment in ICT with joint ventures with the private sector. One immediate impact of growth in ICT investments in LMICs was seen in the spurt in foreign direct investment (FDI) inflows, which led to a rise in local employment and use of local resources, aside from transfer of technology and upgrading of skills. The ICT is now perceived as a basic infrastructural sector for advancing growth in different areas, including the financial sector.

Studies, which include Wilson (1993), Radeck, Wenninger, and Orlow (1997), and Freund, König, and Roth (1997), have highlighted benefits the flowing out of ICT in banks and financial markets. Kumar, Kumar, and Patel (2015) studied the role of ICT on economic development in small South Pacific island countries and found that ICT contributed to long-term economic growth in these countries. Aghaei and Rezagholizadeh (2017) and Niebel (2018) also showed that ICT enhances economic growth in Organisation of Islamic Cooperation countries and other emerging, developing, as well as developed countries. Majeed and Ayub (2018) examined a similar relationship in a sample of 149 economies from 1980–2015 and came to a similar conclusion.

### 5.2.4 Trends in India's Remittance Inflows, Financial Inclusion, and Spread of ICT

#### 5.2.4.1 Remittance Inflows

There has been a revival of the upward trend in REM inflows to developing countries both in terms of absolute amounts of US dollars and as percentages of gross domestic products (GDP) after end of the Great Recession (2008–12). In particular, with the rise in the price of crude oil in the first 10 months of 2018, there was a notable rebound in remittance outflows from the Gulf Cooperation Council (GCC) countries and the Russian Federation. The *Development Brief 30* (World Bank Group & KNOMAD, 2018) has estimated REM flows to LMICs at US$528 billion in 2018. In keeping with the general rising trend, REM flows to South Asia,[6] which belongs to the category of LMICs, are reported to have grown to a new high as well. Inflows of REM to South Asia grew 12% to US$131 billion in 2018, exceeding the 6% growth rate in 2017. The upsurge was influenced by stronger growth in 2018 in the United States and a pick-up in oil prices in 2018, benefitting oil-producing countries.

##### *5.2.4.1.1 A Unique Characteristic of REM*

Ready response to call for aid in times of natural disasters caused by cyclones and earthquakes is one of the unique characteristics of remittances, which distinguishes them from other types of capital transfers. The FDI is generally in long-term projects such as construction of factories with production processes of considerable gestation period. Hence, FDI funds are not pulled out in a short span of time. On the other hand, portfolio investments which are in short-term debt and equity instruments tend to get pulled out at short notice, as they are mainly for speculative reasons, and they move in and out at the click of the mouse in search of higher return. On the other hand, remittances belong to a different category,[7] as they are sent by migrants to their families, without expecting anything in return.

### 5.2.5 Remittances as a Stabilizing Factor

In the context of high degree of volatility in oil prices and stagnant exports, India's trade and current account deficits have started to widen in recent years. Steadily increasing inflows of remittances have, however, been seen reducing the current account deficits to manageable levels.[8] In the absence of REM, the current account deficits would have been much higher and pressure on the Indian rupee would have also been larger. Although, short-term portfolio investment flowed in and strengthened the Indian rupee in 2016 and 2017,[9] their well-known fickle-mindedness betrayed their nature.

Rukhaiyar (2018) noted that 2018 was the worst one for the Indian equity market as it witnessed a heavy outflow of foreign portfolio funds. From January 2018 onwards, oil prices began to rise after a gap of three years. Fears of a widening

**TABLE 5.1**

**Capital Transfers: World, Low- and Middle-Income Countries, South Asia: 1990–2017 (US$ billion)**

| | 1980–89 (Ave.) | 1990–99 (Ave.) | 2000–04 (Ave.) | 2005–09 (Ave.) | 2010 | 2011 | 2012 | 2013 | 2014 | 2015 | 2016 | 2017 |
|---|---|---|---|---|---|---|---|---|---|---|---|---|
| **World** | | | | | | | | | | | | |
| Foreign aid | 36.02 | 57.6 | 64.83 | 115.29 | 130.94 | 141.9 | 133.77 | 151.14 | 161.52 | 153.21 | 158.22 | 162.78 |
| FDI | 88.8 | 373.6 | 946.6 | 2130.2 | 1863.6 | 2290.3 | 2118.7 | 2136.9 | 1860.6 | 2411.4 | 2458.3 | 1949.6 |
| Remittances | 42.64 | 92.37 | 162.36 | 338.81 | 418.67 | 470.27 | 495.47 | 525.18 | 559.36 | 564.51 | 549.69 | 580.49 |
| **Low- and Middle-Income Countries** | | | | | | | | | | | | |
| Foreign aid | 33.39 | 54.42 | 63.04 | 115.04 | 130.27 | 141.31 | 133.28 | 150.87 | 161.5 | 153.12 | 157.97 | 162.6 |
| FDI | 11.1 | 82.2 | 163.5 | 453.3 | 610.9 | 722 | 651.6 | 737.6 | 664.9 | 628.5 | 560 | 537.3 |
| Remittances | 20.87 | 48.26 | 103.51 | 234.96 | 301.54 | 339.9 | 366.91 | 387.48 | 416.5 | 431.32 | 415.75 | 442.08 |
| **South Asia** | | | | | | | | | | | | |
| Foreign aid | 0.26 | 5.72 | 5.99 | 11.43 | 15.46 | 16.93 | 14.13 | 13.94 | 15.58 | 15.65 | 13.94 | 14.71 |
| FDI | 4.99 | 2.23 | 5.76 | 31.9 | 31.7 | 40.65 | 27.78 | 33.53 | 40.31 | 49.67 | 50.84 | 47.06 |
| Remittances | 5.69 | 9.9 | 23.9 | 55.47 | 81.98 | 96.4 | 107.97 | 110.82 | 115.81 | 117.61 | 110.41 | 116.70 |
| **India** | | | | | | | | | | | | |
| Foreign Aid | 1.83 | 1.88 | 1.28 | 1.82 | 2.83 | 3.27 | 1.68 | 2.46 | 2.99 | 3.17 | 2.68 | 3.09 |
| FDI | 0.1 | 1.51 | 4.61 | 26.3 | 27.4 | 36.5 | 24 | 28.15 | 34.58 | 44.01 | 44.5 | 39.97 |
| Remittances | 2.49 | 6.39 | 16.53 | 37.37 | 53.48 | 62.5 | 68.82 | 69.97 | 70.39 | 68.91 | 62.7 | 68.97 |

*Sources*: World Bank (2019), World Development Indicators and authors' calculations.

**TABLE 5.2 Capital Transfers: World, Low- and Middle-Income Countries, South Asia: 1990–2017 (% of GDP)**

| | 1980–89 (Ave.) | 1990–99 (Ave.) | 2000–04 (Ave.) | 2005–09 (Ave.) | 2010 | 2011 | 2012 | 2013 | 2014 | 2015 | 2016 | 2017 |
|---|---|---|---|---|---|---|---|---|---|---|---|---|
| **World** | | | | | | | | | | | | |
| Foreign aid | 0.26 | 0.21 | 0.17 | 0.21 | 0.20 | 0.19 | 0.18 | 0.20 | 0.20 | 0.20 | 0.21 | 0.20 |
| FDI | 0.63 | 1.28 | 2.59 | 3.74 | 2.74 | 3.03 | 2.72 | 2.57 | 2.28 | 3.15 | 3.16 | 2.35 |
| Remittances | 0.40 | 0.35 | 0.45 | 0.61 | 0.65 | 0.65 | 0.67 | 0.69 | 0.72 | 0.77 | 0.74 | 0.73 |
| **Low- and Middle-Income Countries** | | | | | | | | | | | | |
| Foreign aid | 1.26 | 1.27 | 0.98 | 0.87 | 0.65 | 0.60 | 0.53 | 0.57 | 0.58 | 0.58 | 0.60 | 0.56 |
| FDI | 0.54 | 1.78 | 2.54 | 3.30 | 3.03 | 3.05 | 2.57 | 2.74 | 2.38 | 2.41 | 2.14 | 1.88 |
| Remittances | NA | 1.20 | 1.62 | 1.74 | 0.65 | 0.60 | 0.53 | 0.57 | 0.58 | 0.58 | 0.60 | 0.56 |
| **South Asia** | | | | | | | | | | | | |
| Foreign aid | 1.69 | 5.72 | 5.99 | 11.43 | 15.46 | 16.93 | 14.13 | 13.94 | 15.58 | 15.65 | 13.94 | 14.71 |
| FDI | 0.08 | 0.45 | 0.81 | 2.23 | 1.55 | 1.79 | 1.21 | 1.42 | 1.56 | 1.84 | 1.75 | 1.43 |
| Remittances | 1.99 | 1.31 | 0.84 | 0.83 | 0.76 | 0.74 | 0.62 | 0.59 | 0.60 | 0.58 | 0.48 | 0.45 |
| **India** | | | | | | | | | | | | |
| Foreign aid | 0.81 | 0.57 | 0.25 | 0.17 | 0.17 | 0.18 | 0.09 | 0.13 | 0.15 | 0.15 | 0.12 | 0.12 |
| FDI | 0.04 | 0.39 | 0.85 | 2.30 | 1.65 | 2.00 | 1.31 | 1.52 | 1.70 | 2.09 | 1.95 | 1.54 |
| Remittances | 1.10 | 1.74 | 3.01 | 3.37 | 3.23 | 3.43 | 3.77 | 3.77 | 3.45 | 3.28 | 2.76 | 2.65 |

*Sources*: World Bank (2019), World Development Indicators and authors' calculations.

current account deficit dented confidence in the Indian economy. Outflows of foreign portfolio funds began to exceed inflows, and consequently, the rupee depreciated rapidly. It depreciated from Rs 63.64 per US dollar in January 2018 to Rs 74.45 in October, 2018. The rupee fell by 14%. Towards the end of 2018, as oil prices fell, the rupee gained to finish on the last market day of 2018 at Rs 69.77 per dollar. The depreciation of the rupee in 2018 was 9.23%, and the Indian currency was declared the worst performer amongst all the Asian currencies. The fall in the rupee value would have been worse but for REM inflows. While foreign aid inflows were on the decline and FDI was averaging at US$32 billion in 2014–15 to 2016–17, annual REM inflows during these years averaged US$65 billion.

### 5.2.6 Financial Inclusion

India's financial sector[10] began to play a major role in economic growth in the 1990s only after the introduction of economic reforms (Mohan & Ray, 2017). However, the spread of ICT in India towards speeding up the progress of FSD was slow and sluggish until 2010. Although India accounts for more than 10% of the global smartphone market, the findings of a *Mobile Technology and Its Social Impact Survey* study by Pew Research Center (2019) reveal that India's smartphone ownership rate is the lowest among the 11 developing countries covered by survey. Nearly 79 per 100 of the population in 2015–16 had mobile phones and the number of Internet users for every 100 persons was only around 26 in 2015–16. According to a private marketing agency's survey findings (The Economist, 2019), the number of Internet users was around 566 million in December 2018, out of which 493 million were reported to be regular users (defined as those who accessed the Internet in the last 30 days); about 293 million were active users in urban areas; and 200 million in rural areas. About 97% of the users used the mobile phone to access Internet. Internet usage is limited as 47% of mobile phone users had only a basic phone that could not connect to the Internet (The Economist, 2019).

The present urban–rural divide is generally attributed to the following: (1) internet use has been more of an urban activity, as urban areas happen to have assured availability of electricity for longer hours; (2) urban areas have a higher proportion of formal sector institutions, with commercial establishments and offices of private and public sector institutions; and (3) urban towns have a large number of white-collar working communities with tertiary educational backgrounds and high school and tertiary-level students, who tend to be quick to adopt modern technology. The indications are that with an assured supply of electricity and greater availability of broadband width and cheap data plans, and increased awareness of government programmes linking to various services, the number of rural Internet users (251 million in 2018) will reach 290 million in 2019 (35% increase) and the number of urban Internet users will be 336 million in 2019, up by 7% from 315 million in 2018 (The Economist, 2019).

### 5.2.7 Government Initiatives

Introduction of government-sponsored schemes to promote financial inclusion, increasingly supported by spread of mobile technology, have given a big boost to inculcation of banking habits. Financial inclusion is expected to enable the financial sector to provide universal access to a far wider range of financial services beyond banking, such as insurance and equity products, by utilizing technological innovations such as ATMs, credit and debit cards, Internet banking, electronic transfer, and usage of mobile phones. It has enabled financial institutions to reduce the costs involved in maintaining records and in the mobilization of deposits. Indeed, the use of and rapid spread of mobile phones has been seen as a welcome development by banking institutions themselves as it allows them to prosper with "no more brick or mortar branches".

India's financial inclusion programme is said to have been triggered over a three-year period (2014–16) by a trio referred to collectively as *JAM*. The acronym is a combination of three measures: one is *Jan Dhan* (J for Jan); a universal biometric identification system, known as *Aaddhaar* card (A for Aadhaar); and rising smartphone penetration (M for mobile). These initiatives were supplemented in July 2017 by a fully online goods and services tax (GST) system change, enabling a formalization process, which would facilitate fast-tracking India's digitization and bring about a higher degree of formalization of the Indian economy (Desai, Agarwal, & Arya, 2017). The major advantage is seen in better credit delivery than ever before, as data particulars on the would-be-borrowers are now available to the lending institutions. Earlier, lack of such information about the would-be-borrowers was preventing sanction of loans to non-corporate or small enterprises or individual members of the rural community. Digital transactions leave a data footprint that lenders can use easily and assess the credit worthiness of the prospective borrowers and process loans faster (Desai et al., 2017).

## 5.3 DATA, METHODOLOGY, AND ESTIMATION

### 5.3.1 Data

Our empirical study on growth and remittance nexus in India covers a period of 28 years (1990–2017). The key variables include: (1) GDP per capita in US$ constant price, represented by $y$; (2) capital stock per capita in US$ constant price, represented by $k$; and (3) nominal REM in US$ as percent of nominal GDP in US$, represented by *REM*. For FSD, we considered nominal broad money (*BM*) in rupees as percent of GDP and nominal bank credit (*BC*) as percent of GDP in rupees. Out of these two FSD indicators, we choose bank credit to private sector (*BC*) for the following reason: Financial inclusion efforts are aimed at the bringing in the hither-to bypassed households in rural and inaccessible areas, who had no opportunities to access banking services, not only for putting their savings and but also borrowing from banks for productive investments. Therefore, a better indicator is *BC*.

For ICT, we chose mobile subscriptions per 100 persons (*MOB*). Relative to Internet use, whose data series are inadequate and patchy, time series on ownership of mobile phones is complete. All the data series were obtained from *World Development Indicators* (2019) except for capital, which is from Penn Tables available on the website of US Federal Reserve St. Louis (2019). Since the capital stock data series are available only up to 2014, we extrapolated for the missing years: 2015

**TABLE 5.3**
**Descriptive Measures of Key Variables**

| | *Y* | *K* | *REM* | *BM* | *MOB* | *BC*MOB* |
|---|---|---|---|---|---|---|
| Mean | 1029.524 | 10148.27 | 2.699362 | 62.72477 | 25.61113 | 3.667961 |
| Median | 876.5995 | 8126.407 | 2.773336 | 64.7715 | 3.838817 | 5.52063 |
| Maximum | 1964.595 | 20093.85 | 4.210553 | 80.14708 | 87.28492 | 19.33605 |
| Minimum | 530.8947 | 5413.852 | 0.752687 | 42.75409 | 0.007983 | −18.2938 |
| Std. Dev. | 441.3213 | 4630.397 | 0.890965 | 14.0361 | 33.02138 | 13.35156 |
| Skewness | 0.670965 | 0.797313 | −0.58927 | −0.16492 | 0.788473 | −0.14645 |
| Kurtosis | 2.200457 | 2.26713 | 2.601972 | 1.38442 | 1.872208 | 1.412158 |
| Jarque-Bera | 2.846721 | 3.593253 | 1.805272 | 3.172033 | 4.385118 | 3.041538 |
| Probability | 0.240903 | 0.165857 | 0.405499 | 0.20474 | 0.111631 | 0.218544 |

*Source*: Authors' calculations.

**TABLE 5.4**
**Correlation Matrix of the Variables**

| | *Y* | *K* | *REM* | *BM* | *MOB* | *BC*MOB* |
|---|---|---|---|---|---|---|
| *Y* | 1 | | | | | |
| *K* | 0.996798 | 1.00E+00 | | | | |
| *REM* | 0.646768 | 0.609535 | 1 | | | |
| *BC* | 0.879632 | 0.853657 | 0.861705 | 1 | | |
| *MOB* | 0.964323 | 0.977502 | 0.548394 | 0.813399 | 1 | |
| *BC*MOB* | 0.911956 | 0.889775 | 0.797702 | 0.984787 | 0.851052 | 1 |

*Source*: Author's calculations.

to 2017. Table 5.3 presents the descriptive measures of the variables employed and Table 5.4 the shows correlation matrix of the variables.

### 5.3.2 The Model

The model employed in the study is along the lines of the Cobb-Douglas production function, adopted in Luintel, Khan, Arestis, and Theodoridis (2008), Kumar et al. (2015), and Rao, Singh, Singh, and Vadlamannati (2008). Taking into account the constant returns to scale and Hicks-neutral technological development, the real per-capita output ($y_t$) equation is written as:

$$y_t = A_t k_t^{\alpha}, \qquad 0 < \alpha < 1 \tag{5.1}$$

where:

$y_t$ = real GDP per capita;

$A_t$ = stock of technology;
$k_t$ = stock of capital per capita;
$\alpha$ = share of capital.

According to Solow framework, the technological progress is represented by:

$$A_t = A_o e^{gt} \tag{5.2}$$

where:
$A_t$ = represents the aggregate technology;
$A_o$ = the opening stock of technology;
$t$ = time;
$g$ = the exogenous growth rate of technical progress.

In addition to the variables just noted, we also employed an interaction term, which is the product of bank credit (*BC*) and mobile subscription (*MOB*). Consequently, it was decided to adopt the following:

$$A_t = f(k_t, REM_t, BC_t, MOB_t, BC * MOB_t) \tag{5.3}$$

Here:
*REM* = inflow of remittances as share of GDP;
*BC* = bank credit as percentage of GDP;
*MOB* = mobile subscription per 100 inhabitants;
*BC*MOB* = interaction of bank credit and MOB.

Hence, the initial Cobb-Douglas function is amended as:

$$y_t = A_0 e^{\alpha_1 REM_t + \alpha_2 BC_t + \alpha_3 MOB_t + \alpha_4 BC*MOB_t} k_t^{\alpha_5} \tag{5.4}$$

The stochastic model in its single logarithmic form for analysis and estimation reason is expressed as:

$$ly_t = \alpha_0 + \alpha_5 lk_t + \alpha_1 lREM_t + \alpha_2 lBC_t + \alpha_3 lMOB_t + \alpha_4 lBC * lMOB + \varepsilon_t \tag{5.5}$$

The hypotheses to be confirmed are:

1. The explanatory variable capital per capita stock (*ly*) is associated with real output per capita (*lk*); hence, the sign of *lk* is positive.
2. The explanatory variable REM (*lREM*) is expected to positively influence real output per capita (*ly*). Therefore, the sign of *lREM* is positive.
3. The FSD indicator (*lBC*) has a direct association with real output per capita (*ly*); hence, the sign of bank credit (*lBC*) is positive.
4. The ICT indicator (*lMOB*) is anticipated to positively facilitate growth in real output per capita; hence, the sign of (lMOB) is positive.

On the other hand, since there cannot be any *a priori* conclusion about the behaviour of the interaction term, $lBC * lMOB_t$, it is considered appropriate not to formulate a hypothesis but to leave it for the investigation to reach a robust conclusion.

If the interaction term emerges with a positive sign and is found significant as well, it would mean that the combined growth effects of the interaction term are enhanced in a deeper financial system, with FSD and ICT displaying a complementary relationship to each other. However, if the interaction variable turns out to have a negative coefficient and is also significant, ICT and FSD are substitutes for each other. If the interaction variable has a negative sign and is found to be non-significant, the two variables are independent of each other.

To examine the existence of a long-run equilibrium relationship among the variables in the specified model, we employed the bounds-testing procedure advanced by Pesaran, Shin, and Smith (2001). The bounds-testing procedure within the autoregressive distributed lag (ARDL) context has some technical advantages. It enables tests for the existence of a co-integrating relationship between variables even if the underlying regressors are I(0) or I(1) (Pesaran & Shin, 1999). It is also considered more suitable than the Johansen-Juselius multivariate approach when the sample size is small (Mah, 2000; Tang & Nair, 2002). The estimators of the long- and short-run factors are consistent (Pesaran & Shin, 1999).

Given that we do not have prior evidence on the course of long-run co-integration among the variables, we formulate the following unrestricted error correction model equations in the ARDL framework:

$$\begin{aligned}\Delta ly_t &= \alpha_0 + \alpha_1 ly_{t-1} + \alpha_2 lk_{t-1} + \alpha_3 lREM_{t-1} + \alpha_4 lBC_{t-1} + \alpha_5 lMOB_{t-1} \\ &+ \alpha_6 lBC * MOB_{t-1} + \sum_{i=1}^{n} \beta_1 \Delta ly_{t-i} + \sum_{i=0}^{n} \beta_2 \Delta lk_{t-i} + \sum_{i=0}^{n} \beta_3 \Delta lREM_{t-i} \\ &+ \sum_{i=0}^{n} \beta_4 \Delta lBC_{t-i} + \sum_{i=0}^{n} \beta_5 \Delta lMOB_{t-i} + \sum_{i=0}^{n} \beta_6 \Delta lBC * MOB_{t-i} + \varepsilon_t\end{aligned} \tag{5.6}$$

Here, $\Delta$ is the first difference operative and signifies short-term dynamics. The coefficients devoted with a period-lagged variable indicate long-term relationships. Specifically, two steps are required in this method. First, we estimate equation 5.6 using the least squares method. Secondly, the confirmation of a long-run relationship is traced by imposing restriction on the coefficients of the lagged-level variables by equating to zero. Henceforth, the bounds test is based on the F-statistics (or Wald statistics) with the null hypothesis of no co-integration ($H_0$: $\beta_{i1} = \beta_{i2} = \beta_{i3} = \beta_{i4} = \beta_{i5} = 0$) and the alternative hypothesis of a long-run co-integration relationship ($H_1$: $\beta_{i1} \neq \beta_{i2} \neq \beta_{i3} \neq \beta_{i4} \neq \beta_{i5} \neq 0$).

The F-statistic of the bounds test is checked with the lower- and upper-band critical values calculated by Pesaran et al. (2001). However, Narayan (2005) contends that because Pesaran's critical values are on the basis of large observation sizes, they cannot be used in a small sample study. Narayan (2005) calculated another set of values for small samples. Accordingly, we employ the critical values of Narayan (2005). When the estimated F-statistic is higher than the upper-bound critical value, then the null hypothesis is rejected. If the F-statistic is lower than the lower-bound critical value, then the null hypothesis cannot be discarded. When the F-statistic falls between the lower- and upper-bound critical values, then the result is not conclusive.

We then proceed to estimate the long-run elasticity measures and short-run error correction model (ECM). The short-run error-correction-model is useful to study short-run dynamics and to ratify the robustness of the estimated parameters of long-run in regards to equation 5.6. The ECM is specified as shown in equation 5.7:

$$\Delta ly_t = \alpha_0 + \sum_{i=1}^{n} \beta_1 \Delta ly_{t-i} + \sum_{i=0}^{n} \beta_2 \Delta lk_{t-i} + \sum_{i=0}^{n} \beta_3 \Delta lREM_{t-i} + \sum_{i=0}^{n} \beta_4 \Delta lBC_{t-i} + \sum_{i=0}^{n} \beta_5 \Delta lMOB_{t-i} + \sum_{i=0}^{n} \beta_6 \Delta lBC * MOB_{t-i} + \lambda ECM_{t-1} + \varepsilon_t \quad (5.7)$$

The ECM represents the error correction. It is subtracted from the long-term estimated parameters in equation 5.6. The error correction term is expected to be significant and negatively associated with the dependent variable.

## 5.4 RESULTS AND DISCUSSION

The results of analysis at various steps are presented in this section. These include the stationary properties of the series, co-integration analysis, and short-run and long-run analyses.

### 5.4.1 Results for Unit Root Tests

The unit root test results by means of a conventional method are provided in Table 5.5. The Augmented Dickey-Fuller (ADF) unit root test show that the null hypothesis that

**TABLE 5.5**
**Unit Root Test Results**

| Variables | ADF $T_{stat}$ | |
|---|---|---|
| | In Level | In First Diff. |
| *Ly* | 0.677 | 8.072* |
| *Lk* | 1.140 | 4.629** |
| *lREM* | 2.371 | 5.529* |
| *lBC* | 1.133 | 3.218** |
| *lMOB* | 0.259 | 4.155* |
| *lBC*IMOB* | 0.101 | 3.520** |

*Source*: Critical values for ADF test are based on Mackinnon (1996).

Notes: The length lag is based on the Akaike Information Criterion (AIC). The null hypothesis is that a series has a unit root (non-stationary). * and ** indicate 1% and 5% level of significance respectively. They denote rejection of the null hypothesis.

the variables are non-stationary could not be rejected in levels. However, in the first difference form, the null hypothesis that the variables are non-stationery is rejected. Consequently, we arrive at the decision that the variables in the model are integrated of order I (1). The results confirm that the maximum order of integration is one.

### 5.4.2 Co-integration Result

The bounds F-test outcomes are stated in Table 5.6. The estimated F-statistics is 8.731 for equation 5.6, which is greater than the upper-band critical values—implying that the null proposition of no co-integration is rejected. In addition, the estimated result is significant at the 1% level, confirming the presence of long-run relation amongst the variables when real per-capita output (*ly*) is taken as the dependent variable.[11]

### 5.4.3 Discussion of Long-run and Short-run Findings

In the long run, the logs of capital stock and output per capita are positively related (Table 5.7). The estimated share of capital stock is 0.376, which is in line with stylized values for developing countries. It is also consistent with empirical results for developing economies (Rao, 2010). As expected, the coefficient of log REM, which is the measure of elasticity of output with respect to REM, emerged with a positive sign and is found significant at the 5% level both in the long and short run. It suggests that a 1% increase in REM inflow run would induce a 0.034% increase in per capita income; in the short run, a 1% increase would result in a 0.027% increase in per

**TABLE 5.6**
**Results of the Bound F-Test**

| **Dependent Variable/Independent Variable** | | **Calculated F-Statistics** |
|---|---|---|
| *Ly* | | 8.731 |
| *Lk* | | 2.938 |
| *lREM* | | 2.964 |
| *lBC* | | 1.393 |
| *lMOB* | | 2.198 |
| *lBC*lMOB* | | 2.191 |
| **Critical Values from Narayan (2004)[a]** | | |
| **Significance level** | **Lower bound** | **Upper bound** |
| 1% | 4.4 | 5.664 |
| 5% | 3.152 | 4.156 |
| 10% | 2.622 | 3.506 |

Note: [a] Critical values for Bounds test are from Narayan (2004), Case D: restricted intercept and no trend.

**TABLE 5.7**
**Estimated Long-run Coefficients: Dependent Variable (*ly*)**

| Independent Variable | Coefficient | Standard Error | T-Ratio |
|---|---|---|---|
| *Lk* | 0.369 | 0.153 | 2.407** |
| *lREM* | 0.034 | 0.014 | 2.383** |
| *lBC* | 0.231 | 0.058 | 3.969* |
| *lMOB* | 0.321 | 0.062 | 5.132* |
| *lBC×lLMOB* | 0.087 | 0.015 | 5.532* |
| Constant | −2.373 | 0.605 | −3.918** |

$X^2$sc: $X^2(1) = 0.811$, $X^2$ff: $X^2(1) = 0.087$, $X^2$n: $X^2(1) = 0.173$, $X^2$hc: $X^2(1) = 0.995$;

R-square= 0.91, DW – stats. = 2.27, SER = 0.009, AIC = 54.35.

Note: *, and ** designates 1%, and 5% level of statistical significances respectively.

capita income. The finding that REM has been supporting economic growth in India is consistent with findings of similar studies in other LMICs (Guha, 2013; Giuliano & Ruiz-Arranz, 2009).

The results also confirm the hypothesis that an increase in bank credit leads to an increase in real output per capita. The estimated elasticity of output with respect to bank credit is 0.231, which implies that a 1% increase in bank credit raises per capita income by 0.231% in the long run. The effect of MOB representing ICT on per capita output is positive and significant in the long run. Our finding is in line with that of Kumar et al. (2015) and Niebel (2018) who have shown that there is a positive relationship between ICT and economic growth. In the short run, the sign of *lMOB* is positive and statistically significant. The relatively large magnitude of the coefficient of *lMOB* in the long run indicates its predominant role in economic growth.

The interactive term of *lBC* (representing FSD) and *lMOB* (representing ICT) has a positive sign and is statistically significant, showing that they complement each other. By mutually supporting each other, they boost the per capita income in the long run.

The error correction term ($ECM_{t-1}$), which reflects the speed of adjustment towards original equilibrium has the correct, negative sign and is statistically significant at the 1% level. The estimated coefficient (−0.63) in Table 5.8 indicates that deviation from the long-run original equilibrium as a result of any shocks in the current period will be adjusted by around 63% in the next time period. Thus, the model implies a relatively rapid adjustment to the long-run equilibrium relationship.

### 5.4.4 Diagnostic Test Results

We examined the diagnostic test results for the stability of the specified model. Here we took into consideration the following: (1) Lagrange multiplier test of serial correlation ($X^2$sc), (2) Ramsey's RESET test for correct functional form ($X^2$ff) using

**TABLE 5.8**
**Estimated Short-run Coefficients: Dependent Variable ($\Delta ly$)**

| Independent Variable | Coefficient | Standard Error | T-Ratio |
|---|---|---|---|
| $\Delta ly_{t-1}$ | 0.535 | 0.158 | 3.380* |
| $\Delta lk_t$ | 0.305 | 0.153 | 1.983*** |
| $\Delta lREM_t$ | 0.027 | 0.012 | 2.237** |
| $\Delta lBC_t$ | 0.012 | 0.005 | 2.394** |
| $\Delta lMOB_{t-1}$ | 0.013 | 0.006 | 1.986*** |
| $\Delta lBC \times lMOB_t$ | 0.050 | 0.015 | 3.260* |
| $ECM_{t-1}$ | −0.630 | 0.153 | −4.112* |
| *Constant* | −2.373 | 0.605 | −3.198* |

R-square = 0.801, $\bar{X}\Delta ly = 0.044$, $\hat{\sigma}\Delta ly = 0.021$, F-stat. (8, 26) = 12.607 (0.001).

Note: *, **, and *** designate 1%, 5%, and 10% levels of statistical significance respectively.

square of the fitted values, (3) Jarque-Bera's normality ($X^2n$) test, and (4) test for heteroscedasticity ($X^2hc$) using the regression of squared residuals on squared fitted values. The findings of the aforementioned tests are reported in Table 5.7. It indicates that the specified model does not experience any classical econometric problem. There is no serial correlation. The test results also confirm that the sample is normally distributed. The functional form is correct and the presence of homoscedasticity cannot be rejected.

## 5.5. CONCLUSIONS WITH POLICY IMPLICATIONS

This chapter undertook an empirical investigation of India's economic growth and remittances nexus with specific focus on ICT as a contingency factor during the 28-year period of 1990 to 2017. The study adopted the Solow framework with a Cobb-Douglas production function with per capita real GDP ($y$) as a dependent variable and per capita capital stock ($k$) as a fundamental, independent variable along with the chosen variables for the study, acting as shift variables: remittances as percent of GDP (*REM*); bank credit as percent of GDP (*BC*), representing financial sector development; and mobile phone subscriptions per 100 inhabitants (*MOB*) as a proxy for ICT. We also added an interaction variable, that is, the product of *BC* and *MOB*, to check whether financial sector development and ICT were acting as complements to or substitutes for each other.

The bounds-testing methodology within the ARDL framework (Pesaran et al., 2001) which has been found to be consistent and suitable for relatively small sample size was employed. Following the unit root tests, which showed that all variables were of integrated order of one, long-run and short-run models were estimated. The tests showed that all the variables were found to be co-integrated and the causality linkage ran only from logs of $k$, *REM*, *BC*, *MOBS* and the interaction variable to log of $y$. All the coefficients, which are elasticity estimates of respective variables, emerged with positive signs, which were also found to be statistically significant. A 1% rise in *REM* was found to result in a 0.034% increase in per capita real GDP; a 1% rise in *BC* led to a 0.231% increase in per capita real GDP; and a 1% rise in *MOB* gave rise to a 0.321% increase in per capita real GDP. The interaction variable was found to have a positive sign, which confirmed the existence of a complementary relationship between FSD and ICT.

From the policy perspective, it is clear REM and ICT are essential drivers of output growth. It is obvious that the advent of ICT has broken a new ground. Mobile and Internet banking have contributed to making banking operations less expensive, without any additions to brick and mortar branches, and requiring fewer visits by customers to banks. The costs of banking operations have come down as the maintenance costs of records have been drastically reduced. Payments procedures through use of mobile devices have also lessened the need for holding cash by consumers and entrepreneurs. Above all, savings from REM by the recipient families are now more swiftly deposited in banks as deposits, which were once frittered away on needless consumption. Young and ambitious entrepreneurs in the rural areas, who were hitherto denied of banking facilities and hence had no opportunity to put their savings and borrow from them, are now able to have access to bank credit.

The ICT revolution, which is still unfinished in India as the digital divide between urban and rural areas is still wide, should be carried on with greater vigour. First and foremost is the need for improving telecommunication and electricity infrastructure. Inadequacies in these two fundamentals hamper a widespread use of ICT in all economic activities. Improving the accessibility and affordability of ICT services in Indian economy would not only reduce the present digital divide but also enhance the scale-up effect of ICT use. The efficient use of present-generation phone technologies like 3G and 4G is necessary.

In catching-up with the rest of the world, ICT should be promoted more effectively in the key areas of the economy with main focus in banking and finance, which will increase efficiency and productivity, and enhance long-term economic growth. The urban areas are with better off and educated groups absorb ICT culture, while the rural folks tend to get less involved as their capacity to fully benefit from ICTs is limited.

The impact of ICT is dependent on a host of factors relating to user characteristics and environment. The spread of ICT is primarily influenced by an enabling policy and regulatory environment and through investment in infrastructure and improved digital literacy. The digital literacy, in turn, depends on the quality of education now offered in primary and secondary schools in rural areas. Again, the general urban–rural divide continues to dominate the economic scene in terms of investment in infrastructure and equipment in educational institutions in rural areas, just as there are inequities in the provision of health and sanitation services between cities and villages. The whole picture then changes into one of balanced development, bridging the geographical inequalities aside from economic inequities.

Furthermore, improvement in ICT would not only promote economic growth, but if properly accomplished, it could become a crucial enabler of business innovation, job creation, and new services and industries. To this end, decision makers may consider policy incentives such as tax exemption for ICT industries and duty-free importation of technological devices.

## NOTES

1 The criteria observed by the World Bank (2017b) for classifying the countries on the basis of gross national income per capita are as follows:
Low-income countries: GNI less than $1,025 per capita.
Lower middle-income countries : GNI between $1,026 per capita and $4,035 per capita
Upper middle-income countries: GNI between $4,036 per capita and $12,475 per capita
High-income countries: GNI greater than $12,475 per capita

2 The 2014 Intermedia Financial Inclusion Insight Survey (IFIIS) of 45,000 Indian adults conducted by the Consultative Group to Assist the Poor (CGAP) found that 0.3% of adults used mobile money, compared to 76% in Kenya, 48% in Tanzania, 43% in Uganda, and 22% in Bangladesh.

3 Successful stories have been reported in Indian newspapers and in social media. These include: Ghosh (2017) and Hindustan Times (2017).

4 The World Bank Group and KNOMAD (2018) have estimated that in 2018 remittances were US$528 billion, FDI US$473, and foreign aid US$150 billion, while portfolio private investment in short-term debt and equity instruments was US$156 billion.

5 These included setting up postal savings banks and rural and urban cooperative banks. These efforts culminated in the nationalization of private banks in two tranches, one in 1969 and another in 1989, as part of "quit poverty strategy goals".

6 In South Asia, Nepal is a low-income country, with Bangladesh, Bhutan, India, Pakistan, and Sri Lanka assigned to the group of low middle-income countries.

7 Remittances were once called *unrequited transfers*, because they do not have any *quid pro quo* element. There are no expectations of financial return in terms of interest or dividend, unlike the long-term FDI or portfolio investments of short-term nature.

8 India's foreign reserves touched their highest level ever at US$426 billion in April 2018 due to inflows of hot moneys, as the expectations of speculators were then high. However, export earnings from goods and services had been falling ($317.7 billion in 2014, $266.2 billion in 2015, and $261.9 billion in 2016). With higher imports, annual current account deficits began to widen. Steadily rising remittances however reduced the current account deficits: $26.8 billion in 2014, $22.1 billion in 2015, and $15.2 billion in 2016, or in percentages of GDP, –1.3%, 1%, and –0.7% respectively. Remittances have been a major support, rendering the current account deficits smaller and manageable.

9 The net inflows into investment and debt instruments were $42.2 billion in 2014–15. They dived to a negative figure when outflows surpassed inflows of –$4.1billion in 2015–16. They climbed again to $7.6 billion by 2016–17 (Jayaraman, 2018).

10 The financial sector institutions comprise 93 scheduled banks, of which 27 are public and 21 are private; the rest are owned by foreign interests. The other institutions include 95,000 cooperative banks, 56 regional rural banks, post office banks, and 53 insurance companies.

11 Similar specification is tested using other variables where; $lk_t$, $lREM_t$, $lBC_t$, $lMOB_t$, and $lBC*lMOB_t$ were treated as dependent variables to estimate the F-statistics.

## REFERENCES

Aghaei, M., & Rezagholizadeh, M. (2017). The impact of information and communication technology (ICT) on economic growth in the OIC countries. *Economic and Environmental Studies*, *17*(42), 255–276.

Chakrabarty, K. C. (2011, October 24). *Financial inclusion and banks: Issues and perspectives at the FICCI, speech delivered at the UNDP Seminar on Financial Inclusion.* New Delhi. Retrieved April 19, 2018, from https://rbi.org.in/scripts/BS_SpeechesView.aspx?Id=608

Desai, R., Agarwal, A., & Arya, C. (2017, October 20). How digitisation can push India's growth per capita incomes in the coming decade. *Economic Times*. Retrieved from https://economictimes.indiatimes.com/news/economy/indicators/how-digitisation-can-push-indias-growth-and-per-capita-incomes-in-the-coming-decade/articleshow/61158292.cms?from=mdr

The Economist (2019, January 27). Mukesh Ambani aims to become India's first internet Tycoon. *The Economist*. Retrieved from https://www.economist.com/business/2019/01/26/mukesh-ambani-wants-to-be-indias-first-internet-tycoon

Edwards, A. C., & Ureta, M. (2003). International migration, remittances and schooling: Evidence from Salvador. *Journal of Development Economics*, *51*, 387–411.

Freund, B., König, H., & Roth, N. (1997). Impact of information technologies on manufacturing. *International Journal of Technology Management*, *13*(3), 215–228.

Giuliano, P., & Ruiz-Arranz, M. (2009). Remittances, financial development and growth. *Journal of Development Economics*, *90*(1), 144–152.

Ghosh, M. (2017). *Why this small Gujarati village is not affected by demonetisation – Three live examples of a cashless economy*. Retrieved from http://trak.in/tags/business/2016/11/28/digital-village-akodara-cashless-economy/

Guha, P. (2013). Macroeconomic effects of international remittances: The case of developing economies. *Economic Modelling*, *33*, 292–305.

Hildebrandt, N., & McKenzie, D. J. (2005). *The effects of migration on child health in Mexico* (Policy Research Working Paper Series 3573). Washington, DC: World Bank.

Hindustan Times. (2017, November 12). Amid banknotes chaos, 'digital' village that turned cashless is an oasis of calm. *Hindustan Times*. Retrieved from https://www.hindustantimes.com/india-news/amid-banknotes-chaos-digital-village-that-turned-cashless-is-an-oasis-of-calm/story-BRMwBCXMR7MGMiZs8dG8bM.html

Jayaraman, T. K. (2018, 21 August). What remittances do for the economy. *Deccan Herald*, p. 1.

Jayaraman, T. K., Choong, C. K., & Kumar, R. (2012). Role of remittances in India's economic growth. *Global Business and Economic Review*, *14*(3), 159–177.

Kumar, K., & Radcliffe, D. (2015, January 15). 2015 Set to be big year for digital financial inclusion in India. Retrieved from CGAP: http://www.cgap.org/blog/2015-set-be-big- year-digital-financial-inclusion-India

Kumar, R. R., Kumar, R. D., & Patel, A. (2015). Accounting for telecommunications contribution to economic growth: A study of small pacific Island state. *Telecommunications Policy*, *39*(3), 284–295.

Leon-Ledesma, M., & Piracha, M. (2001). *International Migration and the role of Remittances in Eastern Europe* (Studies in Economics, No. 0113). Department of Economics, University of Kent.

Luintel, K. B., Khan, M., Arestis, P., & Theodoridis, K. (2008). Financial structure and economic growth. *Journal of Development Economics*, *86*(11), 181–200.

MacKinnon, J. G. (1996), Numerical distribution functions for unit root and cointegration tests. *Journal of Applied Econometrics*, *11*, 601–618.

Mah, J. S. (2000). An empirical examination of the disaggregated import demand of Korea – The case of information technology products. *Journal of Asian Economics*, *11*, 237–244.

Majeed, T. M., & Ayub, T. (2018). Information and communication technology and economic growth nexus: A comparative global analysis. *Pakistan Journal of Commerce and Social Science*, *12*(2), 443–476.

Mashayekhi, M. (2014). *Remittances and financial inclusion*. Paper presented at the 13th Coordination Meeting on International Migration, New York, 12–13 February 2015, Geneva: UNCTAD.

Mohan, R., & Ray, P. (2017). *Indian financial sector: Structure, trends and turns* (IMF Working Paper 17/7). Washington, DC: International Monetary Fund.

Narayan, P. K. (2004). Fiji's tourism demand: The ARDL Approach to Cointegration. *Tourism Economics*, *10*(2), 193–206. Retrieved from SSRN: https://ssrn.com/abstract=2079394

Narayan, P. K. (2005). The saving and investment nexus for China: Evidence from cointegration tests. *Applied Economics*, *37*(17), 1979–1990.

Niebel, T. (2018). ICT and economic growth – Comparing developing, emerging and developed countries. *World Development*, *104*(C), 197–211.

Page, J., & Adams, R. H. Jr. (2003). *International migration, remittances and poverty in developing countries* (World Bank Policy Research Working Paper No. 3179). Washington, DC: World Bank.

Pesaran, M. H., & Shin, Y. (1999). An autoregressive distributed lag modeling approach to cointegration analysis. In S. Strom, A. Holly, & P. Diamond (Eds.), *Centennial volume of rangar frisch*. Cambridge: Cambridge University Press.

Pesaran, M. H., Shin, Y., & Smith, R. J. (2001). Bounds testing approaches to the analysis of level relationships. *Journal of Applied Economics*, *16*, 289–326.

Pew Research Center. (2019). Mobile technology and its social impact. Retrieved March 10, 2019, from www.pewresearch.org/fact-tank/2019/03/07/7-key-findings-about-mobile-phone-and-social-media-use-in-emerging-economies/

Radeck, L., Wenninger, J., & Orlow, D. K. (1997). Industry structure: Electronic delivery potential effects on retail banking. *Journal of Retail Banking Services*, *XIX*(4).

Rao, B. B. (2010). Estimates of the steady state growth rates for selected Asian countries with an extended Solow model. *Economic Modelling*, *27*, 46–53.

Rao, B., Singh, T. A., Singh, R., & Vadlamannati, K. C. (2008). *Financial developments and the rate of growth of output: An alternative approach* (MPAR Paper No. 8605). Retrieved from http://mpra.ub.uni-muenchen.de/8605

Reserve Bank of India. (2008). *The banking sector in India: Emerging issues and challenges*. Mumbai: RBI.

Rukhaiyar, A. (2018, December 25). Foreign fund outflows: Highest since 2008. *The Hindu*. Retrieved from https://www.thehindu.com/business/Industry/fpi-outflow/article25828495.ece

Siddique, A., Selvanathan, E. A., & Selvanathan, S. (2010). Remittances and economic growth: Empirical evidence from Bangladesh, India and Sri Lanka. *The Journal of Development Studies*, *48*(8),1045–1062.

Stahl, C. W., & Habib, A. (1989). The impact of overseas workers remittances on indigenous industries: Evidence from Bangladesh. *The Developing Economies*, *27*, 269–285.

Tang, T. C., & Nair, M. (2002). A cointegration analysis of the Malaysian import demand function: Reassessment from the bounds test. *Applied Economics Letters*, *9*, 293–296.

United Nations. (2015). *Millennium development goals and post-2015 development Agenda*. Retrieved August 25, 2016, from www.un.org/en/ecosoc/about/mdg.shtml/post2015developmentframework.shtml

United Nations, Department of Economic and Social Affairs, Population Division. (2017). *World population prospects: The 2017 revision, key findings and advance tables* (Working Paper No. ESA/P/WP/248).

Wilson, D. (1993). Assessing the impact of information technology on organizational performance. In R. Banker, R. Kanffan, & M. Mahmood (Eds.), *Hanbury: Strategic information technology management idea group* (pp. 471–514).

World Bank. (2017). *Migration and remittances: Recent developments and outlook*. Retrieved June 20, 2017, from www.worldbank.org/en/topic/migrationremittancesdiasporaissues

World Bank. (2018). *Press note on release on remittances brief 30*. Retrieved April 19, 2019, from www.worldbank.org/en/news/press-release/2019/04/08/record-high-remittances-sent-globally-in-2018

World Bank. (2019). *World development indicators*. Washington, DC: World Bank, Retrieved May 27, 2019, from https://databank.worldbank.org/data/reports.aspx?source=world-development-indicators#

Yang, D. (2008). International migration, remittances and household investment: Evidence from Philippines. *Economic Journal*, *118*(528), 591–630.

# 6 Effects of Capital Adequacy on Operational Efficiency of Banks: Evidence from Bangladesh

*Md. Nur Alam Siddik and Sajal Kabiraj*

CONTENTS



## 6.1 INTRODUCTION

The main task of a commercial bank is to extend loans and the greater portion of its total assets is shaped by loans. This core function can be well performed by banks only when they run in a more efficient manner. In order to induce a sound and safe financial system, regulators require banks to keep adequate capital to meet any losses incurred and reduce moral hazard behavior (Fungáčová, Solanko, & Weill, 2014).

Minimum capital requirements form the base of contemporary banking regulation, and holding such capital comes with a cost, such as trading off financial stability for less liquidity (and efficiency), and inducement of banks to optimize their risk taking. In regard to this, Blum (1999) found evidence that a bank may value an additional unit of equity tomorrow more when there are minimum capital requirements than when such requirements are non-existent. Mitchell (1984) argues that capital forms two functions in a bank, namely, financing purchase of assets and protecting creditors. Banks argue that loan loss reserves should be included in defining bank capital because these accounts perform some of the functions of capital for banks. The Basel Committee on Banking Supervision endorsed that banks should maintain capital at a level so as to diminish the chance of bank failures. This is termed as *capital adequacy requisite*, and it requires banks to maintain a minimum capital-to-assets ratio to remain in operation. This requirement of more capital would eventually make the banks safer, though it will increase their cost of capital. The objectives of the constraint can result in either preventing the banks from taking more risk to enhance their profits or promoting financial stability that provides a safeguard against systemic crises. The Basel Accord was mainly devised as an instrument to control and monitor banks' risk-taking behavior. The reducing chance of investments through increasing capital adequacy and will be insolvent. The lower the profitability of banks in terms of higher the risk-weighted capital adequacy ratios.

In order to reduce insolvency and promote stability, recently bank supervisors advised banks to hold minimum regulatory capital levels (Aggarwal & Jacques, 2001). This supervisory stress conveys disciplinary guidance to bank managers. Capital adequacy is also considered as an indispensable instrument to safeguard banks' creditworthiness and profitability. This is because of the existence of probable information asymmetry between banks and parties that may end in default of loans. Banks are required to have sufficient capital, not only to stay solvent, but also to withstand catastrophes in the financial system and to ensure operational efficiency (Aggarwal & Jacques, 2001).

Allen and Rai (1996) stated that to bring operational efficiency banks must provide quality banking services to the customers at the lowest possible cost. More specifically, banks operate efficiently by supplying loans to those customers who have been well screened and have a good record of repayment (Athanasoglou, Brissimis, & Delis, 2008). Once the loan is provided, careful monitoring of borrowers increases operational efficiency of banks by reducing the probability of default. Technical efficiency of banks, such as adoption of new technologies while providing services to customers and innovation of products tailored to customers, also helps banks to achieve operational efficiency.

There is dispute and contradicting views on the issue of whether the regulatory requirement for capital promotes bank efficiency and performance worldwide. Some researchers have observed that capital adequacy has positive effects on bank performance, whereas others argue that bank capital regulation has a negative relationship with performance and efficiency of banks.

This research adds to the literature by providing empirical evidence of the impacts of the capital adequacy requirement on the operating efficiency of banks operating in Bangladesh. In Bangladesh, commercial banks' financial intermediation processes are characterized by challenges emanating from high business deal costs arising from rising interest rates; high information asymmetry between banks, investors, and borrowers, which can give rise to adverse selection and moral hazards; low liquidity owing to little savings as compared to consumption by a mass of households; and problems in delegated monitoring before and after credit competence is advanced. The level of capital is crucial the same as riskiness of bank deposits in worried. A bank with inadequate capital is more likely to go bankrupt in the face of unfavourable growth on its asset than a satisfactorily capitalized one. Capital, being an important managerial conclusion variable, has theoretically been seen to influence a bank's capital structure and the loan policy for the function of credit creation and overall wealth maximization. This has implications on the performance of banks as financial intermediaries, and hence for the allotment of real resources within the economy. This research therefore sought to address these gaps and analyze in detail the relationship between capital adequacy and operational efficiency of Bangladeshi banks.

## 6.2 LITERATURE REVIEW

The capital base of a financial institution helps it to absorb shocks in the financial market. It also signals that the institution will continue to respect its obligations. Bichsel and Blum (2005) found that capital regulations help in reducing negative externalities (e.g. general loss of confidence in the banking system) in addition to boosting GDP. A minimum quantity of capital is essential to protect the bank and to assure its customers of its financial stability.

The overall capital adequacy ratio (CAR) measures the amount of a bank's core capital stated as a ratio of its weighted credit exposures. Adequate CAR helps banks to absorb unexpected shocks and also indicates that the financial institution will honor its obligations. Capital adequacy eventually determines how well financial institutions can handle shocks to their balance sheets (Haron, 2004).

### 6.2.1 Previous Empirical Studies

A number of studies have been conducted that are relevant to this study. Thakor (1996) argued that an increase in capital regulations stimulates banks to reexamine their inside operations policies with regard to the issue of corporate governance, risk valuation techniques, the credit appraisal processes, the hiring of a more competent and trained workforce, and better internal control measures. According to Thakor (1996), banks with more capital are able to invest in more profitable projects, magnify operations, and manage estimated levels of risks. Thus, the capital adequacy requirement is estimated to have a positive effect on bank efficiency.

Applying cross-county data of 72 countries over the period of 1999–2007, Barth, Lin, Ma, Seade, and Song (2013) examined the effects of bank supervision on operation efficiency of banks and observed that banks of those countries which had adopted stringent capital adequacy requirements were found to be more efficient compared to those banks from countries which had adopted flexible capital adequacy requirements. Chortareas, Girardone, and Ventouri (2012) employed data from 2000 to 2008 on 22 EU countries and found that capital requirements had significant effects on the operating efficiency of banks. In contrast, Altunbas, Carbo, Gardener, and Molyneux (2007) conducted a study on banks in 15 European countries for the period of 1992–2000 and found significant negative impacts of bank capital adequacy on operating efficiency of banks.

Hahn (2002) studied the effects of Basel I on credit growth of 750 universal banks in Austria during the period 1996–2000 using a panel-econometric approach. To define the impacts of the opening of Basel I from other shocks, the author proscribed the impacts caused by loan demand shocks by counting several variables, such as the collective output gap and the collateral value of real estate. The findings showed that minimum capital holding had a negative impact on credit creation in that country. The author also provided proof that amount of obtainable bank capital may work as a binding restraint on liquidity and credit creation.

Diamond and Rajan (2000) also observed that an upturn in the capital adequacy ratio can result in a credit crisis for the poor and can possibly lessen the debt weight of the rich as more security has an unfavourable distributional penalty. Marvin, Lawrence, Angelo, and Tereso (2012), using capital adequacy, management quality, asset quality, earnings performance, and liquidity (CAMEL framework), conducted empirical tests to evaluate the possible impact of economic, regulatory, and bank-specific characteristics on bank intermediation and credit creation in the ASEAN+3 region. Data for the period 2006–2010 exposed, among other things, that bank equity matters in net interest margin but not in the purpose of net loans and regulations do not have uniform effects. In terms of the effects of regulatory variables, the amplification of the reserve requirement reduces the capability of banks to make loans.

Bikker and Hu (2012) employed data on banks of 26 countries and found no support for the credit crisis research. As banks naturally hold the least amount of capital necessary, they found that capital requirements do not seem to be compulsory restrictions on advance supply.

Furfine (2001) incorporated explanations into a hypothetical model that used real US bank data to replicate bank reactions to changes in capital requirements. The author found that collective advancing in the US fell in the early 1990s for the following reasons: (1) greater regulatory scrutiny, (2) less demand for loans due to the Great Recession, and (3) higher capital requirements mandated by Basel I. The author states that "some form of regulatory participation, either raising capital requirements or increasing regulatory monitoring, was a necessary supplier to the credit crisis. That is, the experiential portfolio adjustment undertaken in the early 1990s could not have been merely the result of altering economic circumstances or worldly change" (Furfine, 2001).

Honda (2002) examined credit creation by Japanese banks and found that execution of Basel I decreased aggregate bank lending by a significant amount. Using New England data, Peek and Rosengren (1995) found that credit availability was not connected to episodes of disintermediation but rather due to banks facing binding capital constraints, an experience they termed "capital crunch". They found that it was hard to divide the diminishing in demand for loans that occurred in the collapse from the diminished supply of loans. To alleviate this, they used cross-section data on New England banks facing a similar local economic downturn and found that banking institutions facing a capital crisis regularly modified their balance sheets by either issuing new securities (to raise capital) or regularly switching to assets that required less equity, from the ones that required more, and therefore reduced loan ease of use to businesses, exacerbating the critical situation (Peek & Rosengren, 1995). Mwega (2009) found that capital requirements minimize the likelihood that banks will go bankrupt if sudden shocks happen. He distinguished that the higher the risk-weighted capital adequacy ratio, the lower the probability that commercial banks will be exposed to the risk of insolvency, and that therefore a negative connection exists between the risk-weighted adequacy ratio and insolvency of commercial banks.

## 6.3 RESEARCH METHODOLOGY

### 6.3.1 Sample and Data Collection

Out of 41 private commercial banks, 23 commercial banks were undertaken on the basis of data availability. This research used secondary data from audited annual reports over a five-year period of 2013 to 2017.

### 6.3.2 Econometric Model

$$OPEF_{it} = \alpha_0 + \beta_1 CAR_{it} + \beta_2 ROA_{it} + \beta_3 LQD_{it} + \beta_4 SZ_{it} + \beta_5 NPL_{it} + \varepsilon_{it} \quad (6.1)$$

In the model, $\alpha_0$ is the constant term. *OPEF* refers to the operational efficiency of bank *i* in period *t*. *CAR* is the capital adequacy ratio. *ROA* signifies the return on asset, *LQD* refers to liquidity, *SZ* is the size of the bank, and NPL refers to non-performing loans, $\beta_{1-5}$ are the regression coefficients of respective independent variables, and $\varepsilon$ is the error term.

### 6.3.3 Definition of Variables

#### 6.3.3.1 Dependent Variable

The operating efficiency ratio (OPEF) refers to how much expenditure is occurred than operating income to run smoothly operating activities that affect bank's performance. This ratio is determined as follows:

$$OPEF = \frac{\text{Operating expense}}{\text{Operating income}}$$

#### 6.3.3.2 Independent Variables

Capital adequacy ratio (*CAR*): The total capital of the bank articulated as a percentage of its total risk-weighted assets. The formula for CAR is as follows:

$$CAR = \frac{\text{Total capital funds}}{\text{Total risk-weighted assets}}$$

Return on asset (*ROA*): The returns on assets of the bank measured by the ratio of the bank's profits over the bank's total assets. This ratio shows the effectiveness of management in the use of shareholders' funds. We calculate *ROA* as follows:

$$ROA = \frac{\text{Net income}}{\text{Total assets}}$$

Liquidity ratio (*LQD*): Liquidity ratio refers the ability of probable investment as loans using total deposits. The relationship of liquidity is positive with return on investments. The formula we used is as follows:

$$LQD = \frac{\text{Net loans}}{\text{Total deposits}}$$

Bank's size (SZ): This is measured by taking the natural logarithm of the total assets of a bank.

Non-performing loan (*NPL*): This is a gauge of credit risk management that explains how banks cope with their credit risk and is measured by the share of bad loans over the total assets of the bank.

## 6.4 RESULTS AND DISCUSSIONS

From Table 6.1, we find that operating efficiency averaged 52% with minimum and maximum values of 28.64 and 94.01 respectively. This means that typically the operational expenses of any bank will encompass only 52% of the bank's total income. Capital adequacy has a mean of 13.4508 and minimum and maximum values of 6.62 and 80.43 respectively. This implies that most of the banks in Bangladesh maintain the minimum capital adequacy ratio. Bangladesh imposes a greater capital requirement to establish a new bank. Then, commercial bank will fall liquidity crisis for investments, instant demands of clients. Liquidity has a mean of 72.9342 with minimum and maximum values of 2.07 and 99.65 respectively. It indicates that banks do not depend on other sources of funds.

The correlation matrix of the variable included in the model is presented in Table 6.2. The correlation matrix shows that the data are random, implying that the model is reliable and stable. Lower correlation implies that there is low multicollinearity, and as such not an issue in our analysis.

### 6.4.1 Regression Results

In determining whether to use fixed-effect analyses and random-effect analyses, we used the Hausman test, which examines whether the null hypothesis displays a

**TABLE 6.1**
**Descriptive Statistics**

| Variable | Obs. | Mean | Std. Dev. | Min. | Max. |
|---|---|---|---|---|---|
| OPEF | 115 | 52.8009 | 3.3680 | 28.64 | 94.01 |
| CAR | 115 | 13.4508 | 5.0945 | 6.62 | 80.43 |
| ROA | 115 | 1.1148 | 0.5424 | 0.01 | 3.81 |
| LDR | 115 | 72.9342 | 12.4889 | 2.07 | 99.65 |
| SZ | 115 | 12.1906 | 0.8973 | 8.97 | 14.55 |
| NPL | 115 | 1.4517 | 2.3971 | 0.00 | 8.68 |

**TABLE 6.2**
**Correlation**

| | OPEF | CAR | ROA | LDR | SZ | NPL |
|---|---|---|---|---|---|---|
| OPEF | 1.0000 | | | | | |
| CAR | 0.1264 | 1.0000 | | | | |
| ROA | −0.3942 | 0.1838 | 1.0000 | | | |
| LQD | −0.1283 | −0.2136 | −0.2136 | 1.0000 | | |
| SZ | 0.2614 | −0.5406 | −0.2927 | 0.1568 | 1.0000 | |
| NPL | 0.0278 | −0.0606 | −0.308 | 0.3025 | −0.0015 | 1.0000 |

**TABLE 6.3**
**Fixed Effect Regression Results**

| Explanatory Variable | Dependent Variable: OPEF | | | |
|---|---|---|---|---|
| | Coefficient | Robust SE | *t* | *P* > \| *t* \| |
| CAR | 0.200327 | .0620083 | 3.23 | 0.002*** |
| ROA | −2.12365 | 1.066076 | −4.81 | 0.000*** |
| LQD | −0.07638 | .0600035 | −2.27 | 0.018** |
| SZ | 0.947103 | 1.476455 | 0.64 | 0.524 |
| NPL | −.2312179 | .3983501 | −2.33 | 0.088* |
| _cons | 6.74438 | 1.25426 | 2.88 | 0.006*** |
| No. of obs. | 115 | | | |
| R-squared | 0.4169 | | | |
| F-statistic | 11.08*** | | | |

Note: * signifies variable significant at 10%; ** signifies variable significant at 5%; and *** signifies variable significant at 1%.

statistical dissimilarity (Hausman, 1978). If so, the fixed-effect regression is a better fit to the analysis. Otherwise, the random-effect regression model should be used. Based on the results of the Hausman test, we employed fixed-effects regression analysis.

Table 6.3 shows that the capital adequacy ratio (*CAR*) has significant positive effects on operating efficiency of banks, which implies that strict capital regulation improves banks' operational efficiency. This finding is similar to the findings of Das and Ghosh (2006). The main implication of this finding is that it will stimulate regulators to adopt bank capital regulations and at the same time banks will have the impetus to maintain the CAR because it improves their operational efficiency. Table 6.3 also shows that return on asset (*ROA*), liquidity (*LQD*), and non-performing loan (*NPL*) have significant negative impacts on the operating efficiency of banks. Meanwhile, we found bank size had a positive but insignificant impact on operating efficiency of the banks.

## 6.5 CONCLUSION

The banking sector constitutes a main component of financial trade. The creation of credit forms the core business of every bank by utilizing 85% of deposits available (Saunders, Cornett, & McGraw, 2006). Financial institutions have a great role in financing economic growth, and this is one reason that they are highly monitored through various regulatory measures. The capital adequacy requirements may have played a major role in several bank mergers, acquisitions, conversions, and liquidations that occurred in Bangladesh for compliance purposes. It has also been proved that in this new competitive environment large banks will survive and small banks will only survive if they specialized in a few of their activities. This study aimed at establishing the effect of capital adequacy requirements on operational efficiency of commercial banks in Bangladesh. Employing data of 23 Bangladeshi commercial banks over the period of 2013–17, fixed-effects regression results indicate that the capital adequacy ratio has a positive impact on the operational efficiency of banks. From the research findings, we conclude that there is significant positive rapport between capital adequacy requirements and operating efficiency of commercial banks in Bangladesh. This study was limited only to the factors that originate from capital adequacy requirements; it did not examine interest rate shocks, variations in demand for credit, and other macroeconomic shocks that are equally vital.

## REFERENCES

Aggarwal, R., & Jacques, K. T. (2001). The impact of FDICIA and prompt corrective action on bank capital and risk: Estimates using a simultaneous equations model. *Journal of Banking & Finance*, *25*(6), 1139–1160.

Allen, L., & Rai, A. (1996). Operational efficiency in banking: An international comparison. *Journal of Banking & Finance*, *20*(4), 655–672.

Altunbas, Y., Carbo, S., Gardener, E. P., & Molyneux, P. (2007). Examining the relationships between capital, risk and efficiency in European banking. *European Financial Management*, *13*(1), 49–70.

Athanasoglou, P. P., Brissimis, S. N., & Delis, M. D. (2008). Bank-specific, industry-specific and macroeconomic determinants of bank profitability. *Journal of International Financial Markets, Institutions and Money*, *18*(2), 121–136.

Barth, J. R., Lin, C., Ma, Y., Seade, J., & Song, F. M. (2013). Do bank regulation, supervision and monitoring enhance or impede bank efficiency? *Journal of Banking & Finance*, *37*(8), 2879–2892.

Bichsel, R., & Blum, J. (2005). Capital regulation of banks: Where do we stand and where are we going? *Swiss National Bank, Quarterly Bulletin*, *4*, 42–51.

Bikker, J. A., & Hu, H. (2012). Cyclical patterns in profits, provisioning and lending of banks and procyclicality of the new Basel capital requirements. *PSL Quarterly Review*, *55*(221).

Blum, J. (1999). Do capital adequacy requirements reduce risks in banking? *Journal of Banking & Finance*, *23*(5), 755–771.

Chortareas, G. E., Girardone, C., & Ventouri, A. (2012). Bank supervision, regulation, and efficiency: Evidence from the European Union. *Journal of Financial Stability*, *8*(4), 292–302.

Das, A., & Ghosh, S. (2006). Financial deregulation and efficiency: An empirical analysis of Indian banks during the post reform period. *Review of Financial Economics*, *15*(3), 193–221.

Diamond, D. W., & Rajan, R. G. (2000). A theory of bank capital. *The Journal of Finance*, *55*(6), 2431–2465.

Fungáčová, Z., Solanko, L., & Weill, L. (2014). Does competition influence the bank lending channel in the euro area? *Journal of Banking & Finance*, *49*, 356–366.

Furfine, C. (2001). Bank portfolio allocation: The impact of capital requirements, regulatory monitoring, and economic conditions. *Journal of Financial Services Research*, *20*(1), 33–56.

Hahn, F. R. (2002). *The effects of bank capital on bank credit creation: Panel evidence from Austria* (No. 188). WIFO Working Papers.

Haron, S. (2004). Determinants of Islamic bank profitability. *Global Journal of Finance and Economics*, *1*(1), 11–33.

Hausman, J. A. (1978). Specification tests in econometrics. *Econometrica: Journal of the Econometric Society*, *46*(6), 1251–1271.

Honda, Y. (2002). The effects of the Basel accord on bank credit: The case of Japan. *Applied Economics*, *34*(10), 1233–1239.

Jensen, M. C., & Meckling, W. H. (1976). Theory of the firm: Managerial behavior, agency costs and ownership structure. *Journal of Financial Economics*, *3*(4), 305–360.

Marvin, R. F., Lawrence, B. D., Angelo, B. T., & Tereso, S. T. (2012). Research report. *The roles and functions of the banking sector in the financial system of the ASEAN + 3 region*. Manila, Philippines: De La Salle University – Angelo King Institute (DLSU-AKI). Retrieved January 20, 2019, from https://www.asean.org/uploads/2012/10/Roles%20and%20Functions%20of%20the%20Banking%20Region_Angelo%20King%20Institute_Final.pdf

Mitchell, K. (1984). Capital adequacy at commercial banks. *Economic Review*, 69, 17–30.

Mwega, F. M. (2009). *Global financial crisis discussion series-Kenya* (No. 7). London: Overseas Development Institute.

Peek, J., & Rosengren, E. (1995). Bank regulation and the credit crunch. *Journal of Banking & Finance*, *19*(3–4), 679–692.

Saunders, A., Cornett, M. M., & McGraw, P. A. (2006). *Financial institutions management: A risk management approach*. New York: McGraw-Hill.

Thakor, A. V. (1996). Capital requirements, monetary policy, and aggregate bank lending: Theory and empirical evidence. *The Journal of Finance*, *51*(1), 279–324.

# 7 Time and Frequency Analysis Using the ARMA Model: Evidence from the Indian Stock Market

*Nikita Chopra*

CONTENTS



## 7.1 INTRODUCTION

The stock market is a complex yet a sophisticated mechanism to channel savings into investments. However, with the increasing volume of trade, prediction of stock prices has become even more challenging. Thus, some investors profit from these fluctuations while others lose ample amounts of investments. An investor wants guaranteed returns with minimum risk. This has inspired researchers to evolve and develop predictive models. Researchers have made predictions based on daily returns and monthly returns, where it has been observed that these fluctuations are merely corrective movements of the stock market. However, the existence of the predictive capability of the stock market has been an inconclusive study. Thus, prediction of the financial market needs to be researched in order to empower effective investment strategies for investors and professionals.

Practitioners and academicians have proposed various models based on modern and traditional theories of investments such as random walk theory, the capital asset pricing model, and technical and fundamental analyses to give a more accurate forecast. The fundamental analysis entails an in-depth analysis of the economy, industry, and company, assuming that the trading price of the company depends on its intrinsic value and the expected return of the investors. In contrast, the technical analysis assesses past stock price trends and uses price, volume, and open interest statistical charts to predict future prices with correctional changes. The philosophy behind the technical analysis is that any factor can affect the market at any given point of time. It may be internal or external (Mendelsohn, 2000).

Over the years, several scientific fields have merged with finance to provide better input for making predictions. Disciplines such as psychology have suggested the behaviour of market participants as a possible explanation of this fluctuation, whereas the field of economics has empirically proved the influence of macroeconomic factors such as inflation to be the cause of the stock movement. Even the field of artificial neural networks has contributed to the study and recording of the pattern from the data. Statistical tools such as regression, exponential smoothing, and generalized autoregressive conditional heteroscedasticity (GARCH) have been used to analyze stock market dynamics. Yet, the objective of all these statistical, econometric, and social sciences is the same, that is, to predict the movement of the stock market. Thus, ARMA and ARIMA models are amongst the most prominent models of prediction.

The autoregressive moving average (ARMA) and autoregressive integrated moving average (ARIMA) models are considered to be the most robust and efficient models for time series forecasting. These models estimate autocorrelation and partial autocorrelation, which help in identifying existing patterns in stock market movements. Fluctuations in the financial market have an adverse effect on the confidence and psychology of investors as many retail investors incur losses. The identification of these patterns can help in predicting another financial crisis or any underlying economic problem to take corrective action. Thus, the objective of this study was to conduct a time and frequency analysis of the Indian stock market. Furthermore, it sought to test the presence or absence of the random walk theory in the Indian stock market. This study also sought to address the predictive capability of stock price movements.

## 7.2 THEORETICAL BACKGROUND

Many traditional, modern, and behavioural theories have been developed to explain the movement of the stock market. However, the pattern of stock price movement can be explained by the random walk theory. The premise of the random walk theory begins with the basic assumption that stock exchange markets are efficient. Efficient markets are markets where all the information is available to the public. It further assumes that consecutive price changes of the stock market are independent and random. The efficient market hypothesis has three levels of markets, namely the weak form, semi-strong form, and strong form. The weak form assumes that all information is reflected in past prices. Thus, the random walk theory implies that events cannot be predicted, which is the basis of the weak form of the efficient market hypothesis theory. Similarly, in the semi-strong form, the available public information mirrors stock prices, making it impossible to benefit from the fundamental analysis. Finally, the third level of efficient market is the strong form, which suggests that both public and private information are reflected in the current prices. Thus, the application of the random walk theory and efficient market hypothesis suggests that it is impossible to predict stock prices. Consequently, the theory also proposes that it is not possible to earn more returns than the market returns as the stock prices reflect all available information. Additionally, it suggests that new information supplied to the market is random. However, the principle of technical analysis, which concludes that a stock's future price movement can be forecasted by observing historical prices through charts and indicators, is invalidated by this theory.

In research undertaken to predict stock price movements, the efficient market hypothesis and random walk theory have been used interchangeably. However, the efficient market hypothesis bases its inference on the impact of information asymmetry on stock price predictability, whereas the random walk theory addresses the existence of the pattern of stock prices. These theories further suggest that even the fundamental analysis, which is a study of the overall financial health of the economy, industry, and the business of the company, assumes that an investor cannot outperform the market. Accordingly, any return over and above the market return may be credited to the skill set of the financial analyst and entail the study of behavioural finance.

## 7.3 LITERATURE REVIEW

Fama (1965, 1991) made a notable contribution to stock price forecasting in the form of the random walk theory. Yet, there is a vast difference in the price movement of the stock markets in developed and developing countries. Thus, the theory may not hold true for all the stock markets. Many prominent researchers have attempted to detect the presence of random walk in the Indian stock market. Historically, Rao and Mukherjee (1971) and Sharma and Kennedy (1977) used the run test and spectral density to prove the existence of random walk theory in the monthly return of the Bombay Stock Exchange (BSE). Furthermore, using autocorrelation and run tests, it has been found that the weak form of market was inefficient (Chaudhuri, 1991; Poshakwale (1996).

Mallikarjunappa and Afsal (2007) established that stocks were highly correlated, suggesting an inefficient weak market from 2003 to 2006. However, there are studies that have concluded that there is neither the presence of random walk nor a weak form of market (Khan et al., 2011). Gupta and Siddiqui (2010) used Kolmogorov-Smirnov test (K-S) test along with run test and autocorrelation test to disprove the presence of the weak form of market. Guidi, Gupta, and Suneerl (2011), while testing market efficiency, concluded that it was indeed possible to earn above-the-market returns by analyzing past stock prices. Thus, the existing literature provides inconclusive evidence with respect to the random walk theory.

The literature demonstrates the use of autocorrelation as an econometric tool to detect the presence of a pattern in the stock market used to study random walk theory. The autoregressive moving average (ARMA) and autoregressive integrated moving average (ARIMA) models have been applied for forecasting stock prices and testing the random walk theory. In this section, selected studies have been mentioned. In earlier studies, evidence of the presence of random walk behaviour in Indian stock market returns was found (Gupta, 1985). Furthermore, Amanulla and Kamaiah (1998) examined the distribution pattern of increments of stock market returns on BSE indices using the ARIMA model to conclude that the equity market was efficient in weak form. However, there is evidence against the random walk theory, as found by Poshakwale (2002). The ARIMA model has been used to forecast the future stock indices, both for Indian and foreign stock exchanges. Banerjee (2014) applied the ARIMA model to forecast the future prices of the Indian stock exchange. Similarly, Jin, Wang, Yan, and Zhu (2015) demonstrated the predictive capability of the ARIMA model to model the closing value of the Shanghai Composite Stock Price Index. In recent studies, Ashik and Kannan (2017) applied the Box-Jenkins approach to examine the Nifty 50 index to model the closing stock price of the index. The research further concluded that Nifty 50 displayed a decrease in fluctuation for the year 2015. Pradesh, Venkataramanaiah, and Campus (2018) compared the forecasted and actual data for a 10-year study on BSE Sensex using the ARIMA model to conclude that the investors chose as per the expected returns. Thus, no conclusive results can be drawn and the accurate prediction of stock market is an ongoing area of research.

## 7.4 RESEARCH METHODOLOGY

This study attempted to examine the pattern of stock movement of the Indian stock market. The S&P BSE 500 index was selected for the study as it represents 93% of market capitalization, covering all major industries in the Indian economy. The variable used for the study is the daily closing value of S&P BSE 500 from 1 February 1999 to 31 August 2018. The closing prices reflect all the activities of the index on that particular day. The stationarity of time series data has been assessed using the augmented Dickey-Fuller test. Furthermore, the normality of the time series has been tested using Jarque-Bera test. In the study, the ARMA method has been used to develop a model for stock forecasting. The ARMA model has been developed using the EVIEWS software version 10. The automatic ARIMA forecasting helps in determining the appropriate ARIMAX specification. The series $y^t$ follows an ARIMAX ($p$, $d$, $q$) model if:

$$\begin{gathered} D\left(y^t, d\right) = \beta X_t + v_t \\ v_t = \rho_1 v_{t-1} + \rho_2 v_{t-2} + \ldots + \rho_p v_{t-p} + \theta_1 \epsilon_{t-1} + \theta_2 \epsilon_{t-2} + \ldots + \theta_q \epsilon_{t-q} \end{gathered} \tag{7.1}$$

The exogenous variable $X$ is a constant and trend term. The non-seasonal ARIMA models use parameters $p$, $d$, $q$; which refer to the autoregressive, differencing, and moving average terms for the non-seasonal component of the ARIMA model. Thus, the selection of the levels of difference, the number of AR and MR terms, is based on a model selection criterion such as the Akaike information criterion (AIC) and the Schwarz information criterion (SIC or BIC). ARIMA modeling is also named the Box-Jenkins methodology, as it comprises a series of activities such as identifying, estimating, and diagnosing ARIMA models of a time series.

## 7.5 RESULTS AND DISCUSSION

### 7.5.1 Graphic Analysis of Closing Value of the S&P BSE 500

The spike graph in Figure 7.1 clearly shows that although the closing value of the index has been rising, there is no specific pattern in the closing price of the BSE 500 index. It can be observed that there is a sudden drop in the year 2008–2009 due to the global financial crisis. From the year 2009, volatility and investor response to the aftermath of the financial crisis is evident in the graphic presentation of the closing value of the index.

### 7.5.2 Normality Test

The time series was subject to normality testing using the Jarque-Bera test as shown in Figure 7.2. The results suggest that the time series rejects the assumption of normal distribution which further confirms that the closing values of the index do not follow a normal pattern. The average closing price of the S&P BSE 500 price index from its inception until the 1st October 2018 has been Rs.5888.817 and the maximum closing price on a particular day was Rs.15846.20.

The objective of the research was to capture and predict the movement of the Indian stock market. As per the National Bureau of Economic Research, the recession

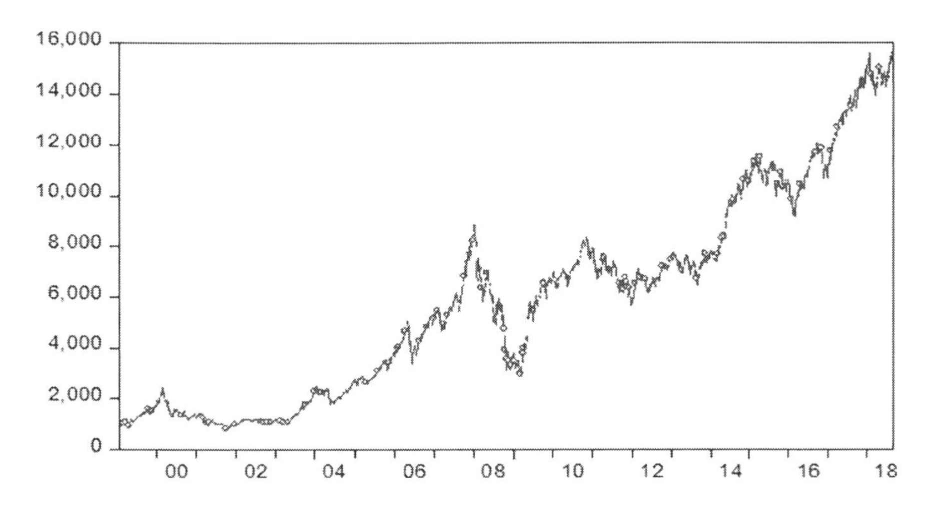


**FIGURE 7.1** Graph of closing value of the S&P BSE 500 from 01-02-1999 to 01-10-2018.

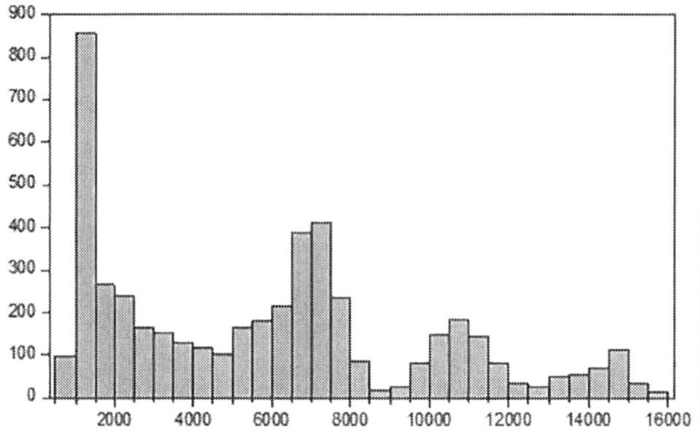


**FIGURE 7.2** Jacque-Bera normality test for the closing value of the S&P BSE 500 index.

**TABLE 7.1**
**Results of Augmented Dickey-Fuller Test**

| Time Series | T-Statistics | Probability | Level |
|---|---|---|---|
| Pre-recession | –42.93215 | 0.0000 | I (1) |
| Post-recession | –44.18608 | 0.0001 | I (1) |

ended in June 2009. Pandey, Patnaik, and Shah (2017) further proved that the recession for the Indian economy ended at the beginning of the third quarter of 2009. Thus, the data have been bifurcated accordingly. The data have been segregated into two sets of data: (1) 01-02-1999 to 01-06-2009 and (2) 01-06-2009 to 01-10-2018.

### 7.5.3 Stationarity Test

The time series, namely, pre-recession and post-recession, were subject to a stationarity test. The augmented Dickey-Fuller test results, as shown in Table 7.1, suggest that the daily closing prices of the S&P BSE 500 are stationary at first-order difference.

### 7.5.4 Autoregressive Moving Average Model Selection

Using the ARMA forecasting, EVIEWS 10 software was utilized for the selection of the appropriate dependent variable and model criteria. A number of iterations using the Box-Jenkins method were computed in the software to select the appropriate model. The ARMA model was popularized by Box-Jenkins as follows:

$$X_t = c + \epsilon_t + \sum_{i=1}^{p} \rho_i X_{t-1} + \sum_{i=1}^{q} \theta_t \epsilon_{t-1} \tag{7.2}$$

Table 7.2 represents the dependent variable selected for the analysis of the movement of the Indian stock market. Pre-recession, the closing value of the S&P

**TABLE 7.2**
**ARMA Model Selection Criterion**

| Sample No. | Sample | Dependent Variable | No. of Obs. | Estimated ARMA Models | Selected ARMA Model | AIC Value |
|---|---|---|---|---|---|---|
| Pre-Recession | 01-02-1999 to 01-06-2009 | D(CLOSE) | 2578 | 25 | (4,4), (0,0) | 11.21368 |
| Post-Recession | 01-06-2009 to 01-10-2018 | DLOG(CLOSE) | 2319 | 25 | (1,0), (0,0) | −6.39364 |

*Source:* Author's calculations.

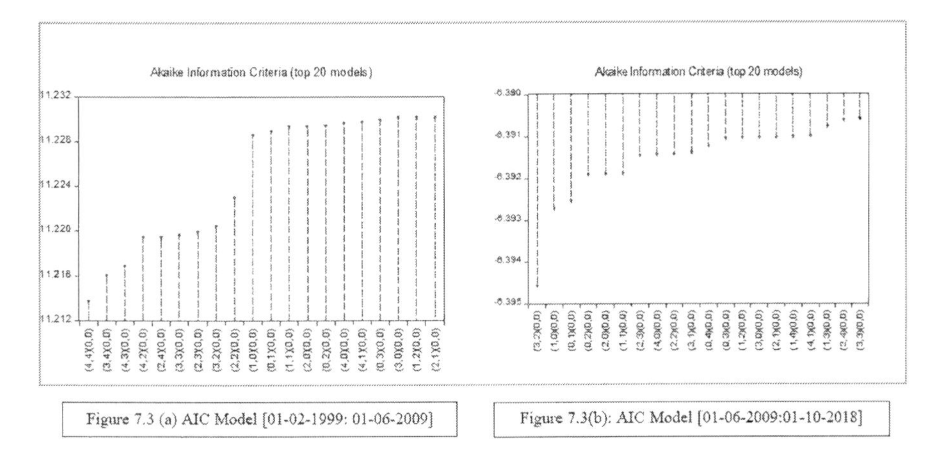


**FIGURE 7.3** (a) AIC model: 01-02-1999 to 01-06-2009. (b) AIC model: 01-06-2009 to 01-10-2018.

BSE 500 was stationary at the first-level difference. Thus, the automatic iterations selected the first-level difference, D (CLOSE), as the dependent variable. While estimating 25 models for 2578 observations, the model with the lowest Akaike information criterion value, 11.21368, was selected for pre-recession. Similarly, for the period after the recession (01-06-2009), due to increased volatility the automatic ARIMA forecasting model selected DLOG (CLOSE) as the dependent model, suggesting that the closing value of the index was stationary after taking the logarithm of the first-level difference. Twenty-five iterations of the model consisting of 2319 observations were calculated and the lowest AIC value of −6.39364 was selected. Figures 7.3(a) and 7.3(b) show the iterations of the respective models of pre-recession and post-recession.

In Figures 7.4(a) and 7.4(b), the AIC (Akaike information criteria), LOGL, BIC (Bayesian or Schwarz information criterion), and HQ values of the prospective models of sample pre-recession and sample post-recession have been calculated. Using the AIC values, the models of the respective samples have been selected. For sample

Dependent Variable: D (CLOSE)

| Model | LogL | AIC* | BIC | HQ |
|---|---|---|---|---|
| (4,4)(0,0) | -14444.435386 | 11.213681 | 11.236392 | 11.221914 |
| (3,4)(0,0) | -14448.433914 | 11.216008 | 11.236447 | 11.223417 |
| (4,3)(0,0) | -14449.584231 | 11.216900 | 11.237340 | 11.224309 |
| (4,2)(0,0) | -14453.830790 | 11.219419 | 11.237587 | 11.226004 |
| (2,4)(0,0) | -14453.847919 | 11.219432 | 11.237600 | 11.226018 |
| (3,3)(0,0) | -14454.132551 | 11.219653 | 11.237821 | 11.226239 |
| (2,3)(0,0) | -14455.449669 | 11.219899 | 11.235796 | 11.225661 |
| (3,2)(0,0) | -14456.085734 | 11.220392 | 11.236290 | 11.226155 |
| (2,2)(0,0) | -14460.409865 | 11.222971 | 11.236597 | 11.227910 |
| (1,0)(0,0) | -14470.624216 | 11.228568 | 11.235381 | 11.231038 |
| (0,1)(0,0) | -14471.050700 | 11.228899 | 11.235712 | 11.231369 |
| (1,1)(0,0) | -14470.623847 | 11.229344 | 11.238428 | 11.232636 |
| (2,0)(0,0) | -14470.623867 | 11.229344 | 11.238428 | 11.232636 |
| (0,2)(0,0) | -14470.687252 | 11.229393 | 11.238477 | 11.232686 |
| (4,0)(0,0) | -14468.969794 | 11.229612 | 11.243238 | 11.234551 |
| (4,1)(0,0) | -14468.105671 | 11.229717 | 11.245615 | 11.235480 |
| (0,3)(0,0) | -14470.364235 | 11.229918 | 11.241273 | 11.234034 |
| (3,0)(0,0) | -14470.603987 | 11.230104 | 11.241459 | 11.234220 |
| (1,2)(0,0) | -14470.620463 | 11.230117 | 11.241472 | 11.234233 |
| (2,1)(0,0) | -14470.623264 | 11.230119 | 11.241474 | 11.234235 |
| (0,4)(0,0) | -14469.779155 | 11.230240 | 11.243866 | 11.235179 |
| (1,4)(0,0) | -14468.824820 | 11.230275 | 11.246173 | 11.236038 |
| (1,3)(0,0) | -14470.166746 | 11.230541 | 11.244167 | 11.235480 |
| (3,1)(0,0) | -14470.259165 | 11.230612 | 11.244239 | 11.235552 |
| (0,0)(0,0) | -14495.255841 | 11.246901 | 11.251443 | 11.248548 |

Figure 7.4 (a) Iterations for Pre-Recession

Dependent Variable: DLOG (CLOSE)

| Model | LogL | AIC* | BIC | HQ |
|---|---|---|---|---|
| (3,2)(0,0) | 7363.970104 | -6.394585 | -6.377120 | -6.388218 |
| (1,0)(0,0) | 7357.815020 | -6.392712 | -6.385227 | -6.389983 |
| (0,1)(0,0) | 7357.626732 | -6.392548 | -6.385063 | -6.389819 |
| (0,2)(0,0) | 7357.879034 | -6.391898 | -6.381918 | -6.388260 |
| (2,0)(0,0) | 7357.878763 | -6.391898 | -6.381918 | -6.388260 |
| (1,1)(0,0) | 7357.871854 | -6.391892 | -6.381912 | -6.388254 |
| (2,3)(0,0) | 7360.379061 | -6.391464 | -6.373998 | -6.385096 |
| (4,0)(0,0) | 7359.342010 | -6.391432 | -6.376461 | -6.385974 |
| (2,2)(0,0) | 7359.336971 | -6.391427 | -6.376457 | -6.385969 |
| (3,1)(0,0) | 7359.303839 | -6.391398 | -6.376428 | -6.385941 |
| (0,4)(0,0) | 7359.101015 | -6.391222 | -6.376252 | -6.385764 |
| (0,3)(0,0) | 7357.922201 | -6.391067 | -6.378591 | -6.386518 |
| (1,2)(0,0) | 7357.886790 | -6.391036 | -6.378561 | -6.386488 |
| (3,0)(0,0) | 7357.884547 | -6.391034 | -6.378559 | -6.386486 |
| (2,1)(0,0) | 7357.879319 | -6.391029 | -6.378554 | -6.386481 |
| (1,4)(0,0) | 7359.864064 | -6.391016 | -6.373551 | -6.384649 |
| (4,1)(0,0) | 7359.836383 | -6.390992 | -6.373527 | -6.384625 |
| (1,3)(0,0) | 7358.579486 | -6.390769 | -6.375799 | -6.385311 |
| (2,4)(0,0) | 7360.412349 | -6.390624 | -6.370663 | -6.383346 |
| (3,3)(0,0) | 7360.370916 | -6.390587 | -6.370627 | -6.383310 |
| (4,2)(0,0) | 7360.088188 | -6.390342 | -6.370381 | -6.383065 |
| (3,4)(0,0) | 7360.339859 | -6.389691 | -6.367236 | -6.381505 |
| (4,3)(0,0) | 7360.338931 | -6.389691 | -6.367235 | -6.381504 |
| (4,4)(0,0) | 7360.339298 | -6.388822 | -6.363871 | -6.379725 |
| (0,0)(0,0) | 7349.121952 | -6.386025 | -6.381035 | -6.384206 |

Figure 7.4(b): Iterations for Post-Recession

**FIGURE 7.4** (a) Iterations: Pre-recession. (b) Iterations: Post-recession.

pre-recession (01-02-1999 to 01-06-2009); the AIC value was 11.21368 and the BIC value was 11.236392 for the ARMA model (4, 4) (0, 0). The selection criteria for post-recession (01-06-2009 to 01-10-2018) shows the value of AIC was −6.39364 and the BIC value of −6.386207 for the ARMA model (3, 2) (0, 0).

### 7.5.5 Equation Analysis

The ARMA equation analysis is based on the transformation of the dependent variable by taking the natural log, selecting the level of difference, and then selecting exogenous regressors and ARMA terms. It can be noted from Figures 7.5(a) and 7.5(b) that the variable selected for the period before the recession (01-06-2009) is first-level difference of the closing value of the index. However, for the period after 01-06-2009, log value with first-level difference of the closing price has been selected. Natural logs of variables are often selected when there is an exponential growth rate in the change in increase and decrease of the variables. Thus, it can be concluded that due to the increased volume and volatility there has been an exponential change in the closing value of the selected index.

In Figures 7.5(a) and 7.5(b) the ARMA equation has been tested for autoregressive (AR) terms 1 to 4 and moving average (MA) terms 1 to 4. The optimization method of Berndt-Hal-Hal-Hausman algorithm (BHHH) has been applied to the maximum likelihood test. For the sample I, 2578 observations with 103 iterations, it can be

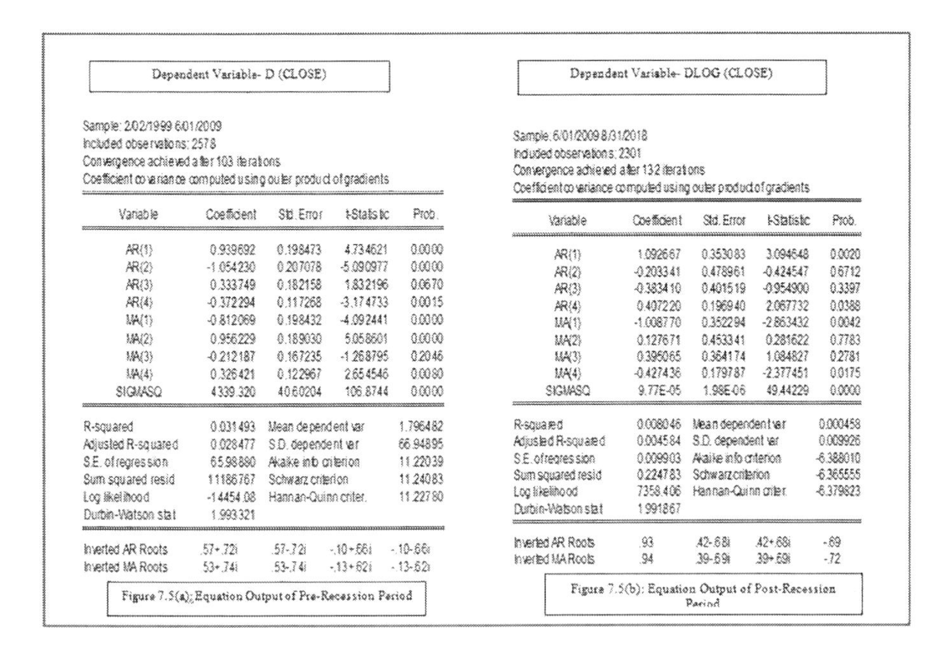

Dependent Variable- D (CLOSE)

Sample: 2/02/1999 6/01/2009
Included observations: 2578
Convergence achieved after 103 iterations
Coefficient covariance computed using outer product of gradients

| Variable | Coefficient | Std. Error | t-Statistic | Prob. |
|---|---|---|---|---|
| AR(1) | 0.939692 | 0.198473 | 4.734621 | 0.0000 |
| AR(2) | -1.054230 | 0.207078 | -5.090977 | 0.0000 |
| AR(3) | 0.333749 | 0.182158 | 1.832196 | 0.0670 |
| AR(4) | -0.372294 | 0.117268 | -3.174733 | 0.0015 |
| MA(1) | -0.812069 | 0.198432 | -4.092441 | 0.0000 |
| MA(2) | 0.956229 | 0.189030 | 5.058601 | 0.0000 |
| MA(3) | -0.212187 | 0.167235 | -1.268795 | 0.2046 |
| MA(4) | 0.326421 | 0.122967 | 2.654546 | 0.0080 |
| SIGMASQ | 4339.320 | 40.60204 | 106.8744 | 0.0000 |

| | | | |
|---|---|---|---|
| R-squared | 0.031493 | Mean dependent var | 1.796482 |
| Adjusted R-squared | 0.028477 | S.D. dependent var | 66.94895 |
| S.E. of regression | 65.98880 | Akaike info criterion | 11.22039 |
| Sum squared resid | 11186767 | Schwarz criterion | 11.24083 |
| Log likelihood | -14454.08 | Hannan-Quinn criter. | 11.22780 |
| Durbin-Watson stat | 1.993321 | | |

| | | | | |
|---|---|---|---|---|
| Inverted AR Roots | .57+.72i | .57-.72i | -.10+.66i | -.10-.66i |
| Inverted MA Roots | .53+.74i | .53-.74i | -.13+.62i | -.13-.62i |

Figure 7.5(a): Equation Output of Pre-Recession Period

Dependent Variable- DLOG (CLOSE)

Sample: 6/01/2009 8/31/2018
Included observations: 2301
Convergence achieved after 132 iterations
Coefficient covariance computed using outer product of gradients

| Variable | Coefficient | Std. Error | t-Statistic | Prob. |
|---|---|---|---|---|
| AR(1) | 1.092667 | 0.353083 | 3.094648 | 0.0020 |
| AR(2) | -0.203341 | 0.478961 | -0.424547 | 0.6712 |
| AR(3) | -0.383410 | 0.401519 | -0.954900 | 0.3397 |
| AR(4) | 0.407220 | 0.196940 | 2.067732 | 0.0388 |
| MA(1) | -1.008770 | 0.352294 | -2.863432 | 0.0042 |
| MA(2) | 0.127671 | 0.453341 | 0.281622 | 0.7783 |
| MA(3) | 0.395065 | 0.364174 | 1.084827 | 0.2781 |
| MA(4) | -0.427436 | 0.179787 | -2.377451 | 0.0175 |
| SIGMASQ | 9.77E-05 | 1.98E-06 | 49.44229 | 0.0000 |

| | | | |
|---|---|---|---|
| R-squared | 0.008046 | Mean dependent var | 0.000458 |
| Adjusted R-squared | 0.004584 | S.D. dependent var | 0.009926 |
| S.E. of regression | 0.009903 | Akaike info criterion | -6.388010 |
| Sum squared resid | 0.224783 | Schwarz criterion | -6.365555 |
| Log likelihood | 7358.406 | Hannan-Quinn criter. | -6.379823 |
| Durbin-Watson stat | 1.991867 | | |

| | | | | |
|---|---|---|---|---|
| Inverted AR Roots | .93 | .42-.68i | .42+.68i | -.69 |
| Inverted MA Roots | .94 | .39-.69i | .39+.69i | -.72 |

Figure 7.5(b): Equation Output of Post-Recession Period

**FIGURE 7.5** (a) Equation output: Pre-recession period. (b) Equation output: Post-recession period.

observed that AR terms 1, 2, and 4 are significant and MA terms 1, 2, and 4 are significant, which was suggested by the automatic ARIMA forecasting. However, for post-recession, 2301 observations and 132 iterations, it can be observed that only AR terms 1 and 4 and MA terms 1 and 4 are significant. The maximum likelihood variable, SIGMASQ, is also statistically significant.

### 7.5.6 Impulse Response

The impulse response function traces the one-time shock in the innovation (reaction of the previous stock prices). Figures 7.6(a) and 7.6(b) reflect the shocks associated with the stock markets for the time period before and after 01–06–2009. In sample pre-recession, it can be observed that there are multiple shock waves in the time series, which reflect a highly volatile market with multiple corrections in stock prices.

Any correction beyond 30–40% in stock prices is considered to be a crash in the market. Thus, the impulse response of sample I clearly reflects the financial crisis. In the post-recession period, it can be observed that there are shocks in the financial market but the correctional changes in the stock prices adjust the movements faster. It can further be deduced that the model for the post-recession series is stationary as the impulse response is asymptote to zero and the accumulated response asymptote to its long-run value.

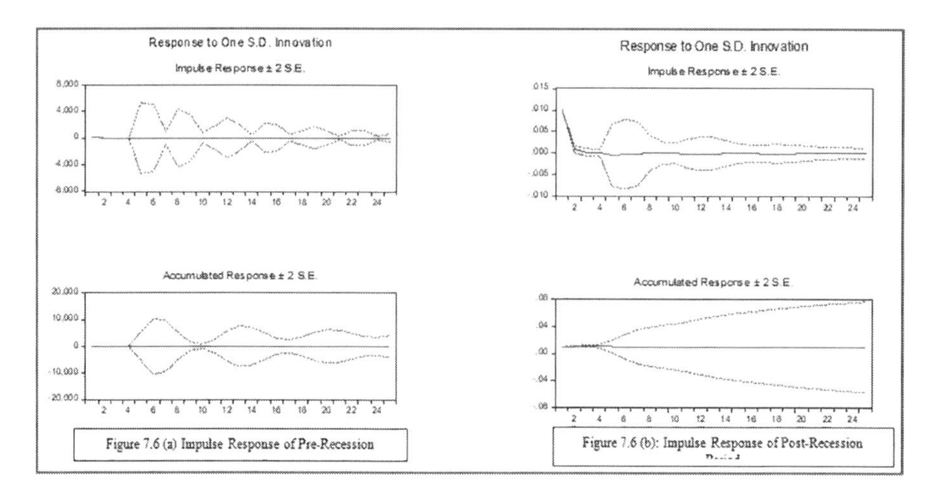


**FIGURE 7.6** (a) Impulse response: Pre-recession. (b) Impulse response: Post-recession.

### 7.5.7 Autocorrelation and Partial Autocorrelation

A correlogram compares the autocorrelation pattern of the structural residuals with that of the estimated model for a specified period of time. Figures 7.7(a) and 7.7(b) have specified a graphical comparison over 24 periods/lags. Since the models are stationary, both the actual and second sample movements are depicted in the graphs. It can be noted from Figure 7.7(a) that in the period before 01-06-2009 only models could be fitted only till period/ lag 11, while in Figure 7.7(b) it can be observed that the longer the period (more than period/lag 7), the more inaccurate the predictions become. It can be inferred that the ARMA models are more accurate for short-term predictions of stock prices.

### 7.5.8 Frequency Analysis

The ARMA frequency spectrum shows the spectrum of estimated ARMA terms in the frequency domain, rather than the time domain. The spectrum of ARMA is a function of its frequency $\lambda$, where $\lambda$ is measured in radians, taking value from $\pi$ to $-\pi$. If a series has white noise, then the frequency spectrum is a flat horizontal line. Figures 7.8(a) and 7.8(b) depict the frequency spectrum of the pre-recession and post-recession periods. When the ARMA model has a strong AR component, the frequency graphs have high cyclic frequencies, as displayed in Figures 7.8(a) and 7.8(b). It can further be observed from the figure that post-recession does not have a very strong AR component, yet has a cyclic component. It can be understood that due to a more aware investor, the fluctuation can be reduced. As interpreted before, the ability of the market to adjust according to the prevalent market has improved after the financial crisis. Thus, the pre-recession period has a smaller $\pi$ value.

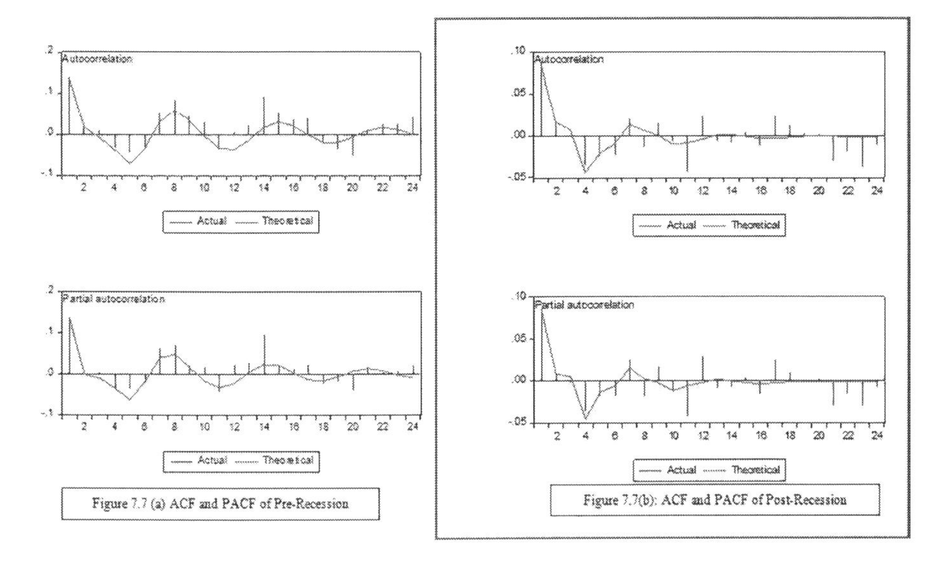


**FIGURE 7.7** (a) ACF and PACF: Pre-recession. (b) ACF and PACF: Post-recession.

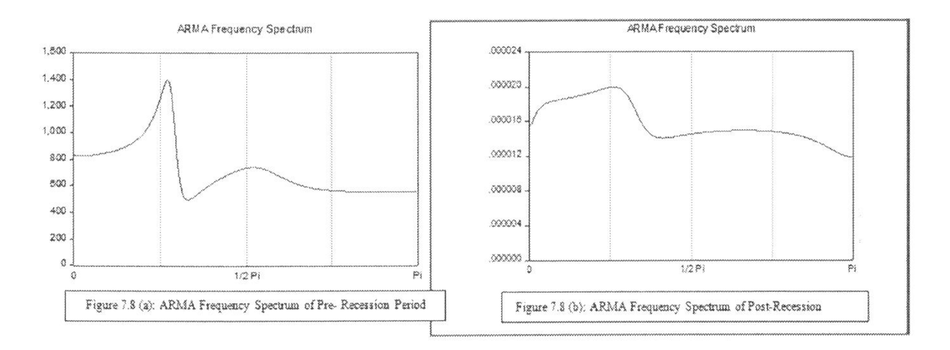


**FIGURE 7.8** (a) ARMA frequency spectrum: Pre-recession. (b) ARMA frequency spectrum: Post-recession.

### 7.5.9 Forecast Analysis

The forecast analysis entails the forecasting of the ARMA equation to create a dynamic forecasted series named **CLOSEF** with a margin of error of ± 2 Standard Error for the pre-recession and post-recession periods. In Figure 7.9, it can be noted that the forecasted series is almost similar to the actual values until 2003. However, from the 2003, the financial bubble started developing to the extent that the actual values exceeded the error margin. Finally, the actual value started falling drastically from the 2007 until the period 01-06-2009.

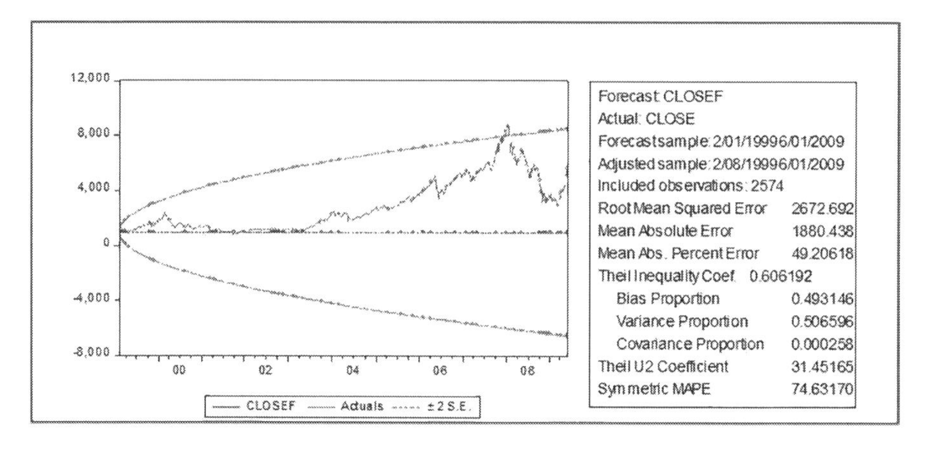


**FIGURE 7.9** Forecast analysis: Pre-recession period.

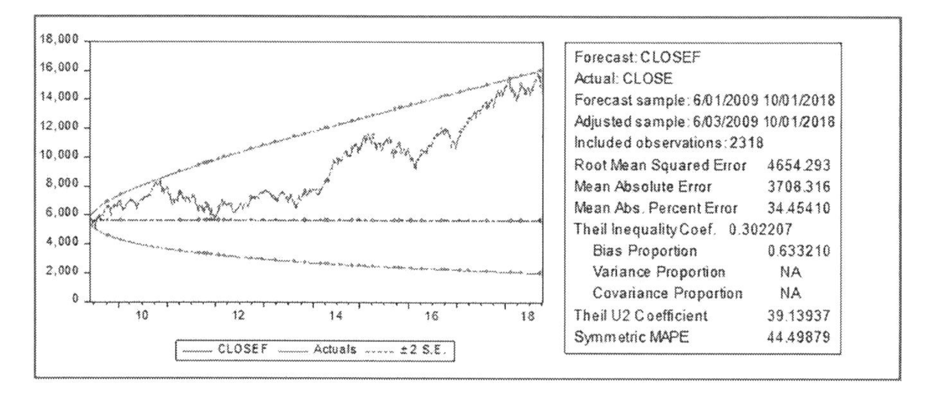


**FIGURE 7.10** Forecast analysis: Post-recession period.

In Figure 7.10, it can be observed that there is a difference between the actual and the forecasted series, yet it falls within the standard error margin. Moreover, the symmetric mean average absolute error percentage (MAPE) is lower for the post-recession period (44.49) as compared to MAPE of the pre-recession period (74.63). It can further be noticed that due to small and immediate corrections in the period after 01-06-2009, even with a highly volatile market, the economy has not started the cycle of another recession. Thus, the results of the autocorrelation partially reject the random walk theory for a shorter duration.

## 7.6 CONCLUSION

The Indian stock market is a highly sophisticated market with the best compliance and prudence norms issued by Security Exchange Board of India (SEBI). Over the years, the volume of stock trade has increased. With the increasing volume, the volatility of stock prices has increased. Thus, it is important to study the movements of stock prices as there is a lot more at stake as the number of retail investors continues to grow. This study examined the time and frequency of the movement of share prices. The study used the ARMA method to study the actual and forecasted values. The time series selected for the study was the daily closing value of the S&P BSE 500 index from 01-02-1999 to 31-08-2018. This time series was bifurcated into two series from 01-01-1999 to 01-06-2009 and 01-06-2009 to 31-08-2018 as the National Bureau of Economic Research suggests that the recession officially ended in June 2009. The time series were not normally distributed, which confirms the volatile movement of the market. The study utilized autoregressive terms 1 to 4 and moving average terms 1 to 4 to analyse the ARMA structure. With the of correlogram and impulse function it can be suggested that the model is fit for a smaller time period. The longer the time period, the more inaccurate the predictions. The frequency spectrum graph suggested that there is strong component of autoregressive (AR) movement in the period before the recession while after the recession due to frequent correctional changes the AR movement did not have high peaks.

The time series suggests that both sample periods were highly volatile, but due to the first-time exposure to the financial crisis the actual and forecasted values differed to a large extent. It can be concluded that small corrective pricing from the year 2009 to the current period have enabled a highly volatile market, yet it is unlikely that any chain of recession events will take place. It can also be associated with the increased investor awareness and financial literacy. Furthermore, the role and accountability of financial intermediaries has improved. The conclusion of the study entails the acknowledgement of the fact that the Indian stock market does not solely function on the previous prices, but that external factors play an equally important role in determining prices. The Indian stock market is a sentimental market where rumors can fuel the movement of stock prices in any direction. Thus, it becomes difficult to predict stock prices. The limitation of the study is that better models of neural network analysis can be used to better forecast stock prices. Yet, it is further proposed to conduct event studies to conclude which particular factor impacts stock movements more.

## REFERENCES

Amanulla, S., & Kamaiah, B. (1998). Asset price behaviour in Indian stock market: Is the "CAPM" still relevant? *Journal of Financial Management & Analysis*, *11*(1), 32.

Ashik, A. M., & Kannan, K. S. (2017). Forecasting national stock price using ARIMA model. *Glob Stoch Anal*, *4*(1), 77–81. Retrieved From: https://www.mukpublications.com/resources/vol._3_no._3/8_MOHAMED_ASHIK%20(1).pdf.

Banerjee, D. (2014, January). *Forecasting of Indian stock market using time-series ARIMA model*. 2nd International Conference on Business and Information Management (ICBIM), IEEE, pp. 131–135.

Chaudhuri, S. K. (1991). Short-run share price behaviour: New evidence on weak form of market efficiency. *Vikalpa*, *16*(4), 17–21.

Fama, E. F. (1965). The behavior of stock-market prices. *The Journal of Business*, *38*(1), 34–105.

Fama, E. F. (1991). Efficient capital markets: II. *The Journal of Finance*, *46*(5), 1575–1617.

Guidi, F., Gupta, R., & Suneerl, M. (2011). Weak-form market efficiency and calendar anomalies for Eastern Europe equity markets. *Journal of Emerging Market Finance, 10*(3), 337–389.

Gupta, O. P. (1985). Behaviour of Share Prices in India — A Test of Market Efficiency, New Delhi: National Publishing House.

Gupta, P. K., & Siddiqui, S. (2010). *Weak form of market efficiency: Evidences from selected NSE indices*. Retrieved from SSRN http://ssrn.com/abstract=1355103 or http://dx.doi.org/10.2139/ssrn.1355103

Jin, R., Wang, S., Yan, F., & Zhu, J. (2015). The application of ARIMA model in 2014 Shanghai composite stock price index. *Science Journal of Applied Mathematics and Statistics*, *3*(4), 199–203.

Khan, A. Q., Ikram, S., & Mehtab, M. (2011). Testing weak form market efficiency of Indian capital market: A case of national stock exchange (NSE) and Bombay stock exchange (BSE). *African Journal of Marketing Management*, *3*(6), 115–127.

Mallikarjunappa, T., & Afsal, E. M. (2007). Random walk hypothesis: A test on Sensex stocks. *Contemporary Management Research*, *1*(2), 14–39.

Mendelsohn, L. B. (2000). *Trend forecasting with technical analysis*. Marketplace Books. Retrieved from https://nseguide.com/images/ebooks/Trend_Forecasting_with_Technical_Analysis.pdf

Pandey, R., Patnaik, I., & Shah, A. (2017). Dating business cycles in India. *Indian Growth and Development Review*, *10*(1), 32–61.

Poshakwale, S. (1996). Evidence on weak form efficiency and day of the week effect in the Indian stock market. *Finance India*, *10*(3), 605–616.

Poshakwale, S. (2002). The random walk hypothesis in the emerging Indian stock market. *Journal of Business Finance & Accounting*, *29*(9–10), 1275–1299.

Pradesh, A., Venkataramanaiah, M., & Campus, G. V. I. (2018). Forecasting time series stock returns using Arima: Evidence from S&P BSE Sensex. *International Journal of Pure and Applied Mathematics*, *118*(24).

Rao, K. N., & Mukherjee, K. (1971). Random walk hypothesis: An empirical study. *Arthaniti*, *14*(1–2), 53–58.

Sharma, J. L., & Kennedy, R. E. (1977). A comparative analysis of stock price behavior on the Bombay, London, and New York stock exchanges. *Journal of Financial and Quantitative Analysis*, *12*(3), 391–413.

# 8 Behavioural Biases and Trading Volume: Empirical Evidence from the Indian Stock Market

*Sunaina Kanojia, Deepti Singh, and Ashutosh Goswami*

## CONTENTS



## 8.1 INTRODUCTION

Financial analysts have long puzzled over the behavior of investors towards trading in the stock market. Several attempts have been made to understand the driving forces behind investors' trading behavior. According to behavioral theorists, two behavioral biases increase stock market trading volume. These are overconfidence and the disposition effect. Overconfidence is overestimating one's actual ability to predict stock performance relative to others (Moore et al., 2007). Overconfident investors believe that they perform better than others in the stock market. According to Shiller (1997), "Overconfidence is associated with people in their own judgments; these individuals underestimate the margins of error likely to be committed". Odean (1998) and Gervais (2001) found that overconfident investors trade more than normal investors and lose the most. High market returns make the investors overconfident, and as a consequence, these investors trade more than others.

The disposition effect is another factor that influences trading by individual investors in the stock market. Statman (1985) was the first to analyze the disposition effect using the prospect theory and found that investors treat unrealized losses and gains in different ways. They prefer to hold on to losers, but to sell the winners, which, in turn, increases the trading volume of the stock, whose price increases. Previous studies have indicated that overconfidence and disposition affect the trading volume in the stock market (Statman, Vorkink, & Thorley, 2006; Chuang & Lee, 2006; Glaser & Weber, 2007; Adel & Mariem, 2013). However, the presence of these two biases has yet to be documented in the emerging markets. This study attempts to capture the impact of these two biases on the trading volume in the Indian stock market.

## 8.2 REVIEW OF THE LITERATURE

A number of studies have been done to gauge the relationship between stock returns and trading volume in different market scenarios. Weber et al. (1998) conducted a controlled laboratory experiment of buying and selling of six hypothetical stocks in the normal course of 14 trading rounds and found that subjects are about 50% more likely to realize gains as compared to losses. Further, Ezzeddine et al. (1998) analyzed the monthly return and trading volume in the Tunisian market using the vector autoregressive model (VAR) and associated impulse response functions and found evidence of overconfidence bias in the Tunisian market. Billings et al. (2001) used gender as a dependent variable to study the impact of overconfidence bias. The study analyzed the common stock investments of men and women between 1991 and 1997 for 35,000 households at the New York Stock Exchange and found that men are more overconfident than women. They trade 45% more than women (based on monthly portfolio turnover), and the returns of men are reduced by 2.65 percentage points a year as opposed to 1.72 percentage points for women due to excess trading rather than poor selection of stock. Siwar (2006) used time series regression, VAR, and GARCH models to examine the presence of overconfidence and disposition effect and found a strong relationship between trading volume and past stock returns in the French stock market. Weber et al. (2007) conducted an online survey to measure different facets of overconfidence such as miscalibration, volatility, and better-than-average effect and found that investors who think that they are above average in terms of investment skills or past performance trade more than others. Zaine (2013) used VAR and associated impulse response functions to analyze the monthly return and trading volume between 2000 and 2006 and found evidence of overconfidence bias in the Chinese stock market. However, no evidence of disposition effect was found in the Chinese market. On the basis of the extant literature, this study used VAR and the impulse response function to understand the impact of overconfidence and disposition on the Indian stock market.

## 8.3 OBJECTIVES OF THE STUDY

This study had three main objectives:

1. To capture the impact of overconfidence on trading volume in the Indian equity market.
2. To capture the impact of the disposition effect on trading volume in the Indian equity market.
3. To segregate the impact of overconfidence and the disposition effect on the volume of transactions in the Indian equity market.

## 8.4 HYPOTHESES

In order to examine the relationship between the trading activity and the lagged returns, the following sub-hypotheses have been framed:

1. $H_{01}$: The trading volume is not positively related to lagged market returns.
   $H_{a1}$: The trading volume is positively related to lagged market returns.
2. $H_{02}$: Individual security turnover is not positively related with the both lagged market and lagged security returns.
   $H_{a2}$: Individual security turnover is positively related with the both lagged market and lagged security returns.
3. $H_{03}$: The positive relationship doesn't exist between the trading volume and volatility.
   $H_{a3}$: The positive relationship exists between the trading volume and volatility.

## 8.5 DATA AND METHODOLOGY

This study examined the impact of overconfidence and the disposition effect on trading volume. The data source and research methodology are as follows.

### 8.5.1 Data Source

The study sought to identify and analyze the impact of overconfidence and the disposition effect on trading volume in the Indian stock market using monthly observations of Nifty 50 stocks for a period of nine years from 1 April 2009 to 31 March 2018 collected from the Centre for Monitoring Indian Economy (CMIE) prowess database. Following Odean (1998), Ezzeddine et al. (1998), Lo and Wang (2000), Statman et al. (2006), and Zaine (2013), the study focused on monthly observations under the perspective that monthly observations reflect the changes in investor overconfidence and the disposition effect. Trading activities were measured by turnover (shares traded divided by outstanding shares) and aggregate security turnover into a market turnover on a value-weighted rather than equal-weighted basis.

### 8.5.2 Definition of the Variables Used in the Study

The study used both exogenous and endogenous variables, which are defined as follows:

**M sig:** Monthly temporal volatility of market return based on daily market returns within the month, correcting for realized autocorrelation, as specified in French, Schwert, and Stambaugh (1987).[1] The volatility control variable is similar to the mean absolute deviation (MAD) measure in the trading volume study of Bessembinder, Chan, and Seguin (1996) but based on Karpoff's (1987) survey of research on the contemporaneous volume–volatility relationship. According to French, Schwert, and Stambaugh (1987), non-synchronous trading of securities causes daily portfolio returns to be autocorrelated, particularly at lag one. However, the negative sign of variance in case of some individual securities leads us to use the approximation of Duffe (1995): $\left(\mathrm{Msig}^2 = \sum_{t=1}^{T} r_t^2\right)$. In fact, French, Schwert, and Stambaugh (1987) approximation result in a negative variance estimate if the first-order autocorrelation of daily returns in a given month is ≤ (−0.5).

**Disp:** Cross-sectional standard deviation of returns for all stocks in month t:

$$\mathrm{Disp} = \sum_{i=1}^{N} W_i \sigma_{i,t}$$

where $W_i$ is the weight of stock i in the market portfolio, and $\sigma_{it}$ is the standard deviation of the returns of stock i in month t.

According to Statman et al. (2006), return dispersion is included as a control variable to account for potential trading activity associated with portfolio rebalancing. For example, large spreads between the individual stock returns might lead to trading activity among investors seeking to maintain fixed portfolio weights.

**M ret:** Monthly stock market return along with dividends.

**Ret:** Return of security at month t.

**M trading:** Monthly market turnover (shares traded divided by outstanding shares).

**Trading:** Monthly turnover of an individual security.

### 8.5.3 Empirical Methodology

The vector autoregression model (VAR) and impulse response function (IRF) were used to capture the impact of overconfidence and the disposition effect on the trading volume in the Indian stock market. The VAR model was applied on the market-wide transaction volume and market returns to investigate investors' overconfidence. It was then applied on security-wide transaction volume, security returns, and market return to capture and segregate the impact of overconfidence and the disposition effect. The IRF was used to verify the validity of the VAR model.

1. **Market-wide VAR model to investigate investors' overconfidence:** Here the endogenous variables are market turnover and market return and the exogenous variables are dispersion and market volatility

$$\begin{bmatrix} mtrading_t \\ mret_t \end{bmatrix} = \begin{bmatrix} \alpha_{mtrading} \\ \alpha_{mret} \end{bmatrix} + \sum_{k=1}^{3} A_k \begin{bmatrix} mtrading_{t-k} \\ mret_{t-k} \end{bmatrix} + \sum_{l=0}^{\square} B_l \begin{bmatrix} msig_{t-1} \\ Disp_{t-1} \end{bmatrix} + \begin{bmatrix} e_{mtrading,t} \\ e_{mret,t} \end{bmatrix} \quad (8.1)$$

$A_k$ is the matrix that measures how trading proxy and returns react to their lags.
$B_l$ is the matrix that measures how trading proxy and returns react to month (t–1) realizations of exogenous variables.
$e_t$ is a (N × 1) residual vector. It captures the contemporaneous correlation between endogenous variables.
K and l are the numbers of endogenous and exogenous observations. K and l are chosen based on the Akaike (1974) (AIC) and Schwartz (SC) information criteria. In this study, the SIC leads to k = 3 and l = 0.

In equation 8.1, any change in the residuals, say $e_{mtrading,t}$, will not only bring change in the current value of market trading, but also affect the future values of market trading and market return because the lagged values of market trading appear in both the equations through the coefficient matrix $A_k$. To test for overconfidence, we shock the market return by one sample standard deviation and track how market trading activity responds over time to the market return residual.

2. **Security-wide VAR Model to investigate and segregate the impact of the disposition effect and overconfidence:** The existing literature suggests that the trading volume of the market increases post positive portfolio returns due to overconfidence among the investors (Statman et al., 2006; Abaoub et al., 2009). However, the disposition effect suggests that an increase in the return of an individual security increases the trading volume of the individual stock because investors enjoy realizing paper gains on individual securities (Weber et al., 1998; Sengupta et al., 2013). Therefore, overconfidence bias increases the trading volume in general, whereas the disposition effect increases the trading volume of individual securities. Following Statman et al. (2006) and Zaine (2013), the trivariate VAR model for individual securities is used, which is as follows:

$$\begin{bmatrix} trading_t \\ return_t \\ mret_t \end{bmatrix} = \begin{bmatrix} \alpha_{trading} \\ \alpha_{return} \\ \alpha_{mret} \end{bmatrix} + \sum_{k=1}^{3} A_k \begin{bmatrix} trading_{t-k} \\ return_{t-k} \\ mret_{t-k} \end{bmatrix} + \sum_{l=0}^{\square} B_l sig_{t-1} + \begin{bmatrix} e_{trading,t} \\ e_{return,t} \\ e_{mret,t} \end{bmatrix} \quad (8.2)$$

In the aforementioned model, the security trading, security return, and market return are the endogenous variables and security volatility is the exogenous variable. Here, the first and foremost concern is how the security turnover responds to shocks in security return and market return. Following Hirshleifer et al. (1998), the study also examined the response of security returns to shocks in security-wide turnover to test the proposition of overconfidence. The market return (mret) was introduced into the model in order to check if the disposition effect explains high trading volume as well as the overconfidence hypothesis. If overconfidence explains volume along with the disposition effect, we should find a positive relationship between the securities trading volume and past market return even when lagged security returns are introduced in the model.

3. **Impulse response function:** An impulse response function traces the effect of one standard deviation shock (measured within the sample) in one residual to current and future values of the dependent variables through the dynamic structure of the VAR. It helps to illustrate how the endogenous variables are related to each other over time. In this study, an IRF was applied along with the market-wide model and security-wide model to investigate the presence of overconfidence and the disposition effect in the Indian stock market.

## 8.6 RESULTS AND DISCUSSION

This section presents the results and discussions of the impact of overconfidence and the disposition effect on the trading volume in the Indian stock market using the VAR model and IRF.

### 8.6.1 Market Turnover

Figure 8.1 presents market turnover from 1 April 2009 to 31 March 2018. There is a wide range of fluctuation, but there is no particular trend found in the turnover. The

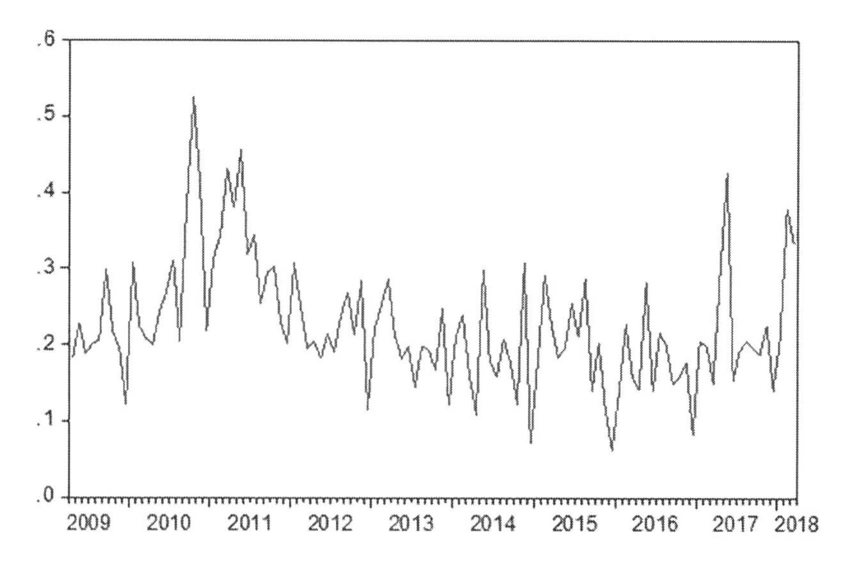


**FIGURE 8.1** Monthly market turnover for the Nifty 50 index.

changes in investors' overconfidence induced by realized portfolio returns are one probable reason for the long-term fluctuations in trading activity.

### 8.6.2 Market VAR Estimation and Test Results

Table 8.1 gives the brief summary of the bivariate VAR model of market trading and market return for equation 8.1.

Table 8.1a exhibits that market trading is autocorrelated, with a highly significant first and third lag coefficient. However, lag observation of market trading is autocorrelated to market return with a significant second lag coefficient.

Table 8.1b depicts the association between the market trading and lagged market returns. The results of the study exhibit that market trading is positively related to lag market returns with the significant second lag coefficient (Hypothesis 1 of the study). The result is consistent with the previous empirical studies of the overconfidence hypothesis

**TABLE 8.1**
**Market VAR Estimation (m trading = turnover)**

**TABLE 8.1A**
**Relations with Lagged Market Trading**

| | | M Trading (t-1) | M Trading (t-2) | M Trading (t-3) |
|---|---|---|---|---|
| M trading (t) | Coefficient | 0.267240 | 0.060801 | 0.290943 |
| | Std. error | (0.06985) | (0.07104) | (0.06547) |
| | P-value | 0.00** | 0.39 | 0.00** |
| M ret (t) | Coefficient | −0.094141 | 0.253769 | −0.022195 |
| | Std. error | (0.11425) | (0.11191) | (0.11212) |
| | P-value | 0.40 | 0.02* | 0.84 |

* Significant at 5% level; ** significant at 1% level.

**TABLE 8.1B**
**Relations with the Lagged Market Return**

| | | M ret (t-1) | M ret (t-2) | M ret (t-3) |
|---|---|---|---|---|
| M trading (t) | Coefficient | 0.022468 | 0.086875 | −0.034724 |
| | Std. error | (0.04212) | (0.04283) | (0.03947) |
| | P-value | 0.59 | 0.04* | 0.37 |
| M ret (t) | Coefficient | −0.003214 | 0.037564 | 0.070043 |
| | Std. error | (0.06888) | (0.06747) | (0.06760) |
| | P-value | 0.96 | 0.57 | 0.30 |

* Significant at 5% level; ** significant at 1% level.

**TABLE 8.1C**
**Relations with Lagged Volatility**

| | | Constant e | M sig (t) | Disp (t) |
|---|---|---|---|---|
| M trading (t) | Coefficient | 0.068783 | 0.026117 | 3.424683 |
| | Std. error | (0.01965) | (0.02106) | (2.40974) |
| | P-value | 0.00** | 0.21 | 0.15 |
| M ret (t) | Coefficient | 0.010746 | −0.071397 | 2.984638 |
| | Std. error | (0.01185) | (0.01270) | (1.45290) |
| | P-value | 0.36 | 0.00** | 0.03* |

* Significant at 5% level; ** significant at 1% level.

(Statman et al., 2006; Chuang & Lee, 2006; Glaser & Weber, 2007; Adel & Mariem, 2013). According to Glaser and Weber (2007), high market returns make investors overconfident in the sense that they underestimate the variance of stock returns.

Table 8.1c presents the relation between endogenous and exogenous variables (msig and disp). The result indicates that there is non-significant association between the trading volume and volatility (Hypothesis 3 of the study). Further, the association between dispersion and trading volume is also non-significant.

### 8.6.3 Market Impulse Response Function

The individual VAR coefficient estimates do not contain the full impact of the independent variable observations. An IRF uses all the VAR coefficient estimates to outline the full impact of a residual shock that is one sample standard deviation from zero. Figure 8.2 contains the four possible IRF graphs using bivariate VAR estimation.

In Figure 8.2 the vertical axis measures the percentage increase in mtrading, that is, the volume, and the horizontal axis depicts the number of months. Further, the blue line is the IRF and the red lines are the 95% confidence intervals.

Figures 2.1 and 2.2 represents responses of mtrading to one standard deviation of mtrading and mret along with confidence bands spaced out at two standard errors. Figure 2.1 depicts a large and persistent response in mtrading to an mtrading shock. Figure 2.2 indicates a positive response in mtrading to mret shock, the key finding of this study.

Figures 2.3 and 2.4 represent responses of mret to one standard deviation of mtrading and mret along with confidence bands spaced out at two standard errors. Figure 2.3, indicates that a one standard deviation shock causes mtrading to increase slightly. In Figure 2.4 the IRF indicates that the impact of the mret shock is positive and persistent.

### 8.6.4 Security VAR Estimation and Test Results

The study found strong empirical evidence for the overconfidence hypothesis. Market-wide trading is positively related to previous market returns in the Indian

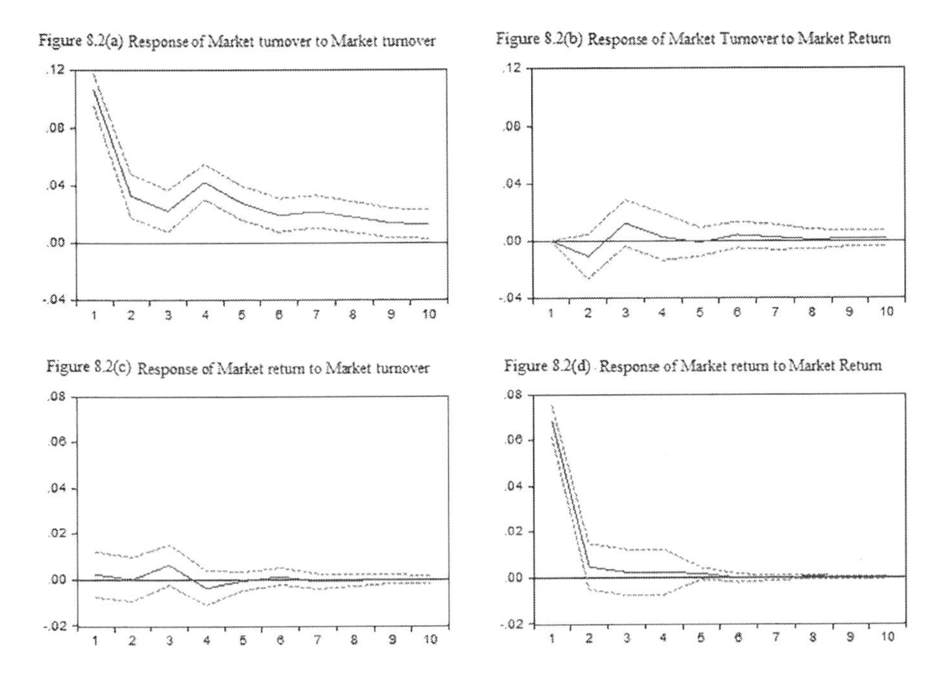


**FIGURE 8.2** Response to Cholesky one S.D. innovations ±2 S.E.

stock market. As explained by Statman et al. (2006), investors start trading more because they feel more confident after having high market returns or simply because they enjoy realizing paper gains on individual securities by selling them, commonly known as the disposition effect.

Gervais and Odean (2001) stated that overconfidence bias increases trading by investors in general, whereas Shefrin and Statman (1985) documented that the disposition effect is an act by investors to sell specific securities in their portfolio to realize the paper gain. In this study, an attempt was made to separate the two theories by examining the trading activity of individual stocks. If overconfidence bias affects the trading activity along with the disposition effect, we should find positive coefficients in regressions of security turnover on lagged market returns even when lagged returns on that stock are included. It is difficult to separate the impact of overconfidence and the disposition effect on the stock market. One could easily raise an argument that when investors sell securities to realize a paper gain they raise cash that could be used to purchase the security in question, which, in turn, increases the trading volume in the stock market. Despite this, an examination of individual security turnover ensures that the documented market-wide turnover patterns are not a simple summation of direct disposition effect evidence in individual security VAR.

The trivariate VAR was estimated in equation 8.2 for each security and the mean coefficient estimates are reported in Table 8.2. The three dependent variables (security trading, security return, and market return) and one independent variable (volatility of the securities) were included in the security VAR model. The security trading and the security returns were similar to the dependent market variables in the bivariate market VAR model. On the other hand, volatility of securities was similar to market volatility in the bivariate market VAR model. The optimal selection criteria based on SIC varied across firms, but to ensure consistency in the results the lag specification used in the market VAR were k = 3 and l = 0 for all individual firms. It was also found that the specification of k = 3 and l = 0 gave the minimum average SIC across all securities.

Table 8.2 provides the brief summary of the security VAR model of market trading and market return for equation 8.2.

Table 8.2a depicts that security turnover is autocorrelated, with a highly significant first, second, and third lag coefficient. However, lag observation of market trading is not correlated to market return with any significant lag coefficient.

**TABLE 8.2**
**Security VAR Estimation (m trading = turnover)**

**TABLE 8.2A**
**Relations with Lagged Security Trading**

| | | Trading (t-1) | Trading (t-2) | Trading (t-3) |
|---|---|---|---|---|
| Trading (t) | Coefficient | 0.242356 | 0.116848 | 0.153686 |
| | Std. error | (0.018652) | (0.017172) | (0.01641) |
| | P-value | 0.00** | 0.00** | 0.00** |
| Ret (t) | Coefficient | 0.109847 | −0.00126 | 0.045155 |
| | Std. error | (0.084782) | (0.051686) | (0.063612) |
| | P-value | 0.20 | 0.98 | 0.48 |

* Significant at 5% level; ** significant at 1% level.

**TABLE 8.2B**
**Relations with Lagged Security Return**

| | | Ret (t-1) | Ret (t-2) | Ret (t-3) |
|---|---|---|---|---|
| Trading (t) | Coefficient | 0.003128 | 0.039022 | −0.01593 |
| | Std. error | (0.006262) | (0.006965) | (0.009109) |
| | P-value | 0.62 | 0.00* | 0.08 |
| Ret (t) | Coefficient | −0.06153 | −0.07327 | −0.00428 |
| | Std. error | (0.017128) | (0.019262) | (0.015167) |
| | P-value | 0.00 | 0.00 | 0.77 |

* Significant at 5% level; ** significant at 1% level.

**TABLE 8.2C**
**Relations with the Lagged Market Return**

| | | M ret (t-1) | M ret (t-2) | M ret (t-3) |
|---|---|---|---|---|
| Trading (t) | Coefficient | 0.014276 | 0.024992 | –0.01055 |
| | Std. error | (0.00367) | (0.006088) | (0.005987) |
| | P-value | 0.00** | 0.00** | 0.08** |
| M ret (t) | Coefficient | 0.018923 | –0.03977 | 0.00104 |
| | Std. error | (0.010682) | (0.012395) | (0.008318) |
| | P-value | 0.08 | 0.00 | 0.90 |

* Significant at 5% level; ** significant at 1% level.

Table 8.2b exhibits that security turnover is autocorrelated with a highly significant second lagged coefficient of security return.

Table 8.2c shows that security turnover is autocorrelated with first and second lagged market return.

These results clearly show the dependence of security trading on lagged security returns and market returns, which is consistent with the results of Thorley et al. (2006). Hence, it is to be concluded that individual security turnover is positively related with the both lagged security and lagged market returns in the Indian stock market (Hypothesis 2 of the study). This confirms the presence of the disposition effect. To check robustness, IRF was used to study the dependence of security trading on lagged security returns and market returns. For brevity, figures are not reported but the results confirm that the disposition effect occurs at the monthly horizon in the Indian stock market.

## 8.7 CONCLUSION

Deciphering the impact of investor behavior on trading volume raises several questions with regard to the validity of standard financial models. Trading volume adversely affects stock prices that may not be justified by their price/earnings (P/E) ratio, resulting in overvaluation or undervaluation of stocks. Undervalued stocks are expected to go higher; overvalued stocks are expected to go lower. The results based on VAR models and IRF confirm the impact of overconfidence bias and the disposition effect on trading volume in the Indian stock market. In order to avoid the impact of these behavioral biases on trading activity, investors can prepare a checklist of these biases before trading in the stock market. They can also analyze the impact of these two biases on the P/E ratio before making an investment decision. As future research, it would be interesting to use weekly or daily data to study the impact of overconfidence and the disposition effect on the trading volume in the Indian stock market.

## NOTE

1 Volatility, according to French, Schwert, and Stambaugh (1987), is calculated as follows: $Msig^2 = \sum_{t=1}^{T} r_t^2 + 2\sum_{t=1}^{T} r_t r_{t-1}$, where $r_t$ is day t's return and T is the number of trading days in month t, in order to adjust the first-order autocorrelation of returns.

## REFERENCES

Black, F. (1986). Noise. *The Journal of Finance*, *41*, 528–543.
Daniel, K., Hirshleifer, D., & Subrahmanyam, A. (1998). Investor psychology and security market under- and overreactions. *The Journal of Finance*, *53*(6), 1839–1885.
Dhar, R., & Zhu, N. (2006). Up close and personal: Investor sophistication and the disposition effect. *Management Science*, *52*(5), 726–740.
Gervais, S., & Odean, T. (2001). Learning to be overconfident. *The Review of Financial Studies*, *14*, 1–27.
Glaser, M., & Weber, M. (2007). Overconfidence and trading volume. *The Geneva Risk and Insurance Review*, *32*(1), 1–36.
Grinblatt, M., & Han, B. (2004). Prospect theory, mental accounting, and momentum. *Journal of Financial Economics*, *78*, 311–339.
Karpoff, J. (1987). The relation between price changes and trading volume: A survey. *The Journal of Financial and Quantitative Analysis*, *22*(1), 109–126.
Lo, A. W., & Wang, J. (2000). Trading volume : Definitions, data analysis, and implications of portfolio theory. *Review of Financial Studies*, *13*(2), 257–300.
Moore, D. A., & Healy, P. J. (2008). The trouble with overconfidence. *Psychological Review*, *115*(2), 502–517.
Odean, T. (1998). Volume, volatility, price, and profit when all traders are above average. *The Journal of Finance*, *53*, 1887–1934.
Shefrin, H., & Statman, M. (1985). The disposition to sell winners too early and ride losers too long: Theory and evidence. *The Journal of Finance*, *40*(3), 777–790.
Statman, M., Thorley, S., & Vorkink, K. (2006). Investor overconfidence and trading volume. *Review of Financial Studies*, *19*(4), 1531–1565.
Taylor, S. E., & Brown, J. D. (1988). Illusion and well-being: A social psychological perspective on mental health. *Psychological Bulletin*, *103*(2), 193–210.
Zaiane, S., & Abaoub, E. (2009). Investor overconfidence and trading volume: The case of an emergent market. *International Review of Business Research Papers*, *5*(2), 213–222.
Zaine, S. (2013). Overconfidence, trading volume, and the disposition effect: Evidence from the Shenzhen stock market of China. *Business Management and Economics*, *1*(7), 163–175.

# 9 Leveraging Tacit Knowledge for Strategizing: Impact on Long-Term Performance

*Krishna Raj Bhandari and Sachin Kumar Raut*

## CONTENTS



## 9.1 INTRODUCTION

The early years of strategic management saw authors like *Chandler, Ansoff, and Porter*—as well as iconic frameworks such as the BCG matrix, the diversification matrix, and the five forces – impose themselves upon our consciousness and spread into other business disciplines. Strategic management was focused and exuberant: a creature of will. More recently, we have witnessed an increase in what we might call "the wit" applied to figuring out what strategy really is and where it comes from. We have seen more attention paid to the creation of strategy and the practices that shape it: different viewpoints have been explored, and multiple schools of thought defined. The fields of strategy and organization studies have spilled over into one another, and the focus on the noun *strategy* has shifted towards an interest in the verb *strategizing*. Many other fields in management now use strategy as an adjective (e.g. strategic marketing, strategic HR), and people at all organizational levels are encouraged to think and act strategically. Esoteric debates about the "true" nature of strategy, often impenetrable to most practitioners, have emerged.

And this "wit" has gradually obscured the focus prevalent at the outset—there are now so many varied views of strategy that it has become difficult to be sure of what we mean when we use the term. Some have suggested that this denotes a "crisis", and that "strategic management should have grown up" years ago. We do not think so; rather, to extend Peter McKiernan's metaphor from his article where he characterized the field as "scrambling from adolescence to adulthood", we would argue that (as might be typical for one in their 30s) strategic management has exercised its faculties to fully explore its possibilities.

Strategy is defined "as a situated, socially accomplished activity, while *strategizing* comprises those actions, interactions and negotiations of multiple actors and the situated practices that they draw upon in accomplishing that activity" (Jarzabkowski, 2005, p.07). There is an understanding that the three elements in strategy-as-practice (SAP) are denoted by three Ps (practice, practitioner, and praxis) (Johnson, Melin, and Whittington, 2003; Wolf and Floyd, 2017). The other way of representing these three pillars is through examining the "who", "what", and "how" of strategy. In general, the intersection of the three elements is called *strategizing* which is measured through *tacit* knowledge, the major contribution of this chapter.

This study goes deeper into the three pillars of *strategizing*. The first pillar of *strategizing* is *practitioners*. Strategy practitioners are those who are directly involved in making strategy, such as managers and consultants, and those with indirect influence, such as policymakers, the media, the gurus, and the business schools, who shape legitimate praxis and practices (Jarzabkowski and Whittington, 2008; Brown and Thompson, 2013). The second pillar of *strategizing* is *praxis*. Praxis "indicate[s] both an embedded concept that may be operationalized at different levels from the institutional to the micro, and dynamic, shifting fluidly through the interactions between levels" (Jarzabkowski et al., 2007, p. 101).

The third pillar of *strategizing* is *practices*. Practices "involve the various routines, discourses, concepts and technologies through which this strategy labor is made possible—eg: academic and consulting tools such as (Porterian analysis, hypothesis testing etc.) and in more material technologies and artefacts" (Jarzabkowski and Whittington, 2008, p. 101). For the present study, these three elements were combined in the stream of literature by developing a measure based on the interaction between praxis, practices, and practitioners (the three Ps). Using this approach, a curvilinear relationship between strategy and long-term performance can be expected. A summary of existing theories in *strategizing* identified four theoretical approaches: situated learning approach; sense-making and organization routines; institutionalist theories; and actor-network theory (Johnson, Langley, Melin, and Whittington, 2007; Collins, 2018). A clear text analysis approach by Bingham and Kahl (2013) is most relevant for this paper. Qualitative approaches are often recommended when relatively little is known about an area of study or when a fresh perspective is needed (Eisenhardt, 1989). The authors began this exploration by arguing that in-depth and largely quantitative data are a central requirement for developing the strategy-as-practice perspective.

## 9.2 BACKGROUND OF THE STUDY

Though there have been multiple approaches for operationalization of *tacit*-knowledge through causal mapping, self-Q, and storytelling, the authors follow epitomes of knowledge to measure *tacit* knowledge, as developed by Haldin and Herrgard (2004). The "epitomes of *tacit* knowledge" introduced by Haldin and Herrgard (2003) to measure *strategizing* as *tacit* knowledge is the major approach followed in this chapter.

Keywords from both appendices from Haldin and Herrgard (2003) were used in computer-aided text analysis (CATA) to model *tacit* knowledge and the intersection of the three Ps in SAP. The authors are the first to use CATA as a method for the SAP literature.

The study examined discourse in annual reports over a 10-year period to determine whether there is a direct relationship between *strategizing* and *long-term* performance. The research design was developed by using CATA to generate keyword counts to be used in panel regression with Tobin's Q. Such an approach enables a longitudinal study which is not possible with cross-sectional survey-based research design. Apart from the methodological contribution, one novel theoretical contribution of this approach is its examination of the influence of *tacit* knowledge on *strategizing* and its impact in long-term performance. The "practice-based view" (PBV) of a firm was proposed by Bromiley and Rau (2014) who argue for the study of specific, actual managerial techniques rather than the macro-level perspective prevalent in existing strategy discourse.

Resource-based view (RBV) (Barney, 1991) argues for resource immobility or inimitability in contrast to Penrosian thinking where fungibility of resource is more valuable which can be transferred across firms as the key performance differential. Thus, the PBV differs from the RBV in two aspects: the dependent variable and the isolating mechanisms. In the RBV the dependent variable is the sustained competitive advantage, whereas in the PBV it is the business unit or the firm performance. Therefore, mixing these two dependent variables is not a good idea. The isolating mechanisms prevent managers from sharing the resource to outsiders, as per the RBV. Thus, it enables sustained competitive advantage.

The other school of thought is anchored in the RBV rather than the PBV (Jarzabkowski, Kaplan, Seidl, and Whittington, 2016a). In a very rare thinking they suggest that *strategizing* is not stand-alone phenomena (Bromiley and Rau, 2014), who use the word "Practice-Based View of Strategy" (PBV). PBV has countered the RBV in generating differential performance. However, Jarzabkowski et al. (2016b) argue that the PBV focuses only on one P, practices, ignoring people and praxis. Our study considers all three Ps while testing the hypothesis using "epitomes of *tacit* knowledge" as a measure.

The theoretical focus of the SAP field seems to be predominantly at the micro level. More specifically, according to Johnson et al. (2007), four main approaches have been developed that link SAP with *tacit* knowledge. The first, the situated learning approach, argues that learning emerges through the activity of those actors involved in the learning process. Therefore, knowledge is not static, but rather evolves dynamically through activities. The second approach is sense-making and organizational routines. Sense-making is a process through which actors work to understand organizational actions and make sense of them. It is a social activity where individuals share, retain, and preserve the plausible stories of organizational events (Johnson et al., 2007). Organizational routines are the repetitive actions of those actors within organization practice. The third approach is institutionalist theories, which view the formal organization as a system of coordinated and controlled activities, including the work embedded within technical relations and boundary-spanning exchange. With that said, institutional theory emphasizes how formal organizations should be through rules and cultural norms (Meyer and Rowan, 1977, p. 3). The fourth approach is actor-network theory. This approach treats objects as a part of social networks. According to actor-network theory (Latour, 1999), "Actors perform with a wide network but not in isolation", and therefore the *tacit* knowledge is spread around to actors to perform organizational tasks.

According to actor-network theory, managers discover their place in the firm through a dynamic knowledge-based activity system. The institutional theory posits that organizational activities are controlled and systematized, and that knowledge creation and dissemination are formalized. As too little or too much of anything is detrimental to performance, the authors hypothesized that too high level of *tacitness* makes it difficult to transfer and thus the related performance impact would not be visible. Too little *tacit* knowledge, on the other hand, means that it is so easy to copy that there is no value of possessing it. Specifically we detect,

Hypothesis 1: SAP has a curvilinear relationship with long-term performance.

Correspondingly, as discussed earlier on the RBV versus PBV preview, too much of *tacitness* which becomes detrimental to transfer will not result into performance. However, too little of *tacitness* which is so easy to copy also makes the knowledge a risky resource. Therefore, there should exist an optimum level of *tacitness* which maximizes the performance in the long-run.

## 9.3 METHOD: COMPUTER-AIDED TEXT ANALYSIS (CATA)

To measure *strategizing*, the authors used keyword analysis of annual reports to measure SAP, adopting the process suggested by Short et al. (2010). The CATA approach makes it possible to conduct a longitudinal analysis. CATA offers several benefits apart from enabling a longitudinal research design. First, the possibility of longitudinal design enables testing of causal relationships (Keil et al., 2017). Second, Krippendorff (2012) suggested that C-ATA is a superior technique over traditional research design, as outlined in Table 9.1.

Annual reports were used as the corpus for the analysis, which provides a firm-level perception of *strategizing* (Maula, Keil, and Zahra, 2013). Use of annual 10-K filings provides numerous advantages over other types of corporate documents (Feldman et al., 2010).

**TABLE 9.1**
**Traditional vs. CATA-Based Research Design**

| | Traditional Approach (Interviews or Surveys) | CATA |
|---|---|---|
| Longitudinal research | Very difficult due to retrospective bias in survey-based research design | Easier to implement |
| Replicability of research | Very difficult | Very easy (Krippendorff, 2012) |
| Reliability | Low | High (Krippendorff, 2012) |
| Obtrusiveness | High | Low (Krippendorff, 2012) |
| Consistency of data across | Low (Eggers & Kaplan, 2009) | High |
| Safety | Low | High (Duriau, Reger, and Pfarrer, 2007). |
| Scalability | Low | High (Duriau, Reger, and Pfarrer, 2007). |
| Cost-effectiveness | Costly | Low cost (Duriau, Reger, and Pfarrer, 2007). |
| Collaboration | Difficult | Easy (Duriau, Reger, and Pfarrer, 2007). |
| Triangulation | Challenging | Possible |

## 9.4 DATA SOURCES AND SAMPLE

The authors collected a unique dataset of 2690 annual reports of Nordic large-cap and mid-cap companies. Authors selected a cross-industry, cross-sector sample of large-cap and mid-cap Nordic companies listed in Nordic NASDAQ index. Therefore, several samples of the most prominent companies from Sweden, Finland, Denmark, and Iceland were included. Norway, although a Nordic country, is not an EU member and is not included in Nordic NASDAQ index calculations. Sweden, Finland, Denmark, and Iceland are very similar to each other in terms of the business environment that the companies are facing (Benito, Grøgaard, and Narula, 2003).

In the Nordic stock exchange, companies with market capitalization of EUR 1 billion or more are considered large-cap companies, whereas those with market capitalization of EUR 150 million or more are considered mid-cap companies. The authors chose to base the sample selection on the new marketing paradigm of service provisioning, as goods manufacturers are also increasingly bundling services in their offerings, making the distinction between manufacturing and service firms less relevant. The authors constructed a panel dataset of the sample companies over the period of 2005 to 2014. Data were collected from the NASDAQ Nordic stock exchange index. As of 2015, 296 firms were listed under the categories of large cap (≥ 1 billion EUR) or mid cap (≥ 150 million EUR). During data collection, the authors identified 27 firms without access to annual reports or substantially missing data. Therefore, the final sample consisted of 269 companies in the NASDAQ Nordic large- and mid-cap indices.

## 9.5 MEASURES

### 9.5.1 Independent Variables

*Strategizing* is based on the dominant view of the RBV where *tacitness* is assumed to fulfill the criteria of a valuable, rare, inimitable, and non-substitutable (VRIN) resource. "*Tacit*-knowledge is deeply rooted in action and in an individual's commitment to a specific context—a craft or a profession, a technology or product market, or the activities of a work group or team" (Nonaka, 1991, p. 165). The multiple approaches for operationalization through causal mapping, self-Q, and storytelling have been used in the past, we embark upon using the epitomes of knowledge to measure *tacit* knowledge, as developed by Haldin and Herrgard (2004).

### 9.5.2 Dependent Variable

The present study used Tobin's Q to measure a firm's long-term performance. Tobin's Q is used as a measure of both short-term and long-term performance (Lubatkin and Shrieves, 1986; Uotila, Maula, Keil, and Zahra, 2009). Various authors have utilized this measure as the market value divided by the book value of assets (Brown and Caylor, 2006); consistent with this approach the authors have utilized this measure in the analysis.

### 9.5.3 Control Variable

Firm size was measured as the logarithm of the number of employees. Year controls and industry controls were used to operationalize as dummy variables at the two-digit SIC code level. The lagged values of the dependent performance variable were included in the models to control for unobserved heterogeneity. Use of 1% Winsorization was taken for all continuous variables.

## 9.6 ANALYSIS METHOD SYSTEM: GMM

The hypothesis was tested by following the latest specification and argumentation of two-step system GMM (Keil et al., 2017). Table 9.2 presents the five key reasons that the system GMM estimator is a robust estimator for the research. First, the type of data demands this method, as the data are panel data with few time periods and many companies. Second, the dependent variable is driven by previous levels of performance. This requires the use of a lagged dependent variable as a control. Third, the panel data are inherent with heteroscedasticity and autocorrelation that needs to be controlled. Fourth, explanatory variables are correlated with past and current realizations of the error term. Fifth, it needs the most prevalent control for unobserved heterogeneity. Lagged dependent variables can be used with time series and panel data where many observations in multiple times are used (Wooldridge, 2009).

**TABLE 9.2**
**Why System GMM?**

| Key Issues in Panel Data Analysis | Does System GMM Handle it? |
|---|---|
| 1. Panel data with few time periods and many companies. | Yes |
| 2. The dependent variable is driven by the previous levels of performance and the need for lagged dependent variable as a control. | Yes |
| 3. The panel data are inherent with heteroscedasticity and autocorrelation that needs to be controlled. | Yes |
| 4. The explanatory variables may be correlated with past and current realizations of the error term. | Yes |
| 5. Need to control for unobserved heterogeneity. | Yes |

Lags refer to time related to other variables. Similarly, in arguing for causality, the second condition, temporal precedence, could be handled through a lagged variable. Two-step system GMM uses lagged variables on the specification itself.

Another issue discussed in the context of panel data is serial correlation or autocorrelation, which occurs when a variable correlate with itself over time (Wooldridge, 2009). The effects of serial correlation of the error term are similar to the effects of heteroscedasticity. Two-step system GMM not only handles first-order serial correlation but goes one step further to handle second-order serial correlation.

System GMM was used as the analysis method to account for the endogeneity problem inherent in the panel data. The Arellano-Bond system GMM estimator was used (Arellano and Bond, 1991; Blundell and Bond, 1998) due to the presence of the highly persistent dependent performance variable. The system GMM estimation was run using the xtabond2 Stata module (Roodman, 2009). This method is good for situations with "small T, large N".

The dire state of endogeneity in strategic management research has been highlighted by Semadeni, Withers, and Certo (2014). Endogeneity biases the ordinary least squares (OLS) regression estimator. One of the statistical methods in the absence of good instruments is GMM and its variant system GMM (Wintoki, Linck, and Netter, 2012). With a panel data of 6000 firms from 1991 to 2003 the findings suggested that there was no causality between board structure and current firm performance. The claim is noteworthy because ruling out the major cause of endogeneity has been done through system GMM as a method. Therefore, to cater to endogeneity, system GMM was used as an analysis method.

## 9.7 RESULTS

Table 9.3 presents the descriptive statistics and correlations. The measure of sustainable performance is Tobin's Q, which had a mean value of 6.30 with a standard deviation of 9.74. The minimum and maximum range was between 0.39 and 67.35. The main independent variable, SAP (measured as *tacit* knowledge), had a mean value

**TABLE 9.3**
**Descriptive Statistics and Correlations**

| | Variable | Mean | S.D. | Min. | Max. | 1 | 2 | 3 | 4 |
|---|---|---|---|---|---|---|---|---|---|
| 1 | Tobin's Q[a] | 6.30 | 9.74 | 0.39 | 67.35 | 1 | | | |
| 3 | *Strategizing* (measured as *tacit* knowledge) | 388.03 | 233.80 | 0.01 | 2007 | 0.21 | 1 | | |
| 5 | R&D intensity (log)[a] | 0.49 | 1.02 | 0.015 | 6.16 | −0.19 | −0.15 | 1 | |
| 4 | Size (log)[a] | 7.48 | 2.17 | 1.61 | 11.72 | −0.18 | 0.39 | −0.08 | 1 |

[a] Winsorized at the 1% level.

of 388.03 and a standard deviation of 233.80. The minimum and maximum range was from 0.01 to 2007. The control variables were the log of R&D intensity and the log of size (number of employees). R&D intensity had a mean of 0.49 and a standard deviation of 1.02 and a minimum and maximum range of 0.015 and 6.16. Similarly, size had a mean value of 7.48 and a standard deviation of 2.17 and a minimum and maximum range of 1.61 and 11.72. The correlation between the variables was in a suitable range, indicating that there was no multicolinearity effect. Therefore, variance inflation factor analysis was not deemed necessary.

Three models were run with system GMM analysis with *strategizing* as the major independent variable and Tobin's Q as a dependent variable (see Table 9.4). Model 1 is the control model. A lagged dependent variable was used as a control to account for unobserved heterogeneity. Similarly, to control the confounding effect of innovation activities in the firm, log of R&D intensity was used as a second control. To avoid spurious effects of firm size, log of number of employees was used as a third control. In Model 2, SAP has been introduced as an independent variable. Though the linear effect is marginally significant, doubts prevail that there is a curvilinear effect in this model. Therefore, model 3 was tested, where the square of SAP is inserted as a second independent variable. The results showed a highly significant curvilinear component. Thus, the hypothesis is strongly supported.

## 9.8 DISCUSSION

As expected in the study hypothesis, the authors found a curvilinear relationship between *strategizing* and sustainable performance in a sample of large-cap and mid-cap companies, suggesting that there is an optimum level of *strategizing*. Before the optimum point is reached there is a positive impact, but at the higher level of *strategizing* there is a negative impact on the performance. Therefore, as demonstrated in the Figure 9.1, the middle range of SAP is best suited from 0.35 to 0.7 range (parameters scaled to 100 for plotting). Our findings support the right practice-based view anchored into the decades of SAP research anchored in the RBV. The results of this study demonstrate that *tacit* knowledge as the measure of interaction between the three Ps of SAP is a good measure to understand the relevance of SAP for a

**TABLE 9.4**
**System GMM Regression of SAP with Tobin's Q**

| Variable | Model 1 (Control) | Model 2 (Linear) | Model 3 (Non-linear) |
|---|---|---|---|
| *Explanatory variables* | | | |
| Strategizing (measured as *tacit* knowledge)$_{(t-1)}$ | | 0.0004 (0.000)$^{+}$ | 0.001 (0.00)** |
| strategizing × strategizing$_{t-1}$ | | | −9.84e-07 (2.55e-07)*** |
| Tobin's Q$_{t-1}$ | 0.211 (0.01)** | 0.78 (0.01)*** | 0.78 (0.01)*** |
| Size, ln$_{t-1}$ | −0.96 (0.23)*** | −0.01 (0.08)ns | −0.03 (0.07)ns |
| R&D intensity, ln$_{t-1}$ | −0.66 (0.19)** | 0.09 (0.06)* | 0.03 (0.06)ns |
| N | 2690 | 2690 | 2690 |
| Wald $\chi^2$ | 1805.14*** | 98425.85*** | 133103.08*** |
| Hansen | 0.05 | 0.47 | 0.54 |
| **$z_1$ or AR$_1$(p values)** | 0.12 | 0.05 | 0.05 |
| **$z_2$ or AR$_2$ (p values)** | 0.54 | 0.28 | 0.29 |
| Number of groups | 254 | 197 | 197 |
| Number of instruments | 76 | 92 | 108 |

**Note**: For each variable beta is reported first followed by the standard error in the parentheses.
$^{+}$ p < 0.1 level, * p < 0.05 level, ** p < 0.01 level, *** p < 0.001 level.

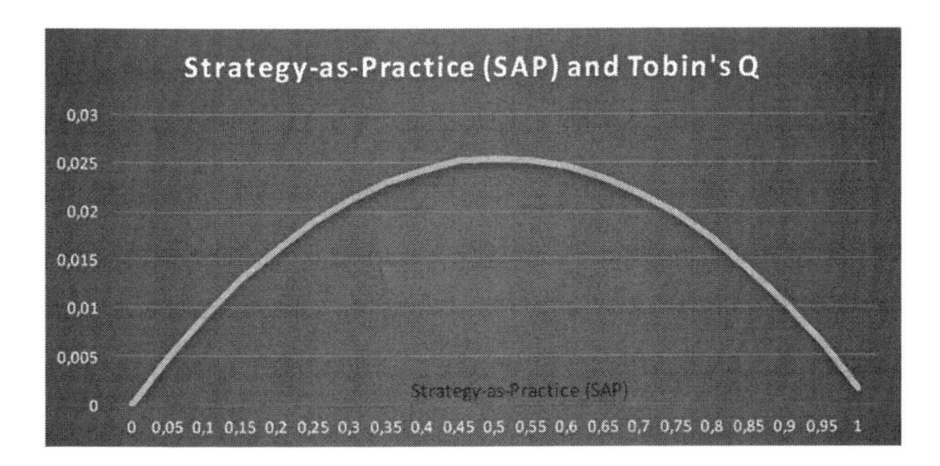


**FIGURE 9.1** Curvilinear relationship between strategy-as-practice (SAP) and sustainable performance.

sustainable performance. Not only the RBV, it also supports the knowledge-based view (KBV) notion of knowledge being the cornerstone of sustainable competitive advantage (Kogut and Zander, 1993). The cornerstone of this approach is the treatment of the firms as a social community whose productive knowledge defines a competitive advantage. This approach shares similarities with, and is yet distinct from, the standard economic treatment of the multinational corporation. A compelling explanation for the determination of the boundaries of the firm has rested on two observations. The first is that a necessary condition for trade among firms and among countries is a comparative advantage: differences in productivity in carrying out economic activities make it desirable for firms and nations to specialize and trade the products and services that reflect their superior capabilities. A second observation is that the hazard or cost of relying upon the market necessitates the "internalization" of trade (or transactions) within the firm. When the *tacitness* of the knowledge is too high it cannot be implemented or transferred inside the companies, resulting into performance decline. Similarly, when the *tacitness* is too low, it is easy to transfer; any outsiders or competitors can copy it.

Buckley and Casson (1976) provided the theory of a multinational enterprise (MNE) that is sufficiently powerful to afford long-term projections of the future growth. The theory emphasizes a "very general form of imperfect competition stemming from the costs of organizing markets". Internalization of such imperfect external markets, when this occurs across national boundaries, leads to the creation of MNEs. In Buckley and Casson's (1976) view, a simultaneous occurrence of five elements lead to the rapid growth of MNE activity: (1) the rise in demand for technology-intensive products; (2) efficiency and scale economy gains in knowledge production; (3) problems associated with organizing external markets for this new knowledge; (4) reductions in international communication costs; and (5) increasing scope for tax reduction through transfer pricing. Buckley and Casson (1976) focused especially on the third factor, the existence of market imperfections, which generates benefits of internalization. Internalization occurs only to the point where the benefits equal the costs. Buckley and Casson (1976) recognized four sets of parameters relevant to the internalization decision: (1) industry-specific factors, (2) region-specific factors, (3) nation-specific factors, and (4) firm-specific factors, focusing on the ability of the management to organize an internal market.

Perhaps the most interesting part of this analysis is the perspective on the MNE as an "international intelligence system for the acquisition and collation of basic knowledge relevant to R&D, and for the exploitation of the commer- cially applicable knowledge generated by R&D" (Buckley and Casson, 1976,p.35)." Here, it is critical to observe that Buckley and Casson (1976) already recognized that both the initial and final stages of R&D should in many cases be decentralized. The former should be located close to the sources of new information, especially basic research institutions, whereas the latter, involving the debugging of new products and processes, as well as adaptation to local market conditions, requires proximate contacts with production and marketing people respectively. Therefore, one needs to find the tactiness level suitable for internalization (Buckley and Casson, 1976) but avoid external leakage.

## 9.9 MANAGERIAL AND POLICY IMPLICATIONS

The major contribution of this research is the novel measurement approach of strategy-as-practice (SAP), and its impact on long-term performance. Managers can avoid managerial myopia driven by the management fad existing today on the value of SAP. In this distorted stream of literature, the present piece of work aims to contribute a unique insight based on a novel measurement and its impact on long-term performance. Policy makers benefit in creating innovation-related policy instruments that enable experimentation, learning, and testing to build SAP as a dynamic capability to maneuver in volatile environments.

## REFERENCES

Ambrosini, V., & Bowman, C. (2001). Tacit knowledge: Some suggestions for operationalization. *Journal of Management Studies*, *38*(6), 811–829.

Arellano, M., & Bond, S. (1991). Some tests of specification for panel data: Monte Carlo evidence and an application to employment equations. *The Review of Economic Studies*, *58*(2), 277–297.

Barney, J. (1991). Firm Resources and sustained competitive advantage. *Journal of management*, 17(1), 99–120.

Benito, G. R., Grøgaard, B., & Narula, R. (2003). Environmental influences on MNE subsidiary roles: Economic integration and the Nordic countries. *Journal of International Business Studies*, *34*(5), 443–456.

Bingham, C. B., & Kahl, S. J. (2013). The process of schema emergence: Assimilation, deconstruction, unitization and the plurality of analogies. *Academy of Management Journal*, *56*(1), 14–34.

Blundell, R., & Bond, S. (1998). Initial conditions and moment restrictions in dynamic panel data models. *Journal of Econometrics*, *87*(1), 115–143.

Bromiley, P., & Rau, D. (2014). Towards a practice-based view of strategy. *Strategic Management Journal*, *35*(8), 1249–1256.

Brown, A. D., & Thompson, E. R. (2013). A narrative approach to strategy-as-practice. *Business History*, *55*(7), 1143–1167.

Brown, L. D., & Caylor, M. L. (2006). Corporate governance and firm valuation. *Journal of Accounting and Public Policy*, *25*(4), 409–434.

Buckley, P. J., & Casson, M. (1976). *The future of the multinational enterprise*. New York: The McMillan Company.

Collins, H. (2018). *Creative research: The theory and practice of research for the creative industries*. UK: Bloomsbury Publishing.

Duriau, V. J., Reger, R. K., & Pfarrer, M. D. (2007). A content analysis of the content analysis literature in organization studies: Research themes, data sources, and methodological refinements. *Organizational Research Methods*, *10*(1), 5–34.

Eggers, J. P., & Kaplan, S. (2009). Cognition and renewal: Comparing CEO and organizational effects on incumbent adaptation to technical change. *Organization Science*, *20*(2), 461–477.

Eisenhardt, K. M. (1989). Building theories from case study research. *Academy of Management Review*, *14*(4), 532–550.

Feldman, R., Govindaraj, S., Livnat, J., & Segal, B. (2010). Management's tone change, post earnings announcement drift and accruals. *Review of Accounting Studies*, *15*(4), 915–953.

Haldin-Herrgard, T. (2003). Mapping tacit knowledge with "Epitomes". *Systèmes d'Information et Management (French Journal of Management Information Systems)*, *8*(2), 93–111.

Haldin-Herrgard, T. (2004). *Diving under the surface of tacit knowledge*. In Fifth European Conference on Organizational Knowledge, Learning, and Capabilities (OLKC).

Jarzabkowski, P. *Strategy as practice: An activity- based approach*. London: Sage, 2005.
Jarzabkowski, P., Balogun, J., & Seidl, D. (2005). Strategizing: The challenges of a practice perspective. *Human Relations*, *60*(1), 5–27.
Jarzabkowski, P., Balogun, J., & Seidl, D. (2007). Strategizing: The challenges of a practice perspective. *Human Relations*, *60*(1), 5–27.
Jarzabkowski, P., Kaplan, S., Seidl, D., & Whittington, R. (2016a). If you aren't talking about practices, don't call it a practice-based view: Rejoinder to Bromiley and Rau in strategic organization. *Strategic Organization*, *14*(3), 270–274.
Jarzabkowski, P., Kaplan, S., Seidl, D., & Whittington, R. (2016b). On the risk of studying practices in isolation: Linking what, who, and how in strategy research. *Strategic Organization*, *14*(3), 248–259.
Jarzabkowski, P., & Whittington, R. (2008). Hard to disagree, mostly. *Strategic Organization*, *6*, 101–106.
Johnson, G., Melin, L., & Whittington, R. (2003). Micro strategy and strategizing: Towards an activity-based view. *Journal of Management Studies*, *40*(1), 3–22.
Johnson, G. A., Langley, A., Melin, L., & Whittington, R. (2007). *Strategy as practice: Research directions and resources*. Cambridge: Cambridge University Press.
Keil, T., Maula, M., & Syrigos, E. (2017). CEO entrepreneurial orientation, entrenchment, and firm value creation. *Entrepreneurship Theory and Practice*, *41*(4), 475–504.
Kogut, B., & Zander, U. (1993). Knowledge of the firm and the evolutionary theory of the multinational corporation. *Journal of International Business Studies*, *24*(4), 625–645.
Krippendorff, K. (2012). *Content analysis: An introduction to its methodology*. Sage. USA
Latour, B. (1999). *We have never been modern*. Translated by C. Porter. London: Harvester Wheatsheaf.
Lubatkin, M., & Shrieves, R. E. (1986). Towards reconciliation of market performance measures to strategic management research. *Academy of Management Review*, *11*(3), 497–512.
Maula, M. V., Keil, T., & Zahra, S. A. (2013). Top management's attention to discontinuous technological change: Corporate venture capital as an alert mechanism. *Organization Science*, *24*(3), 926–947.
Meyer, J. W., & Rowan, B. (1977). Institutionalized organizations: Formal structure as myth and ceremony. *American Journal of Sociology*, *83*(2), 340–363.
Moeini, M., Rahrovani, Y., & Chan, Y. E. (2019). A review of the practical relevance of IS strategy scholarly research. *The Journal of Strategic Information Systems*. Retrieved from www.sciencedirect.com/science/article/abs/pii/S0963868717302135
Nonaka, I. (1991). The knowledge-creating company. *Harvard Business Review*, *69*(6), 96–104.
Penrose, E. T. (1959). *The theory of the growth of the firm*. New York: Sharpe.
Roodman, D. (2009). How to do xtabond2: An introduction to difference and system GMM in stata. *Stata Journal*, *9*(1), 86–136. For Online Publication.
Semadeni, M., Withers, M. C., & Certo, S. T. (2014). The perils of endogeneity and instrumental variables in strategy research: Understanding through simulations. *Strategic Management Journal*, *35*(7), 1070–1079.
Short, J. C., Broberg, J. C., Cogliser, C. C., & Brigham, K. H. (2010). Construct validation using computer-aided text analysis (CATA) an illustration using entrepreneurial orientation. *Organizational Research Methods*, *13*(2), 320–347.
Uotila, J., Maula, M., Keil, T., & Zahra, S. A. (2009). Exploration, exploitation, and financial performance: Analysis of S&P 500 corporations. *Strategic Management Journal*, *30*(2), 221–231.
Wintoki, M. B., Linck, J. S., & Netter, J. M. (2012). Endogeneity and the dynamics of internal corporate governance. *Journal of Financial Economics*, *105*(3), 581–606.
Wolf, C., & Floyd, S. W. (2017). Strategic planning research: Toward a theory-driven agenda. *Journal of Management*, *43*(6), 1754–1788.
Wooldridge, J. M. (2009). *Introductory econometrics: A modern approach* (4th International Student ed., p. 212). Cincinnati: South-Western Cengage Learning and South-Western Cengage Learning.

# 10 Analysis of Investors' Perceptions of Mutual Fund Investment in the Context of the Delhi/NCR Region

*Gurinder Singh, Vikas Garg, and Shalini Srivastav*

## CONTENTS



## 10.1 INTRODUCTION

A mutual fund can be understood as a vehicle that has a large number of investors who pool their money with the common financial objective of earning returns. Anyone who has surplus money after meeting their daily expenses can invest in various types of mutual funds. Different investors have different mutual fund investment objectives and a variety of types of mutual funds have been developed. The fund manager invests the investors' funds in different types of securities, which may include shares in a number of different firms, debentures, and money market instruments, depending on the needs and likes of the customers. The income that is earned by such investments and the increase in capital are shared by the number of units owned by the investors.

Mutual funds are considered a good investment alternative because they allow investors to invest and to diversify their savings into a professionally managed portfolio of securities at a comparatively low cost. In addition, people can also invest small amounts of money into a fund. The investors' money is invested in various asset classes that meet the policies and objectives of the mutual fund. The investment objectives of the mutual fund are based on the investors' decision to contribute money; the fund manager cannot deviate from the fund's specified objectives. Every mutual fund is managed by a mutual fund manager who is responsible for using his or her investment management skills and ensuring that all the required research is performed. The increase in capital and other income earned from various investments is invested in a mix of various stocks and bonds that take advantage of both the growth potential of stocks the reduced risk of bonds. It is also known as hybrid fund.

### 10.1.1 Assets Under Management

The Indian mutual funds industry is a fast-growing one. The Indian mutual fund industry has generated maximum growth in management over the past 10 years. Most of the cash flows into equity funds are from individual investors.

### 10.1.2 Mutual Funds

According to the Association of Mutual Funds in India, individual investors held slightly under 50% of mutual fund assets and corporations held slightly over 50% as of the end March 2007.

## 10.2 LITERATURE REVIEW

Mutual funds are a more recent investment vehicle in the Indian financial market and offer investors the ability to invest smaller amounts of money. As a result, the industry is rapidly growing, increasing the assets of various mutual fund companies. Investments in mutual funds are less risky in comparison to direct investment in the share market (Prabhu, 2015).

All the innovative things are not a part of the system but they all are coming as a new idea of fund management in terms of services and products. The fund management industry includes registrars, distributors, and various underwriters who are concerned about the market. In order to earn profits, all the funds have focused on various recession-free sectors such as FMCG, pharmaceuticals, and technology. Fund assets have been increasing year after year, reaching 1 trillion (Roy, 2012).

Mutual funds offer better capital growth and better income by means of dividends, reinvestment, etc. The portfolio manager guides the mutual fund investors and tries to update them on various investment options. Mutual funds are considered the best investment vehicle for investors looking toward long-term investment strategies. The performance of any mutual fund depends on the relationship between risk and return (Pandey, 2017).

Due to mutual fund investment in the market, the financial markets are becoming more and more elaborate, and investors can select from various financial products in order to develop an innovative portfolio that meets their investment goals. Thus, it is important to understand investors' thinking, ideas, perceptions, and expectations with regards to mutual funds (Vyas, 2012).

Investors' ideas and opinions have been studied and new strategies formulated to launch mutual fund investment schemes that are beneficial and can attract a large number of investors. The industry is working on the deficiencies in different areas of the mutual fund industry and trying to identify those factors that attract investors. The role of financial advisors are brokers is also being redefined in order to fulfill the challenges of the Indian mutual fund industry (Muthukrishnan, 2016).

An investment decision means that the investor has decided where, how, and when to invest into mutual funds and other instruments in order to generate income or increase wealth. Thus, the investment decision is defined as the decision taken by individual investors to invest in mutual funds. Scholars who study behavioral finance note that such decisions are made based on psychological behavior. This will help the investors to take an appropriate decision and also avoid doing mistakes (Kumar, 2014).

The mutual fund industry of India has a department that checks the rise and fall of mutual funds on a quarterly basis known as assets under management (AUM). As of September 2013, the mutual fund industry fell from Rs.8.08 lakh crore to Rs.38,355 crore, which is was .5% less than the previous quarter. But the mutual fund industry revived and then grew by 12% in October 2013 to Rs.8.34 lakh crore. This increase was led by large inflows in terms of cash in October 2013, which is around 55%, which is a record of the highest inflow of funds in a period of six months. All the banks and companies reinvested their surplus funds, and as a result the financial system realized the cyclical cash inflow in the month of October 2013 (Jain, 2013).

In tier 2 and tier 3 cities in India, there has been a tremendous increase in the various investment options for investors. The Indian mutual funds market also has developed a number of tax saving and special schemes such as money markets, income and balanced funds, growth rate funds, etc., that are trying to fulfill the needs and requirements of investors by reducing risk and maximizing returns (Duggimpudi, 2010).

Various analysts has found that in order to understand the investor details, its likes and dislikes, choices and tastes are majorly related to the UTI mutual fund and it is trying to improve its activities by its own performance and services. Researchers have attempted to examine investors' decision-making methods by understanding their interests, perceptions, as well as temperaments. Behaviour makes trading easy and also very risky. As a result, investors try to follow each other, so it is important to understand common behavior, perceptions, and patterns of investing of different groups of investors (Shukla, 2014).

Many researchers state that investors invest in mutual fund for high returns and low risk. As a result, mutual funds capture the attention of many people in many segments of the society. Thus, for this study we include all stakeholders. This chapter is an effort to make people understand investment process in a safe mode which is acceptable by people in Mathura and also to understand their tastes and choices of investment in mutual funds and also the comparative availability of other related investment avenues (Rathnamani, 2013).

## 10.3 RESEARCH METHODOLOGY

Figure 10.1 presents an overview of the research methodology.

### 10.3.1 Type of Research

This research was a descriptive study because we describe mutual fund investment strategies of investors in the Delhi/NCR region.

### 10.3.2 Objectives of Study

The objectives of the study were as follows:

- To analyze investors' perceptions in the context of mutual fund investment.
- To identify and recognize the issues of investors in investing their money in various mutual fund schemes.
- To understand the factors influencing investors' investment in mutual funds.

### 10.3.3 Research Gap

The study area was the Delhi /NCR region of India. The Delhi/NCR region is traditionally considered a business center, but is considered lagging in mutual fund investment. Many government and non-government institutions offer mutual fund investment options but people do not know a lot about mutual funds. Institutions are taking major steps to increase investors' knowledge in mutual funds and share markets. Despite their efforts, they are finding it difficult to increase the number of

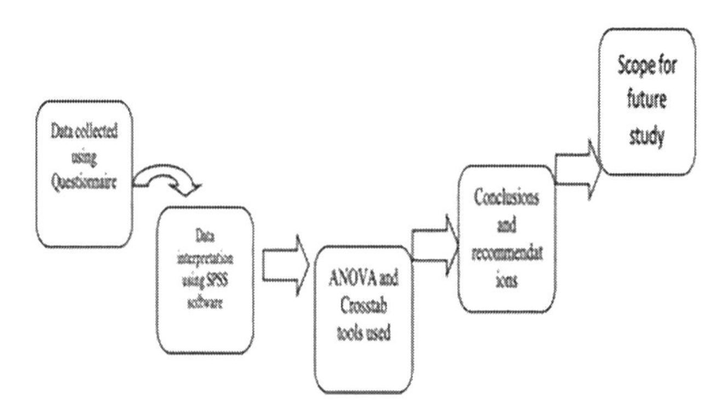


**FIGURE 10.1** Research process.

investors. In view of these problems, this study becomes very important to look into the perceptions of mutual fund investors with reference to some important areas in the Delhi/NCR region of India.

### 10.3.4 Data Collection Tools

Data were classified, tabulated, and processed in an organized manner. ANOVA was used as a statistical tool. ANOVA stands for analysis of variance, which is a statistical tool for analyzing variances among and between the group data.

### 10.3.5 Area of the Study

Study respondents were be randomly selected from the Delhi/NCR region of India.

### 10.3.6 Sampling Technique and Sample Size

The convenience sampling method was used. The sample size was 116.

### 10.3.7 Research Instrument

This study used both primary and secondary data collection methods. A questionnaire was used to collect the primary data. Various magazines, journals, reports, research work from SEBI (Security and Exchange board of India), AMFI (Association of Mutual fund of India), and RBI (Reserve Bank of India) were used as secondary data sources.

## 10.4 DATA COLLECTION AND ANALYSIS

Table 10.1 shows that overall perception about mutual fund investment are more likely to adopt in females more than males. Table 10.2 shows that there is a non-significant relationship between the independent variable (overall perception about mutual fund investment) and dependent variable (gender).

Table 10.3 shows that overall perception about mutual fund investment is a more favorable perception in unmarried people than married people. Table 10.4 shows that

**TABLE 10.1**
**Descriptives: Overall Perception about Mutual Fund Investment by Gender**

| | N | Mean | Std. Deviation | Std. Error | 95% Confidence Interval for Mean | | Minimum | Maximum |
|---|---|---|---|---|---|---|---|---|
| | | | | | Lower Bound | Upper Bound | | |
| Female | 51 | 2.12 | 0.711 | 0.100 | 1.92 | 2.32 | 1 | 3 |
| Male | 66 | 2.12 | 0.645 | 0.079 | 1.96 | 2.28 | 1 | 3 |
| Total | 117 | 2.12 | 0.672 | 0.062 | 2.00 | 2.24 | 1 | 3 |

**TABLE 10.2**
**ANOVA: Overall Perception about Mutual Fund Investment by Gender**

| | Sum of Squares | Df | Mean Square | F | Sig. |
|---|---|---|---|---|---|
| Between Groups | 0.000 | 1 | 0.000 | 0.001 | 0.977 |
| Within Groups | 52.324 | 115 | 0.455 | | |
| Total | 52.325 | 116 | | | |

**TABLE 10.3**
**Descriptives: Overall Perception about Mutual Fund Investment by Marital Status**

| | N | Mean | Std. Deviation | Std. Error | 95% Confidence Interval for Mean | | Minimum | Maximum |
|---|---|---|---|---|---|---|---|---|
| | | | | | Lower Bound | Upper Bound | | |
| Married | 62 | 1.92 | 0.489 | 0.062 | 1.80 | 2.04 | 1 | 3 |
| Unmarried | 55 | 2.35 | 0.775 | 0.105 | 2.14 | 2.55 | 1 | 3 |
| Total | 117 | 2.12 | 0.672 | 0.062 | 2.00 | 2.24 | 1 | 3 |

**TABLE 10.4**
**ANOVA: Overall Perception about Mutual Fund Investment by Marital Status**

| | Sum of Squares | df | Mean Square | F | Sig. |
|---|---|---|---|---|---|
| Between Groups | 5.292 | 1 | 5.292 | 12.939 | 0.000 |
| Within Groups | 47.033 | 115 | 0.409 | | |
| Total | 52.325 | 116 | | | |

there is a significant relationship between the independent variable (overall perception about mutual fund investment) and the dependent variable (marital status).

Table 10.5 shows business professionals are more likely to have a positive view of mutual funds than those who are employed in other areas. Table 10.6 shows that a significant relationship exists between the independent variable (overall perception about mutual fund investment) and the dependent variable (occupation).

Table 10.7 shows that overall perception about mutual fund investment is more likely be accepted by people having income less than Rs 1 lakh. Table 10.8 shows that there is a significant relationship between the independent variable (overall perception about mutual fund investment) and the dependent variable (income).

Table 10.9 shows that overall perception about mutual fund investment is more likely to adopt in people who are majorly graduates. Table 10.10 shows that there is

**TABLE 10.5**
**Descriptives: Overall Perception about Mutual Fund Investment by Occupation**

| | N | Mean | Std. Deviation | Std. Error | 95% Confidence Interval for Mean | | Minimum | Maximum |
|---|---|---|---|---|---|---|---|---|
| | | | | | Lower Bound | Upper Bound | | |
| Business | 53 | 2.28 | 0.769 | 0.106 | 2.07 | 2.49 | 1 | 3 |
| Government | 44 | 2.11 | 0.321 | 0.048 | 2.02 | 2.21 | 2 | 3 |
| Others | 20 | 1.70 | 0.801 | 0.179 | 1.32 | 2.08 | 1 | 3 |
| Total | 117 | 2.12 | 0.672 | 0.062 | 2.00 | 2.24 | 1 | 3 |

**TABLE 10.6**
**ANOVA: Overall Perception about Mutual Fund Investment by Occupation**

| | Sum of Squares | df | Mean Square | F | Sig. |
|---|---|---|---|---|---|
| Between Groups | 4.938 | 2 | 2.469 | 5.940 | 0.004 |
| Within Groups | 47.387 | 114 | 0.416 | | |
| Total | 52.325 | 116 | | | |

**TABLE 10.7**
**Descriptives: Overall Perception about Mutual Fund Investment by Income**

| | N | Mean | Std. Deviation | Std. Error | 95% Confidence Interval for Mean | | Minimum | Maximum |
|---|---|---|---|---|---|---|---|---|
| | | | | | Lower Bound | Upper Bound | | |
| Less than 1 lakh | 34 | 2.15 | 0.359 | 0.062 | 2.02 | 2.27 | 2 | 3 |
| Less than 50,000 | 20 | 1.70 | 0.801 | 0.179 | 1.32 | 2.08 | 1 | 3 |
| More than 1 lakh | 63 | 2.24 | 0.712 | 0.090 | 2.06 | 2.42 | 1 | 3 |
| Total | 117 | 2.12 | 0.672 | 0.062 | 2.00 | 2.24 | 1 | 3 |

**TABLE 10.8**
**ANOVA: Overall Perception about Mutual Fund Investment by Income**

| | Sum of Squares | df | Mean Square | F | Sig. |
|---|---|---|---|---|---|
| Between Groups | 4.432 | 2 | 2.216 | 5.274 | 0.006 |
| Within Groups | 47.893 | 114 | 0.420 | | |
| Total | 52.325 | 116 | | | |

**TABLE 10.9**
**Descriptives: Overall Perception about Mutual Fund Investment by Qualification**

| | N | Mean | Std. Deviation | Std. Error | 95% Confidence Interval for Mean | | Minimum | Maximum |
|---|---|---|---|---|---|---|---|---|
| | | | | | Lower Bound | Upper Bound | | |
| Graduate | 32 | 2.56 | 0.504 | 0.089 | 2.38 | 2.74 | 2 | 3 |
| Others | 26 | 2.35 | 0.797 | 0.156 | 2.02 | 2.67 | 1 | 3 |
| Post graduate | 59 | 1.78 | 0.494 | 0.064 | 1.65 | 1.91 | 1 | 3 |
| Total | 117 | 2.12 | 0.672 | 0.062 | 2.00 | 2.24 | 1 | 3 |

**TABLE 10.10**
**ANOVA: Overall Perception about Mutual Fund Investment by Qualification**

| | Sum of Squares | df | Mean Square | F | Sig. |
|---|---|---|---|---|---|
| Between Groups | 14.430 | 2 | 7.215 | 21.704 | 0.000 |
| Within Groups | 37.895 | 114 | 0.332 | | |
| Total | 52.325 | 116 | | | |

**TABLE 10.11**
**Descriptives: Overall Perception about Mutual Fund Investment by Savings Level**

| | N | Mean | Std. Deviation | Std. Error | 95% Confidence Interval for Mean | | Minimum | Maximum |
|---|---|---|---|---|---|---|---|---|
| | | | | | Lower Bound | Upper Bound | | |
| Above 1,00,000 | 58 | 2.17 | 0.704 | 0.092 | 1.99 | 2.36 | 1 | 3 |
| Above 10,000 | 20 | 1.60 | 0.681 | 0.152 | 1.28 | 1.92 | 1 | 3 |
| Above 50,000 | 39 | 2.31 | 0.468 | 0.075 | 2.16 | 2.46 | 2 | 3 |
| Total | 117 | 2.12 | 0.672 | 0.062 | 2.00 | 2.24 | 1 | 3 |

a significant relationship between the independent variable (overall perception about mutual fund investment) and the dependent variable (qualification).

Table 10.11 shows that overall perception about mutual fund investment is more likely to adopt in people who are having savings above Rs.50,000. Table 10.12 shows that there is a significant relationship between the independent variable (overall perception about mutual fund investment) and the dependent variable (savings).

**TABLE 10.12**
**ANOVA: Overall Perception about Mutual Fund Investment by Savings Level**

| | Sum of Squares | df | Mean Square | F | Sig. |
|---|---|---|---|---|---|
| Between Groups | 6.941 | 2 | 3.471 | 8.718 | 0.000 |
| Within Groups | 45.384 | 114 | 0.398 | | |
| Total | 52.325 | 116 | | | |

## 10.5 MAIN RESULTS

The study found that, in comparison to males, females are more interested in investing into mutual funds, as by nature they try to play a safe game. Moreover, people from business backgrounds and who seek reasonable returns are more interested in investing in mutual funds as a safe investment. Also, investors having a sufficient amount of savings after all expenses are more likely to go for a mutual fund investment.

## 10.6 CONCLUSION AND RECOMMENDATION

Wealth creation is an art, and the securities market is the best game in town to create wealth within the four walls of one's home, but only if the portfolio is regularly monitored and diversified. Putting all one's eggs in a single basket is not advisable, hence monitoring of stocks is essential, in combination with all the other processes of portfolio management.

In the current scenario, investors have various investment options, such as equities, bonds, etc. The behavior of these investors is affected by various demographic factors, including age, gender, and income. Mainly investors aim about less risk with high return and under this approach research factors in terms of investment are income levels, education, occupation etc. Thus, this paper provides a perception which is affected by all these factors of mutual fund investment which tells us about the perception of investors in Delhi/NCR. The data also deal with the middle-class segment of people who prefer mutual funds as a primary investment source to earn a high rate of return.

Young investors are considered to more interested in such jobs because can making their future secure in a more reliable manner. This study is supported by primary data and current facts and figures to make genuine changes in various mutual fund organizations in order to attract the investors.

## 10.7 LIMITATIONS OF THE STUDY AND FUTURE SCOPE OF STUDY

- The sample size could be increased to give a more realistic view.
- The researcher used ANOVA as a tool of analysis, which has its own limitations.

- Respondents could be classified based on their innovation adoption attitude.
- Data collection methods such as the interview method could be used to make the responses more accurate.

## REFERENCES

Duggimpudi, R. A. (2010). An evaluation of equity diversified mutual funds: The case of the Indian market. *Investment Management and Financial Innovations*, 76–84.

Jain, G. A. (2013). Investor's preference towards mutual fund in comparison to other investment avenues. *Journal of Indian Research*, 115–131.

Kumar, S. K. (2014). Influence of risk perception of investors on investment decisions: An empirical analysis. *Journal of Finance and Bank Management*, 15–25.

Muthukrishnan, S. K. (2016). Demographic conditions influencing the investors' perception towards mutual fund investment in Virudhunagar district, Tamil Nadu. *International Journal of Advance Research and Innovative Ideas in Education*, 3789–3794.

Pandey, S. (2017). A comparative study of performance of top mutual fund schemes in India. *Abhinav*, 114–124.

Prabhu, G. (2015). Perception of Indian Investor towards investment in mutual funds with special reference to MIP Funds. *IOSR Journal of Economics and Finance*, 66–74.

Rathnamani, V. R. (2013). Investor's preferences towards mutual fund industry in Trichy. *IOSR Journal of Business and Management*, 48–55.

Roy, J. B. (2012). Customer perception on mutual fund product: A technical analysis. *Interscience Management Review*, 63–66.

Shukla, M. (2014). A literature review on analyzing investors' perceptions towards mutual funds. *AE International Journal of Multidisciplinary Research*, 1–6.

Vyas, R. (2012). Mutual fund investor's behaviour and perception in Indore city. *Researchers World Journal of Arts, Science & Commerce*, 67–76.

# 11 Measuring Customer Brand Equity and Loyalty: A Study of Fuel Retail Outlets in Lucknow

*Shubhendra Singh Parihar and Ankit Mehrotra*

## CONTENTS



## 11.1 INTRODUCTION

Why do firms invest so heavily on developing their brand? There must be some tangible and intangible benefits perceived by both the consumer and seller. The *term* brand originated from the ancient practice of stamping cattle so that the owners could recognize and claim them as their cattle. A brand is basically a differential factor that every seller wishes to create for the product or service offered. In modern trade practices the brand has played a very significant role in businesses.

The concepts of brand culture, brand equity, and brand value describe the benefits a brand offers to customers and the firm. The brand culture is created by the firm itself, its users and their experiences, brand communication by the media and celebrity endorsements, and influencers, who can be from varied backgrounds, such as bloggers, industry opinion leaders, and reviewers.

Brand equity can be positive or negative. David Aaker (1992) outlined that brand equity has components such as brand awareness, perceived quality, brand association, brand personality, brand loyalty, and other assets such as trademarks and patents.

Measuring customer perceptions towards brands and their repeat purchase behavior from the same petrol pump can be used to determine brand equity. This study investigated customer behavior with regards to fuel purchasing decisions.

## 11.2 PETROLEUM PRODUCT AND INDUSTRY OVERVIEW

In India, the marketing of petroleum products for retail customers is performed by two major players. The first is the Public Sector Oil Marketing Companies (OMCs). The OMCs are Indian Oil Corporation Ltd (IOCL), Hindustan Petroleum Corporation Ltd (HPCL), Bharat Petroleum Corporation Ltd (BPCL), Numaligarh Refinery Ltd (NRL), Mangalore Refinery & Petrochemicals Ltd (MRPL), and Bharat Oman Refineries Ltd (BORL). The second includes private companies such as Reliance, Essar, and Shell. As per data available on the website of the Ministry of Petroleum and Natural Gas, 189 liquid petroleum gas (LPG) bottling plants and 320 terminals/depots were operational with 19,136 LPG distributorships as 1 September 2017. As of 1 April 2017, there were 59,595 retail outlets, including those of private companies with 6543 SKO/LDO dealers across nation (http://petroleum.nic.in/marketing/about-marketing).

Considering the intensified competition in the segment, industry players are looking at ways to retain their customer base. This study sought to identify the meaningful factors that impact on brand preference and loyalty.

## 11.3 LITERATURE REVIEW

The concept of brand equity is still under development, because since the 1980s many definitions have been proposed by different authors. Early definitions of *brand equity* ranged from "the net present value of the incremental cash flows attributable to a brand name" (Shocker & Weitz, 1988) to the "set of associations and behavior on the part of the brand's consumers, channel members and parent corporation.

The most accepted definition of brand equity was given by Aaker (1991), as a set of assets and liabilities linked to a brand, its name and symbol that adds to or subtracts from the value provided by a product or a service to a firm and/or to that firm's customers. The American Marketing Association defines it from the consumer's perspective as brand equity is based on consumer's attitude towards the brand and favorable consequences of brand use. There have been many conceptualizations of brand equity given its complexity and multifaceted nature.

Ali and Mukadas (2015) studied the impact of brand equity on brand loyalty with the mediation of customer satisfaction in restaurant sector of Lahore, Pakistan. The study uses seven dimensions of brand equity, which include physical quality, staff behavior, ideal self-congruence, brand identification, life style-congruence, trust and environment.

Srivastava and Shocker (1991) talked about brand value which has long-term effect on cash flows and future profits. Hsin and Ya Ming Liu (2009) explained perceived quality as sustainable and greater preference and purchase intentions in service industry. Brand loyalty is much closer to the concept of overall brand equity than brand awareness associations and perceived quality.

Prasanna and Milan (2013) in their study identified several factors that influenced customers to select a particular petrol retail outlet: speed of delivery in meeting basic needs, value-added services, reliability, and exceptional service. They categorized the four levels of customers' expectations from a petrol retail outlet.

A petrol service station with a forecourt store represents a hybrid retail business that combines franchise and convenience store. A review of franchising research determined that franchising research has mainly focused on franchise structure, consumer exchange (i.e., price and promotion), and strategic intention (Chabowski et al., 2011).

Ali and Mukadas (2015) stated in their study the effect of trust on brand loyalty with mediation of customer satisfaction and offer practical help for managers to train employees which could enhance customer satisfaction and loyalty.

Attri et al. (2012) stated in their study that PSU petrol firms need to work on integrated marketing communication strategies to increase brand awareness. Chahal and Bala (2015) in their confirmatory study on brand equity and brand loyalty: A special look at the impact of attitudinal and behavioral loyalty examined the relationship between brand equity and brand loyalty. Šalčiuviene and Auruškevičiene (2003). Study on dimensions of fuel customer loyalty-multidimensional concept explained the big picture of loyalty is far from its component factors. The loyalty holistically is different from its component factors. Pope et al. (2012) in their study quoted the usefulness of multiple attribute decision model to help resolve such complex decisions. The location is everything in retailing as said by many decision makers.

Kumar and Sahay (2004) underlined the petrol retail outlets not only commodity dispenser but also the service centers for developing loyalty among the customers.

## 11.4 OBJECTIVE OF THE STUDY

This study was undertaken with the objective to measure customer perceptions towards the quality, quantity, and services offered by fuel retail outlets and to identify the major components of retail outlet preference for a customer (retail outlet loyalty).

## 11.5 METHODOLOGY

Lucknow was the geographical area chosen for study. A sample of 35 petrol pumps were chosen from Lucknow. Data were captured from customers visiting a specific pump for their fuel needs. A questionnaire was developed to interview regular customers of fuel stations. To engage customers and to get their valuable time for an interview, respondents were provided a fuel coupon based on funds provided by a petroleum firm once they spent time answering questions and giving honest feedback.

A detailed questionnaire was prepared covering all the factors that could be used to measure the perceived brand equity and customer loyalty. The questionnaires were uploaded onto tablets that were used to help collect the responses and perform data analysis.

The final list of variables was made after a pilot study involving five petrol pumps with a small sample of customers to verify the significance of the variables used to measure customer perceptions towards the quality, quantity, location, and impact of value-added services in customer purchase intentions.

### 11.5.1 Sample Profile

The data were collected using a survey method from 2300 respondents from different petrol pumps in Lucknow. Out of the collected data, 1715 questionnaires were used for analysis, as the rest of the data were found to be incomplete in one or the other parameters.

The questionnaire was put to reliability analysis which came out to be .851 and is appropriate for taking up further analysis.

## 11.6 ANALYSIS AND INTERPRETATION

### 11.6.1 Determination of the Importance Attached to Various Parameters by Customers in Choosing a Petrol Pump

In order to find out the importance attached by customers to various parameters under study for choosing a petrol pump, Kendall's W test was applied. As per Table 11.1, it can be seen that the asymptotic significance for Kendall's W test came out to be significant, with convenient location being given the highest importance by a customer for choosing a petrol pump outlet, followed by fuel quality and quantity. Provision of additional services did not have a significant impact on a customer's choice of an outlet. Strangely, complaint handling was rated lowest by the customer in making a choice for a petrol pump. This can be attributed to the fact that people hardly ever get into arguments with officials and workers of petrol pumps with regards to quantity and quality, as they have accepted that their complaint will not lead to any result. Rather, customers prefer those petrol pumps that are convenient in location to them and that they perceive the quality and quantity supplied to be relatively better than the others.

### 11.6.2 Determination of the Relationship Between Time Duration of Association with a Petrol Pump vs. Importance Attached to Various Parameters

In order to determine the relationship between time duration of association with a petrol pump and the importance attached to a parameter, crosstab analysis with a Chi-square test was performed.

**TABLE 11.1**
**Importance Attached to Parameters**
**Ranks**

| | Mean Rank | Test Statistics | |
|---|---|---|---|
| Convenient Location | 5.78 | N | 1715 |
| Fuel Quality | 5.50 | Kendall's W* | .062 |
| Fuel Quantity | 5.48 | Chi-square | 851.179 |
| Cleanliness Maintenance | 5.10 | df | 8 |
| Behavior Of Staff | 5.02 | Asymp. sig. | .000 |
| Prompt Service | 4.91 | | |
| Basic Amenities | 4.48 | | |
| Additional Services | 4.41 | | |
| Complaint Handling | 4.30 | | |

* Kendall's coefficient of concordance.

#### 11.6.2.1 Time Duration vs. Fuel Quality

The Table 11.2 clearly shows that fuel quality has a significant impact on a customer's choice of visiting and revisiting a petrol pump. It can be seen from Table 11.2 that customers have a direct relationship between visiting a petrol pump vis-à-vis perceived good fuel quality as people who have been visiting a petrol pump for a year or more finding fuel quality to be on the better side (good, very good, and excellent).

#### 11.6.2.2 Time Duration vs. Fuel Quality

It can be seen from Table 11.3 that fuel quantity has a significant impact on a customer's choice to revisit a petrol pump. It can be seen that there is a high percentage of concentration of customers in good (15.1%), very good (18.8%), and excellent (32.0%), especially for the time period of visit being a year or above. This shows that the perceived fulfillment of promised quantity does make an impact on being a loyal customer of a petrol pump.

#### 11.6.2.3 Time Duration vs. Service

The Table 11.4 shows that service delivery time has an impact on customer loyalty and length of association with the petrol pump as customers who stay longer find and rate promptness in service as very good or excellent.

#### 11.6.2.4 Time Duration vs. Behaviour

The Table 11.5 shows that behavior of staff does have an impact on a customer being associated with a petrol pump. It is quite obvious that every human being likes to be treated with respect and dignity, and the behavior of officials and staff make a significant difference for the same. Congenial and cordial behavior of staff even in the situation of argument forces customers to think positively of the petrol pump and urge to stay longer.

**TABLE 11.2**
**Since When You Have Been Taking Fuel vs. Fuel Quality**

| % of Total | | @6aFuelQuality | | | | | | |
|---|---|---|---|---|---|---|---|---|
| | | Poor | Very Bad | Bad | Average | Good | Very Good | Excellent |
| @3_SinceWhen | Last month | .2% | .1% | .1% | .5% | 1.2% | 1.0% | 1.9% |
| | Last six months | .8% | .1% | .1% | .5% | .6% | 2.3% | 3.2% |
| | One year | .2% | .1% | .4% | .8% | 1.7% | 2.3% | 6.9% |
| | More than a year | 2.1% | .6% | 1.0% | 5.4% | 15.8% | 17.2% | 32.8% |
| Total | | 3.4% | .8% | 1.7% | 7.1% | 19.3% | 22.9% | 44.8% |

**Chi-Square Tests**

| | Value | df | Asymp. Sig. (2-sided) |
|---|---|---|---|
| Pearson Chi-Square | 59.699* | 18 | .000 |
| Likelihood Ratio | 53.253 | 18 | .000 |
| Linear-by-Linear Association | 3.671 | 1 | .055 |
| N of Valid Cases | 1715 | | |

* Eight cells (28.6%) have an expected count less than 5. The minimum expected count is .72.

**TABLE 11.3**
**Since When You Have Been Taking Fuel vs. Fuel Quantity**

| | | @6bFuelQuantity | | | | | | |
|---|---|---|---|---|---|---|---|---|
| | | Poor | Very Bad | Bad | Average | Good | Very Good | Excellent |
| @3_SinceWhen | Last month | | .3% | .2% | .6% | .8% | 1.0% | 2.1% |
| | Last six months | .8% | .1% | .2% | .2% | .8% | 2.6% | 2.9% |
| | One year | .2% | .1% | .5% | .6% | 1.7% | 3.3% | 6.0% |
| | More than a year | 2.0% | .8% | 1.1% | 5.1% | 15.1% | 18.8% | 32.0% |
| Total | | 3.0% | 1.2% | 2.0% | 6.6% | 18.5% | 25.7% | 43.0% |

**Chi-Square Tests**

| | Value | df | Asymp. Sig. (2-sided) |
|---|---|---|---|
| Pearson Chi-Square | 72.486* | 18 | .000 |
| Likelihood Ratio | 59.964 | 18 | .000 |
| Linear-by-Linear Association | 3.464 | 1 | .063 |
| N of Valid Cases | 1715 | | |

* Eight cells (28.6%) have an expected count less than 5. The minimum expected count is 1.08.

## TABLE 11.4
## Since When You Have Been Taking Fuel vs. Prompt Service

| | | @6cPromptService | | | | | | |
|---|---|---|---|---|---|---|---|---|
| | | Poor | Very Bad | Bad | Average | Good | Very Good | Excellent |
| @3_SinceWhen | Last month | .1% | .1% | .3% | .7% | .8% | 1.5% | 1.6% |
| | Last six months | .7% | .2% | .2% | .2% | 1.2% | 2.3% | 2.7% |
| | One year | .1% | .2% | .5% | .7% | 2.1% | 4.7% | 4.1% |
| | More than a year | 2.2% | 1.2% | 2.0% | 7.1% | 15.6% | 21.0% | 25.9% |
| **Total** | | 3.1% | 1.7% | 3.0% | 8.7% | 19.7% | 29.5% | 34.4% |

**Chi-Square Tests**

| | Value | df | Asymp. Sig. (2-sided) |
|---|---|---|---|
| **Pearson Chi-Square** | 44.207* | 18 | .001 |
| **Likelihood Ratio** | 40.371 | 18 | .002 |
| **Linear-by-Linear Association** | .896 | 1 | .344 |
| **N of Valid Cases** | 1715 | | |

* Seven cells (25.0%) have an expected count less than 5. The minimum expected count is 1.49.

## TABLE 11.5
## Since When You Have Been Taking Fuel vs. Behavior of Staff

% of Total

| | | @6dBehaviorOfStaff | | | | | | |
|---|---|---|---|---|---|---|---|---|
| | | Poor | Very Bad | Bad | Average | Good | Very Good | Excellent |
| @3_SinceWhen | Last month | .1% | .1% | .2% | .6% | 1.4% | 1.0% | 1.8% |
| | Last six months | .6% | .3% | .1% | .5% | .9% | 2.6% | 2.5% |
| | One year | .2% | .2% | .2% | .6% | 2.0% | 4.8% | 4.3% |
| | More than a year | 2.3% | 1.5% | 1.3% | 6.8% | 15.5% | 20.5% | 27.1% |
| **Total** | | 3.1% | 2.2% | 1.9% | 8.5% | 19.8% | 28.9% | 35.7% |

**Chi-Square Tests**

| | Value | df | Asymp. Sig. (2-sided) |
|---|---|---|---|
| **Pearson Chi-Square** | 43.936* | 18 | .001 |
| **Likelihood Ratio** | 41.616 | 18 | .001 |
| **Linear-by-Linear Association** | .680 | 1 | .410 |
| **N of Valid Cases** | 1715 | | |

* Eight cells (28.6%) have an expected count less than 5. The minimum expected count is 1.64.

### 11.6.2.5 Time Duration vs. Complaint Handling

The Table 11.6 corroborates the findings of Table 11.5 where complaint handling, which is a fallout of behavior of staff, has a significant impact on customers' perceptions and feelings of returning to a petrol pump even if an argument/complaint has occurred.

### 11.6.2.6 Time Duration vs. Additional Services

Table 11.7 shows that presence of additional services does not have a significant impact on a customer's length of stay. This is quite interesting, as one would have expected that additional services like filling of tires with air, cleaning of car windows, and the like would have a positive impact on the minds of customers as they would have seen getting more than what they had paid for.

### 11.6.2.7 Time Duration vs. Basic Amenities, Cleanliness, and Maintenance

As can be seen from Tables 11.8 and 11.9, basic amenities such as provision of drinking water, providing shade, etc., and cleanliness and maintenance of ambience does significantly affect customer association with a petrol pump.

### 11.6.2.8 Time Duration vs. Location

Table 11.10 clearly brings out the fact that location convenience has a significant impact of length of association of a customer with a petrol pump. This corroborates

**TABLE 11.6**
**Since When You Have Been Taking Fuel vs. Complaint Handling**

| | | @6eComplaintHandling | | | | | | |
|---|---|---|---|---|---|---|---|---|
| | | Poor | Very Bad | Bad | Average | Good | Very Good | Excellent |
| @3_SinceWhen | Last month | .1% | .1% | .1% | 1.6% | 1.0% | .8% | 1.4% |
| | Last six months | .4% | .3% | .4% | 1.2% | 1.0% | 2.0% | 2.2% |
| | One year | .2% | .2% | .2% | 1.5% | 2.3% | 4.5% | 3.6% |
| | More than a year | 1.5% | 1.3% | 3.5% | 14.1% | 14.1% | 18.5% | 21.9% |
| Total | | 2.2% | 2.0% | 4.3% | 18.3% | 18.4% | 25.8% | 29.0% |

**Chi-Square Tests**

| | Value | df | Asymp. Sig. (2-sided) |
|---|---|---|---|
| Pearson Chi-Square | 46.35* | 18 | .000 |
| Likelihood Ratio | 43.133 | 18 | .001 |
| Linear-by-Linear Association | .972 | 1 | .324 |
| N of Valid Cases | 1715 | | |

* Seven cells (25.0%) have an expected count less than 5. The minimum expected count is 1.74.

## TABLE 11.7
## Since When You Have Been Taking Fuel vs. Additional Services

| | | @6fAdditionalServices | | | | | | |
|---|---|---|---|---|---|---|---|---|
| | | Poor | Very Bad | Bad | Average | Good | Very Good | Excellent |
| @3_SinceWhen | Last month | .1% | .1% | .2% | .9% | 1.0% | 1.7% | 1.1% |
| | Last six months | .2% | .5% | .6% | .8% | 1.0% | 2.3% | 2.1% |
| | One year | .2% | .2% | .5% | 1.3% | 2.5% | 3.8% | 3.9% |
| | More than a year | 1.8% | 2.2% | 3.7% | 10.3% | 15.9% | 18.7% | 22.3% |
| Total | | 2.3% | 3.0% | 5.0% | 13.2% | 20.4% | 26.6% | 29.4% |

**Chi-Square Tests**

| | Value | df | Asymp. Sig. (2-sided) |
|---|---|---|---|
| Pearson Chi-Square | 26.447* | 18 | .090 |
| Likelihood Ratio | 25.912 | 18 | .102 |
| Linear-by-Linear Association | .034 | 1 | .853 |
| N of Valid Cases | 1715 | | |

* Six cells (21.4%) have expected count less than 5. The minimum expected count is 2.05.

## TABLE 11.8
## Since When You Have Been Taking Fuel vs. Basic Amenities

| | | @6gBasicAmenities | | | | | | |
|---|---|---|---|---|---|---|---|---|
| | | Poor | Very Bad | Bad | Average | Good | Very Good | Excellent |
| @3_SinceWhen | Last month | .2% | .2% | .3% | .8% | 1.0% | 1.0% | 1.6% |
| | Last six months | .2% | .5% | .5% | 1.2% | .9% | 1.9% | 2.4% |
| | One year | | .1% | .4% | 1.4% | 3.0% | 2.9% | 4.5% |
| | More than a year | .8% | 1.4% | 3.4% | 12.8% | 15.7% | 18.4% | 22.4% |
| Total | | 1.1% | 2.1% | 4.7% | 16.2% | 20.7% | 24.3% | 31.0% |

**Chi-Square Tests**

| | Value | Df | Asymp. Sig. (2-sided) |
|---|---|---|---|
| Pearson Chi-Square | 38.724* | 18 | .003 |
| Likelihood Ratio | 37.516 | 18 | .004 |
| Linear-by-Linear Association | .737 | 1 | .391 |
| N of Valid Cases | 1715 | | |

* Seven cells (25.0%) have an expected count less than 5. The minimum expected count is .97.

**TABLE 11.9**
**Since When You Have Been Taking Fuel vs. Cleanliness/Maintenance**

| | | @6hCleanlinessMaintenance | | | | | | |
|---|---|---|---|---|---|---|---|---|
| | | Poor | Very Bad | Bad | Average | Good | Very Good | Excellent |
| @3_ SinceWhen | Last month | | .1% | .4% | .4% | .8% | 2.0% | 1.4% |
| | Last six months | .2% | .3% | .5% | .7% | 1.0% | 2.6% | 2.3% |
| | One year | | .1% | .3% | .9% | 1.8% | 4.3% | 4.9% |
| | More than a year | .5% | .9% | 2.6% | 8.6% | 13.8% | 21.5% | 26.9% |
| Total | | .6% | 1.5% | 3.9% | 10.7% | 17.4% | 30.4% | 35.5% |

**Chi-Square Tests**

| | Value | Df | Asymp. Sig. (2-sided) |
|---|---|---|---|
| Pearson Chi-Square | 42.422* | 18 | .001 |
| Likelihood Ratio | 38.330 | 18 | .004 |
| Linear-by-Linear Association | 1.763 | 1 | .184 |
| N of Valid Cases | 1715 | | |

* Seven cells (25.0%) have an expected count less than 5. The minimum expected count is .56.

**TABLE 11.10**
**Since When You Have Been Taking Fuel vs. Convenient Location**

| | | @6iConvenientLocation | | | | | | |
|---|---|---|---|---|---|---|---|---|
| | | Poor | Very Bad | Bad | Average | Good | Very Good | Excellent |
| @3_SinceWhen | Last month | .1% | .1% | .3% | .4% | .6% | 1.4% | 2.2% |
| | Last six months | .8% | .1% | .3% | .5% | .9% | 1.8% | 3.1% |
| | One year | .2% | | .1% | .5% | 1.3% | 3.7% | 6.5% |
| | More than a year | 1.8% | .6% | 1.5% | 5.2% | 10.3% | 17.0% | 38.4% |
| Total | | 2.9% | .8% | 2.3% | 6.6% | 13.2% | 23.9% | 50.3% |

**Chi-Square Tests**

| | Value | Df | Asymp. Sig. (2-sided) |
|---|---|---|---|
| Pearson Chi-Square | 50.218* | 18 | .000 |
| Likelihood Ratio | 41.535 | 18 | .001 |
| Linear-by-Linear Association | 7.969 | 1 | .005 |
| N of Valid Cases | 1715 | | |

* Eight cells (28.6%) have an expected count less than 5. The minimum expected count is .72.

the findings of Kendall's W test where convenient location was rated as the most important parameter for choosing a petrol pump.

## 11.7 DISCUSSION

Customer perception towards brands varies with the type of product and service. Customers may also be influenced by information that may be true or untrue. Word of mouth and a firm's own marketing communication play a vital role in shaping customer perceptions. If a petroleum firm communicates a message such as "Pure for sure," it means that they are claiming to offer high-quality to customers. However, customer perceptions result from the influence of multiple factors. There are several studies on brand equity. Since the 1990s, a lot of work has been performed on brand equity, with Keller (1993) giving the first brand equity definition from the customer perspective, which has been quoted in literature review.

Brand preference is a component of brand equity from the customer viewpoint in retail marketing of petroleum products, especially for diesel and petrol which are regulated products and the price controlled by the government based upon international crude oil fluctuations and other factors.

This study determined that location is the most important factor in brand preference. Even in retail marketing it is endorsed that location is everything. That why almost every petroleum firm focuses first on location when planning a retail outlet. Quality and quantity of the petroleum product is the second most influential component in purchase decisions. But it is very difficult for customers to measure the quality of a petroleum product. Normally they measure the quality by the mileage they get after fueling their vehicle. But petroleum firms assure that they are offering quality and quantity in every transaction with the customer. In the recent past, several petrol pumps from one of the OMCs was selling compromised quality and quantity petrol. This kind of incident causes a lot of damage to brand equity. But the effect on customer preference hardly make any significant difference because most of the retail outlets do not have effective complaint mechanisms. Even if the company offers a complaint form on its website, customers are not aware of this, so the customer complaint mechanism does not have any significant effect on brand preference. Value-added services such as tire checks and window cleaning have some effect on brand preference, but behavior of staff is not be considered to be a major contributor to brand preference because it is expected by the customer that they will be treated well. Any misbehavior would obviously be considered a major influencing factor.

So, to conclude, the findings of this study suggest that location (convenience), quality and quantity (core product), and value-added services are the major components of brand preference in the oil retail marketing segment. Brand image and repeat purchase contribute to overall brand equity. Thus, oil marketing companies should focus on location and core product offerings. Due to the changing market and intensified competition, factors such as value-added services and complaint redressal mechanisms should be considered.

## REFERENCES

Aaker, D. (1991). Capitalizing the value of a brand name. MB equity.

Aaker, D. (1992). The value of brand equity. *Emerald Insights.*

Al, B. C. (2011). The structure of sustainability research in marketing 1958–2008: A basis for future research. *Journal of the Academy of Marketing Science*, *39*(1).

Ali, F., & Mukadas, S. (2015). The impact of brand equity on brand loyalty: The mediating role of customer satisfaction. *Pakistan Journal of Commerce & Social Sciences*, *9*(3), 890–915.

Attri, R., Urkude, A. M., & Pahwa, M. S. (2011, December). Measuring public sector oil marketing companies brand awareness. *IUP Journal of Brand Management*, *8*(4), 7–24, 18p.

Brahmbhatt, D., & Shah, J. (2017). Determinants of Brand equity from the consumer's perspective: A literature review. *IUP Journal of Brand Management.*

Chahal, H., & Bala, M. (2010, January–June). Confirmatory study on brand equity and brand loyalty: A special look at the attitudinal and behavioral loyalty. *Vision* (09722629), *14*(1–2), 1–12, 13p.

Ferris, P. W., Gregg, E. A., Chinn, B., & Razuri, M. *Brand equity: An overview.* Harvard Business Publishing.

Keller, K. L. (1993). Conceptualizing, measuring, and managing customer-based brand equity. *Journal of Marketing*, *57*(1), 1–22.

Kumar, P., & Sahay. (2003 October–2004 March). Retailing at petrol pumps: From commodity dispensing to customer service. *A Journal of Services Research*, *3*(2), 29–55, 27p.

Liu, H. H. (2009). The impact of brand equity on brand preference and purchase intentions in the service industries. *The Service Industrial Journal*, *29.*

Milan, P. M. (2013). Measuring customers' choices and commitment of petrol retail outlets-comparative study between cosmopolitan city & Tier-2 city. *Journal of Marketing & Communication,* 9(2).

Pope, J. A., Lane, W. R., & Stein, J. (2012). A multiple attribute decision model for store location decisions. *Southern Business Review*, *37*(2), 15–25, 11p. Database: Business Source Complete.

Prasanna, R. M., & Milan, R. D. R. (2013, September–December). *Journal of Marketing & Communication*, *9*(2), 66–72, 7p. Database: Business Source Complete.

Šalčiuviene, L., & Auruškevičiene, V. (2003). Study on dimensions of fuel customer loyalty-multidimensional concept. *Management of Organizations: Systematic Research*, *28*, 163–171, 9p.

Shocker, A. A. (1995). Brand equity is a relational concept. *Journal of Brand Management.*

## WEB LINK

1. http://petroleum.nic.in/marketing/about-marketing

# 12 Random Walk Hypothesis: Evidence from the Top 10 Stock Exchanges Using the Variance Ratio Test

*Madhu Arora, Miklesh Prasad Yadav, and Smita Mishra*

## CONTENTS



## 12.1 INTRODUCTION

Many researchers have examined what determines the price of a stock. When all of the available information is reflected in the stock price, there is efficiency in the market, but the major question is that how quickly it is absorbed into the stock price. Market efficiency has been classified as weak, semi-strong, and strong (Fama, 1965). EMH says that the price of a stock tends to reflect all of the information available such that no investor can enjoy an abnormal profit. It says that the price of a security is independent from its past prices. Investors can easily identify the significance of market efficiency. If the stock market follows the random walk model, no investor can generate an abnormal profit. Hence, an investor can study the market carefully on the basis of it. Arusha and Genertne (2007) determined that foreign investment can be attracted and domestic saving can be encouraged if there is presence of random walk in the stock market. This provides the motivation for the present study. One of the approaches to determining whether the random walk is present is to verify the occurrence of the unit root in security prices. If the security price includes the unit root, then the shock be reflected permanently as the prices will move to a new path. But when prices follow a mean-reverting process, future price movement can be estimated from past values. Ahmad et al. (2006), Alimov et al. (2004), and Gupta Basu (2007) studied efficiency in the stock market. The efficient market hypothesis explains that, statistically, share price changes have a similar distribution and are independent of each other. It means that share price fluctuation has no memory, and thus an investor cannot estimate future market prices based on the past history of the stock price. Fama (1970) identified three forms of market efficiency. The first is the weak form of market efficiency that presumes that the present share price fully reflects all the available market information. It says that the past price and volume of stock are not associated with future prices, and thus one cannot generate excess returns. The second is the strong form of market efficiency that assumes that the current share price adjusts rapidly in response to the release of new public information. The third is the strong form of market efficiency that says that the current share price fully reflects all public and private information. It assumes market, non-market, and inside information are all factored into the share price.

The motivation behind the occurrence of an efficient market is competition amongst investors to gain from new information that becomes available in the marketplace. In a perfect capital market, every investor has homogeneous information and no one can benefit from the shifts because all the investors have similar access to announcements. It means that it is difficult to systematically under- or overvalue stock prices. Amid the great contributions of studies of market efficiency was the determination that there could be anomalies in the short run. Marseguerra (1998) says that there are some assumptions for this market efficiency like there are no taxes, no agency costs, and the all the investors have the similar information. The random walk theory assumes that a rational person is aware of the risks associated with investment and measures that risk premium. It says that fluctuations in stock price are not associated with any exogenous variables, but rather are reflected through investor behaviour based on new information that arrives in the marketplace. In the present study, the stock price of the top 10 countries in the world has been considered for testing the efficiency. An efficient market increases the confidence of investors who have invested in stock market and can motivate others who want to invest as no one can generate abnormal amount of profitability (Marulkar & Muhammadriyaj, 2017). In the current study, the top 10 stock markets in terms of market capitalization have been considered for testing market efficiency.

## 12.2 REVIEW OF THE LITERATURE

In order to find out the research gap, a thorough study on the random walk model was performed (see also Table 12.1). Husain (1997) determined that the random walk does not exist in the Pakistan stock market because of its strong dependence on stock returns. Mustafa and Nishat (2007) found that Middle Eastern markets have a significant correlation with returns, and thus the random walk hypothesis does apply. However, the leading emerging market is not consistent in this aspect. Liu, Song, and Romilly (1997) found the weak form of efficiency in China. Darrat and Zhong (2000) and Lee, Chen, and Rui (2001) found independence of stock returns series for Chinese stock markets. Nevertheless, some conflicting results

**TABLE 12.1**
**Literature Review**

| Author(s) | Year | Methodology | Statistical Tool | Findings |
|---|---|---|---|---|
| Amanulla and Kamaiah | 1998 | BSE Sensex | Rank correlation | Weak form of market efficiency. |
| Poshakwale | 2002 | BSE Sensex | — | Inconclusive. |
| Hoque et al. | 2007 | Eight emerging markets | Variance ratio | Random walk does not exists. |
| Worthington and Higgs | 2009 | Australian Stock Exchange | Correlation test | Random walk does not exist. |
| Abdmoulah | 2010 | Arab stock markets | — | Inefficient. |
| Borges | 2011 | European stock exchanges | Autocorrelation test, run test, variance test | Mixed results. |
| Majumder | 2012 | BRIC stock markets | Serial correlation test | Inefficiency. |
| Vidali Maria | 2013 | London Stock Exchange | Variance ratio test, autocorrelation test | Weak efficiency. |
| Dsouza and Mallikarjunappa | 2015 | Eight emerging stock markets | Autocorrelation test, run test, variance ratio test | Do not follow random walk. |
| Marulkar, Kedar Faniband, and Muhammadriyaj | 2017 | Bombay Stock Exchange (BSE) | Run test, serial correlation test | Efficient in weak form due to which it follows random walk. |
| Tahir et al. | 2017 | Karachi Stock Exchange (KSE) | ADF test, autocorrelation | Not efficient. |

were reported in the same market by Lima and Tabak (2004). The Chinese-A stock and Singapore stock market are weak-form efficient but the Chinese-B shares and Hong Kong market were found to be autocorrelated with stock returns. Gupta (2007) determined that the Indian stock market follows efficiency in the stock market. The researchers noted that market capitalization and liquidity generated conflicting results in the stock market in the same country. Lock (2007), Charles and Darne (2008), and, Fifield and Jetty (2008) found that the Chinese stock market follows the random walk hypothesis.

Worthington and Higgs (2009) concluded that the Australian market rejected the random walk hypothesis because of its strongest dependence in security returns in past trading days. Hoque, Kim, and Pyun (2007) found serial correlation in stock returns in eight emerging markets. Benjelloun and Squalli (2008) studied the market efficiency in different sectors using multiple variance tests. They found inconsistency in the results amongst various sectors and different economies. The efficient market hypothesis has been studied by Borges (2011) in the European stock market. The author tested market efficiency by applying the multiple variance ratio, ADF unit root, autocorrelation, and run test, with mixed results in terms of market efficiency in different countries. Nakamura and Small (2007) applied the small-shuffle surrogate method to test the market efficiency. The study found random walk in US and Japanese stock returns in the given time period.

The majority of the studies on market efficiency have been global in nature, but the results have been inconclusive. In the line with the above contradiction, the present study focused on testing the random walk hypothesis based on top 10 stock exchanges based on market capitalization.

## 12.3 DATA AND RESEARCH METHODOLOGY

### 12.3.1 Data

The daily stock indices of the top 10 stock exchanges by market capitalisation were considered. Stock prices from 1 January 2010 to 25 January 2018 were examined. The adjusted closing prices were collected from the Prowess database of CMIE and Yahoo Finance. Table 12.2 lists the stock market indices examined.

## 12.4 METHODOLOGY

The basic purpose of the study is to examine whether the top 10 stock indices move randomly or not. For the same, the random walk theory is applied. The random walk without drift can be presented as follows:

$$Y_t = Y_{t-1} + e_t \tag{12.1}$$

**TABLE 12.2**
**Stock Market Indices**

| Rank | Market | Country | Stock Indices | No. of Observation |
|---|---|---|---|---|
| 1 | New York Stock Exchange | US | NYSE | 2785 |
| 2 | NASDAQ | US | NASDAQ 100 | 2785 |
| 3 | London Stock Exchange | UK | FTSE 100 | 2797 |
| 4 | Tokyo Stock Exchange | Japan | Nikkei | 913 |
| 5 | Shanghai Stock Exchange | China | SSE | 2696 |
| 6 | Shenzhen Stock Exchange | China | SZSE | 2180 |
| 7 | Toronto Stock Exchange | Canada | TSX | 2778 |
| 8 | Bombay Stock Exchange | India | Sensex | 2117 |
| 9 | Frankfurt Stock Exchange | Germany | FWB | 1904 |
| 10 | Hong Kong Stock Exchange | Hong Kong | HKEX | 2187 |

*Source*: Author's own presentation.

where $Y_t$ is the value in time period t, $Y_{t-1}$ is the value in time period t-1, and $e_t$ is the value of the error term in time period t.

Lo and Mackinaly (1988) developed a variance ratio test that distinguishes alternative stochastic processes; that is, if the stock prices follow a random walk process, then the variance of yearly log price has to be 12 times as large as the variance of monthly stock return. A stochastic process is presented as follow:

$$R_t = \mu + \ln P_t - P_{t-1} + \varepsilon_t \qquad (12.2)$$

where $R_t$ is stock returns, $\mu$ is the drift parameter; ln $P_t$ is the log price at time t, and $P_{t-1}$ is the log price at t-1. The error term is independent identical distribution (iid) and noise disturbances are uncorrelated. Under the random walk hypothesis, the variance of $r_{t+} r_{t-1}$ should be twice the variance of $r_t$. The variance ratio can be shown as follows:

$$\begin{aligned} VR(2) &= \frac{Var[r_t(2)]}{2Var[r_t]} = \frac{Var[r_t + r_{t-1}]}{2Var[r_t]} \\ &= \frac{2Var[r_t] + 2Cov[r_{t,} r_{t-1}]}{2Var[r_t]} \qquad (12.3) \\ VR(2) &= 1 + p(1) \end{aligned}$$

where p(1) is depicted as the first-order autocorrelation coefficient of returns ($r_t$). When VR (2) becomes equal to 1, it means the stock return holds true with regard to the random walk hypothesis. The VR (2) will be applied to any period in the series. If the value of the variance ratio is less than one, then negative autocorrelation arises; if it is more than one, then positive autocorrelation occurs.

The variance ratio of Lo and Mackinlay (1988) tests the individual variance of the ratio at a time t. Therefore, the variance ratio is used to examine the variance of various q values. If the variance ratio of some q period is rejected, the null of random walk is rejected. But the random walk hypothesis needs the variance ratios for entire periods to be equal to one; therefore, the test must be applied jointly for entire q periods. In order to combat this problem, Chow and Denning (1993) developed a multiple variance ratio test in which multiple variance ratios for various periods can be tested to observe whether the ratio is jointly equal to one or not. The null hypothesis in the Lo-Mackinlay test is VR (q) = 1, whereas in the multiple variance ratio test $M_r = (q_i) = VR(q) - 1 = 0$. It can be presented as:

$$\{M_r(q_i) | i = 1, 2 \ldots, m\} \qquad (12.4)$$

Under the random walk hypothesis, the null and alternative hypotheses can be show as follows:

$$H_{oi} = M_r(q_i) = 0 \text{ for } i = 1,2,\ldots,m$$
$$H_{oi} = M_r(q_i)^1 0 \text{ for } i = 1,2,\ldots,m \quad (12.5)$$
$$CD1 = \sqrt{T}maxn|1 \leq i \leq Z(q_i)$$

One of the most often used tests for the market efficiency or random walk model is testing the unit root in stock returns. The logic behind testing for the presence of a unit root is that the presence of a unit root keeps shocks forever. If a stock series has a unit root, it means that a shock never dies; it assumes that shock is permanent. It says that price movements are found because of random shocks, which cannot be estimated, and thus it is impossible to predict the future stock price. But any series that does not have a unit root is stationary, and thus it can be predicted that it will revert back to its natural mean.

## 12.5 RESULTS AND DISCUSSION

Table 12.3 presents the descriptive statistics for the 10 indices. The Hong Kong Exchange (HKEX) had the highest average returns compared to rest of the stock exchanges. The standard deviation of the Frankfurt Stock Exchange was highest due, reflecting its volatility. The returns of seven of the stock exchanges were negatively skewed (returns were flatter to the left of the normal distribution) and three stock exchanges were positively skewed (flatter to the right of normal distribution). The Jarque-Bera (1980) of all the stock exchanges were less than 5%, which confirms that the returns were non-normally distributed.

The series follow random walk when they are nonstationary (Dsouza & Mallikarjunappa, 2015). The ADF test was used to determine the stationarity of the series. The ADF test statistics, probability values, and test critical values of different significance levels are provided in Table 12.4. All the stock exchanges reported negative ADF test statistics and test critical values at 1%, 5%, and 10% significance levels. The probability value was less than 5%, signifying rejection of the null hypothesis (series has unit root); hence, the return series of stock exchanges are stationary. These stationary results are similar to Lock (2007), Charles and Darne (2008), and Dsouza and Mallikarjunappa (2015). Therefore, it can be said that markets are inefficient in all the sample stock exchanges and do not follow random walk.

The Table 12.5 presents the results of the serial correlation LM test. The test says that either there is autocorrelation with its own lag or not. The result of different lags—lag 2, lag 4, lag 8, and lag 16—have been depicted. The *p*-value of all the lags of the 10 stock exchanges were significant except the Tokyo Stock Exchange. This means that there is autocorrelation in nine of the stock exchanges, whereas there is efficiency in the Tokyo Stock Exchange. A technical analyst or fundamental analyst can generate abnormal returns in these nine stock exchanges, but one cannot generate abnormal returns in the Tokyo Stock Exchange. As per the result of the LM

**TABLE 12.3**
**Descriptive Statistics on Returns**

| | BSE | FRANKFURT | HKEX | LSE | NASDAQ | NYSE | SHANGHAI | SZSE | TSE | TSX |
|---|---|---|---|---|---|---|---|---|---|---|
| Mean | 0.000300 | −0.000101 | 0.000498 | 0.000464 | 0.000297 | 3.58E−05 | 0.000127 | 0.000577 | 2.11E−05 | 2.11E−05 |
| Median | 0.000548 | −6.04E−05 | −0.000788 | 0.000000 | 0.001070 | 0.000679 | 0.000940 | 0.002523 | 0.000813 | 0.000813 |
| Maximum | 0.159900 | 0.165094 | 0.182532 | 0.156440 | 0.111594 | 0.115258 | 0.090345 | 0.094686 | 0.093702 | 0.093702 |
| Minimum | −0.116044 | −0.681662 | −0.149532 | −0.138948 | −0.095877 | −0.102321 | −0.092561 | −0.183143 | −0.097879 | −0.097879 |
| Std. Dev. | 0.015716 | 0.027322 | 0.024578 | 0.024338 | 0.014500 | 0.014477 | 0.018369 | 0.020779 | 0.012479 | 0.012479 |
| Skewness | 0.169697 | −7.012258 | 0.757578 | 0.142965 | −0.250508 | −0.390506 | −0.539710 | −0.896225 | −0.673336 | −0.673336 |
| Kurtosis | 11.80970 | 182.1060 | 10.12048 | 9.117242 | 9.828974 | 12.10155 | 6.454345 | 7.673028 | 12.93686 | 12.93686 |
| Jarque-Bera | 7063.352 | 2933048. | 4816.097 | 3408.031 | 4260.757 | 7583.349 | 1190.249 | 2276.425 | 9137.910 | 9137.910 |
| Probability | 0.000000 | 0.000000 | 0.000000 | 0.000000 | 0.000000 | 0.000000 | 0.000000 | 0.000000 | 0.000000 | 0.000000 |
| Sum | 0.654901 | −0.220163 | 1.086225 | 1.012543 | 0.648801 | 0.078070 | 0.276017 | 1.259086 | 0.045987 | 0.045987 |
| Sum Sq. Dev. | 0.538431 | 1.627296 | 1.316892 | 1.291319 | 0.458350 | 0.456914 | 0.735586 | 0.941250 | 0.339501 | 0.339501 |

### TABLE 12.4
### Result of ADF Test of Sample Indices

| Indices | ADF Test Statistic | Prob. Value | Test Critical Value | | |
|---|---|---|---|---|---|
| | | | 1% level | 5% level | 10% level |
| BSE | −48.45659 | 0.0001 | −3.432549 | −2.862397 | −2.567271 |
| FLANKFURT | −53.24689 | 0.0001 | −3.432486 | −2.862369 | −2.567256 |
| HKEX | −47.76131 | 0.0000 | −3.961398 | −3.411450 | −3.127580 |
| LSE | −54.13720 | 0.0000 | −3.961324 | −3.411414 | −3.127559 |
| NASDAQ | −56.81752 | 0.0000 | −3.961338 | −3.411421 | −3.127563 |
| NYSE | −41.22304 | 0.0000 | −3.961339 | −3.411421 | −3.127563 |
| SHANGHAI | −51.01395 | 0.0000 | −3.961447 | −3.411474 | −3.127595 |
| SZSE | −41.87419 | 0.0000 | −3.962256 | −3.411870 | −3.127830 |
| TSE | −54.79891 | 0.0000 | −3.961346 | −3.411425 | −3.127565 |
| TSX | −54.80271 | 0.0001 | −3.432512 | −2.862381 | −2.567262 |

*Source*: Author's calculation.

### TABLE 12.5
### Result of Breusch-Godfrey Serial Correlation LM Test

| Indices | No. of Observations | 2 | 4 | 8 | 16 |
|---|---|---|---|---|---|
| BSE | 2117 | 0.553250 | 0.285241 | 0.132996 | 0.066787 |
| | | (−10.93192) | (−10.09045) | (−8.337059) | (−6.320499) |
| FLANKFURT | 1904 | 0.515937 | 0.248773 | 0.117223 | 0.061416 |
| | | (−3.688610) | (−3.784272) | (−3.710024) | (−3.516740) |
| HKEX | 2187 | 0.543139 | 0.278152 | 0.132338 | 0.068226 |
| | | (−12.41142) | (−11.19569) | (−9.097688) | (−6.986588) |
| LSE | 2797 | 0.473354 | 0.236317 | 0.120199 | 0.058997 |
| | | (−12.72969) | (−10.69044) | (−8.308551) | (−6.226367) |
| NASDAQ | 2785 | 0.484889 | 0.240847 | 0.115942 | 0.055412 |
| | | (−11.44694) | (−9.476565) | (−7.473016) | (−5.575173) |
| NYSE | 2785 | 0.487190 | 0.236375 | 0.113766 | 0.054193 |
| | | (−10.62053) | (−8.883317) | (−6.850697) | (−5.010283) |
| SHANGHAI | 2696 | 0.523655 | 0.235599 | 0.126261 | 0.063186 |
| | | (−13.53578) | (−12.48560) | (−9.765104) | (−7.476904) |
| SZSE | 2180 | 0.568562 | 0.260655 | 0.133929 | 0.068999 |
| | | (−12.97417) | (−12.82527) | (−10.35067) | (−7.892557) |
| TSE | 913 | 1.039940 | 1.075298 | 1.068748 | 1.051043 |
| | | (0.987884) | (1.017748) | (0.587295) | (0.297365) |
| TSX | 2778 | 0.502161 | 0.239970 | 0.117919 | 0.058342 |
| | | (−9.056880) | (−8.150457) | (−6.331297) | (−4.531632) |

*Source*: Author's calculation.

Test, it is clear that these nine stock exchanges are inefficient but the Tokyo Stock Exchange is efficient.

## 12.6 CONCLUSION

The purpose of the present study is to test the weak market efficiency of top 10 stock exchanges. Investors want to generate abnormal returns; hence, understanding the stock market is important to them. The present study helps investors to know which stock exchange can create abnormal returns. The top 10 stock exchanges have been considered in the present study. It has been found that the NYSE, NASDAQ 100, FTSE 100, Nikkei, SSE, SZSE, Sensex, FWB, and HKEX are not efficient, which means that investors can earn abnormal returns. In contrast, the Tokyo Stock Exchange (TSX) is an efficient stock exchange. Efforts to predict returns in the Tokyo Stock Exchange (TSX) will be incorrect.

## REFERENCES

Abdmoulah, W. (2010). Testing the evolving efficiency of Arab stock markets. *International Review of Financial Analysis*, *19*(1), 25–34.

Amanulla, S., & Kamaiah, B. (1998). Indian stock market: Is it informationally efficient? *Prajnan*, *25*(4), 473–485.

Arusha, C., & Guneratne, W. (2007). The efficiency of emerging stock markets: Empirical evidence from the South Asian region. *The Journal of Developing Areas*, *14*(1), 171–183.

Borges, M. R. (2011). Random walk tests for the Lisbon stock market. *Applied Economics*, *43*(5), 631–639.

Charles, A., & Darne, O. (2008). The random walk hypothesis for Chinese stock markets: Evidence from variance ratio tests. *Economic Systems Elsevier*, *33*(2), 117–126.

Darrat, A. F., & Zhong, M. (2000). On testing the random-walk hypothesis: A model comparison approach. *The Financial Review*, *35*(3), 105–124.

Dsouza, J. J., & Mallikarjunappa, T. (2015). Do the stock market indices follow random walk? *Asia-Pacific Journal of Management Research and Innovation*, *11*(4), 251–273.

Fama, E. F. (1965). The behavior of stock market prices. *The Journal of Business*, *38*(1), 34–105.

Fama, E. F. (1970). Efficient capital markets: A review of theory and empirical work. *The Journal of Finance*, *25*(1), 383–417.

Fifield, S. G. M., & Jetty, J. (2008). Further evidence on the efficiency of the Chinese stock markets: A note. *Research International Business Finance*, *22*(3), 351–361.

Gupta, R., & Basu, P. (2007). Weak form efficiency in Indian stock markets. *Journal International Business and Economics.*

Hoque, H. A. A. B., Kim, J. H., & Pyun, C. S. (2007). A comparison of variance ratio tests of random walk: A case of Asian emerging stock markets. *The International Review of Economics & Finance*, *16*(4), 488–502.

Husain, F. (1997). The random walk model in the Pakistani equity market: An examination. *Pakistan Development Review*, *36*(3), 221–240.

Lee, C. F., Chen, G., & Rui, O. M. (2001). Stock returns and volatility on China's stock markets. *Journal of Finance Research*, *24*(4), 523–543.

Lima, E. J. A., & Tabak, B. M. (2004). Tests of the random walk hypothesis for equity markets: Evidence from China, Hong Kong and Singapore. *Applied Economics Letters*, *11*(4), 255–258.

Liu, X., Song, H., & Romilly, P. (1997). Are Chinese stock markets efficient? A cointegration and causality analysis. *Applied Economics Letters*, *4*(8), 511–515.

Lock, D. B. (2007). The China a shares follow random walk but the b shares do not. *Economics Bulletin*, *7*(9), 1–12.

Majumder, D. (2012). When the market becomes inefficient: Comparing BRIC markets with markets in the USA. *International Review of Financial Analysis*, *24*(C), 84–92.

Marulkar, K., & Faniband, M. (2017). *An empirical study of random walk theory: Evidence from Bombay stock exchange.*

Mustafa, K., & Nishat, M. (2007). Testing for market efficiency in emerging markets: A case study of Karachi stock market. *The Lahore Journal of Economics*, *12*(1), 119–140.

Olowe, R. A. (1999). Weak form efficiency of the Nigerian stock market further evidence. *African Development Review*, *11*(1).

Poshakwale, S. (2002). The random walk hypothesis in the emerging Indian stock market. *Journal of Business Finance Account*, *29*(9), 1275–1299.

Vidali, M. (2013). Efficient market hypothesis: The case of the London stock exchange. *Research Journal*, *6*(3), 57–64.

Worthington, A. C., & Higgs, H. (2009). Efficiency in the Australian stock market, 1875–2006: A note on extreme long-run random walk behaviour. *Applied Economics Letters*, *16*, 301–306.

# 13 Priority-Based Time Minimization Transportation Problem with Flow and Mixed Constraints

*Bindu Kaushal and Shalini Arora*

**CONTENTS**



## 13.1 INTRODUCTION

Among several business problems, the transportation problem is the most important application that deals with the distribution of product. In a standard cost-minimization transportation problem (CMTP), the motive is to obtain the minimum cost of transportation between the source destination pairs which works within the given capacity. The CMTP is defined as: let $a_i$ be the availability at given sources $\forall(i = 1 \text{ to } m)$ and $b_j$ be the requirement of given destinations $\forall(j = 1 \text{ to } n)$. For each transportation between the $i$th source and $j$th destination, the per unit cost of each product is $c_{ij}$ and $z_{ij}$ is the quantity transported. Thus, a balanced cost-minimization transportation problem ($\Sigma ai = \Sigma bj$) deals with minimization of total cost.

Mathematically, CMTP is stated as:

$$\min \sum_i \sum_j c_{ij} z_{ij}$$

subject to:

$$\sum_{j=1}^{n} z_{ij} = a_i, \forall i \in I = 1 \text{ to } m$$

$$\sum_{i=1}^{m} z_{ij} = b_j, \forall j \in J = 1 \text{ to } n$$

$$z_{ij} \geq 0, \forall (i, j) \in IXJ$$

where:

$I$ (set of sources) = 1 to $m$
$J$ (set of destinations) =1 to $n$
$a_i$ = availability at each source
$b_j$ = requirement of each destination
$z_{ij}$ = transported quantity to $j$th destination by $i$th source

Initially, CMTP was solved by Hitchcock (1941). Later, Dantzig (Dantzig, 1963; Dantzig & Thapa, 1963) developed the simplex method to solve CMTP. Many researchers widely studied this problem and found several applications in different fields. A variant of CMTP is studied by Appa (1973). A lot of work has been done on CMTP with mixed constraints and flow constraints (Brigden, 1974; Klingman & Russel, 1974; Khanna, Bakshi, & Puri, 1981).

Most real-world problems require different objectives, minimizing the time of transportation instead of cost. For example, in a situation like war, time is the main priority in transportation of various operational needs in the battlefield. Thus, TMTP is the special case of the bottleneck linear programming problem and was initially formulated by Hammer (1969). It deals with the minimization over the convex polytope of concave objective function. Various authors have considered the bottleneck programming problem (Garfinkel & Rao, 1971; Hammer, 1971; Issermann, 1984). In TMTP, the aim is to find out the minimum possible time in which the material can be supplied from the sources to destinations. Several algorithms have been developed by Szwarc (1971, 1996), Bhatia, Swarup, and Puri (1977), Arora et al. (1997), and Nikolic (2007) to find the optimal solution of the TMTP. Some authors have generated the lexicographic solutions of the bottleneck programming problem to declare optimality (Arora & Puri, 1999; Burkard & Rendl, 1991). In lexicographic optimization, our motive is to find that solution vector in which the first component is minimized to as small as possible and, if the first component is minimized, then to minimize the second component; if the first and second components are minimized, then to minimize the third component; and so on. Lexicographic optimal solution of TMTP with mixed constraints was considered by Khanna, Bakshi, and Puri (1983) and Stayaparkash (1982).

Mathematically, TMTP is stated as:

$$\min \left[ \max_{(i,j)\in IXJ} t_{ij}(z_{ij}) \right]$$

subject to:

$$\sum_{j=1}^{n} z_{ij} = a_{i}, \forall i \in I = 1 \text{ to } m$$

$$\sum_{i=1}^{m} z_{ij} = b_{j}, \forall j \in J = 1 \text{ to } n$$

$$z_{ij} \geq 0, \forall (i,j) \in IXJ$$

where:

$I$ (set of sources) = 1 to $m$
$J$ (set of destinations) = 1 to $n$
$a_i$ = availability at each source
$b_j$ = requirement of each destination
$z_{ij}$ = transported quantity to $j$th destination by $i$th source
$t_{ij}$:- transportation time consumed when the destination $j$ is being supplied by the source $i$

$$t_{ij}(z_{ij}) = \begin{cases} t_{ij}, & \text{if } jth \text{ destination receive quantity by } ith \text{ source} \\ 0, & \text{otherwise} \end{cases}$$

Several authors have worked on two-stage TMTP and the assignment problem (Sharma, Verma, Kaur, & Dahiya, 2015; Sonia, 2004, 2008). In the most realistic circumstances, rather than supplying to all the destinations only some destinations fulfill our aim of optimality. This situation is modeled as a priority-based time minimization transportation problem (PBTMTP). This chapter examines a PBTMTP with mixed constraints and a PBTMTP with flow constraints. The discussed problems are the extension of the PBTMTP. Here, lexicographic optimization is used to find the optimal solutions. Section 13.2 shows the mathematical model of the problems. Section 13.3 consists of theoretical results of the problem. Section 13.4 shows the working steps. A numerical illustration for different cases is shown in Section 13.5. Computational details are reported in Section 13.6. Some concluding remarks are given toward the end.

## 13.2 PROBLEM DESCRIPTION

### 13.2.1 PBTMTP with Mixed Constraints

A priority-based time minimization transportation problem (PBTMTP) can be extended to the case with mixed and flow constraints. In PBTMTP some destinations are at a prior level in comparison to the others. Thus, prior destinations are treated as primary destinations and the others are secondary destinations. Consider a TMTP of $m$ sources and $n$ destinations in which some destinations are labeled as priority.

Let $a_i$ and $b_j$ show the availability and the requirement at each source and destination respectively. It is assumed that $\min_{i \in I} a_i > \max_{j \in J} b_j$.

Mathematically, PBTMTP with mixed constraints is defined as:

$$\min_{z_{ij} \in S(Z)} (\max_{IXJ_1} t_{ij}(z_{ij})) + \max_{z_{ij} \in S(\overline{Z})} (t_{ij}(\overline{z_{ij}}))$$

$$S(Z) = \left\{ \begin{array}{l} \sum_{j \in J'} z_{ij} \leq a_i, \forall i \in I_1 \mid \sum_{i \in I} z_{ij} \leq b_j, \forall i \in J_1 \\ \sum_{j \in J'} z_{ij} = a_i, \forall i \in I_2 \mid \sum_{i \in I} z_{ij} = b_j, \forall i \in J_2 \\ \sum_{j \in J'} z_{ij} \geq a_i, \forall i \in I_3 \mid \sum_{i \in I} z_{ij} \geq b_j, \forall i \in J_3 \end{array} \right\} \forall z_{ij} \geq 0, \forall (i,j) \in IXJ'$$

where $\bigcup_{i=1}^{3} I_i = I$ and $\bigcup_{j=1}^{3} J_j = J'$

$$S(\overline{Z}) = \left\{ \begin{array}{l} \sum_{j \in J''} \overline{z_{ij}} \leq a_i', \forall i \in \overline{I_1} \mid \sum_{i \in I} \overline{z_{ij}} \leq b_j, \forall i \in \overline{J_1} \\ \sum_{j \in J''} \overline{z_{ij}} = a_i', \forall i \in \overline{I_2} \mid \sum_{i \in I} \overline{z_{ij}} = b_j, \forall i \in \overline{J_2} \\ \sum_{j \in J''} \overline{z_{ij}} \geq a_i', \forall i \in \overline{I_3} \mid \sum_{i \in I} \overline{z_{ij}} \geq b_j, \forall i \in \overline{J_3} \end{array} \right\} \forall z_{ij} \geq 0, \forall (i,j) \in IXJ''$$

where $\bigcup_{i=1}^{3} \overline{I_i} = \overline{I}$ and $\bigcup_{j=1}^{3} \overline{J_j} = J'$

I (set of sources) = $\{1,2,\ldots,m\}$
$J'$ = Primary destinations
$J''$ = Secondary destinations
$a_i$ = Availability at each source
$b_j$ = Requirement of each destination
$t_{ij}$ = Transportation time when $i$th source supply to $j$th destination
$a_i' = a_i - \sum_{i \in I} x_{ij}, \forall \mathrm{j} \in \mathrm{J}_1$ ; i.e. availability is updated at each source $i$ when all primary destinations are served.
$\overline{I}$ = Set of sources that are available to supply secondary destinations
$J_2 = J - J_1$: Set of Secondary destinations

### 13.2.2 PBTMTP WITH FLOW CONSTRAINTS

It is assumed that $\min_{i \in I} a_i > \max_{j \in J} b_j$.

Mathematically, PBTMTP with flow constraints $\sum_I \sum_J z_{ij} = P$ can be stated as:

$$\min_{z_{ij} \in S(Z)} (\max_{IXJ_1} t_{ij}(z_{ij})) + \max_{z_{ij} \in S(\overline{Z})} (t_{ij}(\overline{z_{ij}}))$$

subject to:

**Case I:** If $P \geq \max(\Sigma a_i, \Sigma b_j)$

$$S(Z) = \{z_{ij} \in R^{m \times n} \mid \sum_{j \in J_1} z_{ij} \geq a_i, \forall i \in I, \sum_{i \in I} z_{ij} \geq b_j, \forall j \in J_1, \sum_I \sum_J z_{ij} = P, z_{ij} \geq 0, \forall (i,j) \in IXJ_1\}$$

$$S(\overline{Z}) = \{\overline{z_{ij}} \in R^{m \times (n-m)} \mid \sum_{j \in J_2} \overline{z_{ij}} \geq a_i', \forall i \in \overline{I}, \sum_{i \in \overline{I}} \overline{z_{ij}} \geq b_j, \forall j \in J_2, \sum_I \sum_J z_{ij} = P, \overline{z_{ij}} \geq 0, \forall (i,j) \in \overline{I}XJ_2\}$$

**Case II:** If $P \leq \min(\Sigma a_i, \Sigma b_j)$

$$S(Z) = \{z_{ij} \in R^{m \times n} \mid \sum_{j \in J_1} z_{ij} \geq a_i, \forall i \in I, \sum_{i \in I} z_{ij} \geq b_j, \forall j \in J_1, \sum_I \sum_J z_{ij} = P, z_{ij} \geq 0, \forall (i,j) \in IXJ_1\}$$

$$S(\overline{Z}) = \{\overline{z_{ij}} \in R^{m \times (n-m)} \mid \sum_{j \in J_2} \overline{z_{ij}} \geq a_i', \forall i \in \overline{I}, \sum_{i \in \overline{I}} \overline{z_{ij}} \geq b_j, \forall j \in J_2, \sum_I \sum_J z_{ij} = P, \overline{z_{ij}} \geq 0, \forall (i,j) \in \overline{I}XJ_2\}$$

**Case III:** If $\min(\Sigma a_i, \Sigma b_j) \leq P \leq \max(\Sigma a_i, \Sigma b_j)$

$$S(Z) = \{z_{ij} \in R^{m \times n} \mid \sum_{j \in J_1} z_{ij} \geq a_i, \forall i \in I, \sum_{i \in I} z_{ij} \geq b_j, \forall j \in J_1, \sum_I \sum_J z_{ij} = P, z_{ij} \geq 0, \forall (i,j) \in IXJ_1\}$$

$$S(\overline{Z}) = \{\overline{z_{ij}} \in R^{m \times (n-m)} \mid \sum_{j \in J_2} \overline{z_{ij}} \geq a_i', \forall i \in \overline{I}, \sum_{i \in \overline{I}} \overline{z_{ij}} \geq b_j, \forall j \in J_2, \sum_I \sum_J z_{ij} = P, \overline{z_{ij}} \geq 0, \forall (i,j) \in \overline{I}XJ_2\}$$

where:

I (set of /sources) = $\{1,2, \ldots, m\}$
J′ = Primary destinations
J″ = Secondary destinations
$a_i$ = Availability at each source
$b_j$ = Requirement of each destination
$t_{ij}$ = Transportation time when $i$th source supply to $j$th destination
$a_i' = a_i - \sum_{i \in I} x_{ij}, \forall j \in J_1$; i.e. availability is updated at each source I when all primary destinations are served.
$\overline{I}$ = Set of sources which are available to supply secondary destinations
$J_2 = J - J_1$: Set of secondary destinations

## 13.3 THEORETICAL DEVELOPMENT

For the corresponding balanced PBTMTP, first partition the whole transportation route $IXJ = \{(i, j)\}$, where $i = 1$ to $m$ and $j = 1$ to $n$ in disjoint sets $M_p$, $p = 1,2, \ldots, q$, according to the corresponding time entries such that $t_1 > t_2 > t_3 > \ldots, \ldots > t_q$. Now, corresponding to each defined set $M_p$, positive weights are attached, say $\lambda_{q-p+1}$ for $p = 1,2, \ldots, q$, where $\lambda_{p+1} >> \lambda_p$, $\forall p = 1,2, \ldots, q-1$. One can refer to Mazzola and Neebe (1993) and Sherali (1982) for computation of values of $\lambda$. These techniques have already been defined by Kaushal and Arora (2018), but for the ready reference it is explained as follows:

1. Sherali's technique
   - The values of $\lambda_{q-p+1}$ are given as:

$$\lambda_{q-p+1} = D^{q-p}, p = 1,2,\ldots,\ldots,q.$$

$$D = 1 + Max[UB(\sum \sum_{(i,j)\in M_p} z_{ij}, l = 1,2,\ldots,\ldots,q)],$$ where UB denote the upper bound.

   - The values of $\lambda_{q-p+1}$ are found as:

$$\lambda_1 = 1$$

$$u_{q-p+1} = u_{q-p} + \lambda_{q-p+1} \sum_{Mp} z_{ij}, p = q-1, q-2,\ldots,\ldots,2,1.$$

This gives

$$u_q = \lambda_q \sum_{(i,j)\in M_1} z_{ij} + \lambda_{q-1} \sum_{(i,j)\in M_2} z_{ij} + \ldots,\ldots, \lambda_1 \sum_{(i,j)\in M_q} z_{ij}$$

2. Mazzola's technique

$$\lambda_1 = 1$$

$$\lambda_p = (m+n-1)\lambda_{p-1} + 1, p = 2,3,\ldots,\ldots,q$$

Here, we proceed using Mazzola's technique (Mazzola & Neebe, 1993) for computation of weights.

### 13.3.1 M-Feasible Solution

A problem $(CP_k)$, defines an M-feasible solution (FS) if $\exists$ a FS which satisfies the following condition: $z_{ij} = 0 \ \forall(i,j) \in IXJ$ if $t_{ij} = M$.

### 13.3.2 Lexicographic Optimal Solution (LOS)

Let $F: S \to R^t$ be the $t$-dimensional function, where $S \subset R^t$ and $f_k$ is the $k$th component of $F$ and suppose $Z \in S$ be the lexicographic feasible solution. In addition to minimizing the function $f_1$, one is also interested in minimizing $f_2$ and, if $f_1$ and $f_2$ are as small as possible, then to minimize $f_3$, and so on. Hence, LOS of any time minimization transportation is obtained by $Lex \min_{Z\in S} F(Z)$. A feasible solution $\overline{Z}$ is lexicographically better than $Z \in S$ for the lexicographic general optimization problem (LGOP) if $\exists\, k \in \{1,2,\ldots,\ldots,t-1\}$, s.t

$$f_s(\overline{Z}) = f_s(Z), s = 1,2,\ldots,\ldots,k$$

$$f_{s+1}(\overline{Z}) < f_{s+1}(Z).$$

It is represented by $F(\overline{Z}) < F(Z)$ or we can say, $\overline{Z}$ is LOS of LGOP if there does not exist $Z \in S$ for which $F(Z) < F(\overline{Z})$, where (LGOP) is $Lex \min_{Z\in S} F(Z)$. The lexicographic TMTP can be defined as:

$$(LTMTP): Lex \min_{Z\in S} F(Z)$$

Any $s$th component of $F(Z)$ is given by $f_s(Z)=\sum\sum_{(i,j)\in M_s} c_{ij}z_{ij}, s=1,2,\ldots,\ldots,q.$

Now, to find the optimal solution we find *Lex* min $[f_1(Z), f_2(Z), \ldots, \ldots, f_q(Z)]$;

i.e. $Lex\min[\sum\sum_{(i,j)\in M_1} c_{ij}z_{ij}, \sum\sum_{(i,j)\in M_2} c_{ij}z_{ij},\ldots,\ldots,\sum\sum_{(i,j)\in M_q} c_{ij}z_{ij}]$.

**Theorem 1:** The "necessary" and "sufficient" condition for $\overline{Z}$ to be LOS of (LGOP) is that $\overline{Z}$ is the optimal solution of $\sum_{s=1}^{q}\lambda_s f_s(Z)$, where $F$ is a non-constant $t$-dimensional vector function and $\lambda_1\,\lambda_2,\ldots,\ldots,\lambda_q$ are positive real numbers such that $\sum_{s=1}^{q}\lambda s\, fs(Z)$ has the sign of $\lambda_i f_i(Z)$ where $i=\min_{s=1,2,\ldots,q}(s: f_s(Z)\neq 0)$.

**Proof.** One can refer to Theorem 1 of Arora and Puri (1997).

**Theorem 2:** An optimal solution of $\min\sum_{s=1}^{q}\lambda_s(\sum\sum_{(i,j)\in M_s} c_{ij}z_{ij})$ is LOS of time minimization transportation problem and conversely.

**Proof.** This theorem pursues the proof from the above theorem.

## 13.4 WORKING STEPS

### 13.4.1 Problem with Mixed Constraints

**Step 1** For a given PBTMTP with mixed constraints, formulate the corresponding balanced transportation problem.

$$I^* = I\cup(m+1)$$
$$J^* = J\cup(n+1)$$
$$a_{m+1}=\sum_{I} b_j+1$$
$$b_{n+1}=\sum_{I} a_i+1$$
$$c_{m+1}=0 \text{ if } j\in J_1$$
$$c_{m+1,j}=\min_{i\in I_3} c_{ij}, \text{ if } j\in J_2\bigcup J_3$$
$$c_{i,n+1}=0, \text{ if } i\in I_1$$
$$c_{i,n+1}=\min_{j\in J_3} c_{ij}, \text{ if } i\in I_2\bigcup I_3$$
$$c_{m+1,n+1}=0, \text{ if } i\in I_1$$

**Step 2** Generate pairs $T^r$ and find $[T_r, T_{r'}]$ ; i.e. the time corresponding to primary and secondary destinations respectively using working steps of Kaushal and Arora (2018; see Section 13.4.2). Find $f_r=[T_r+T_{r'}]$.

**Step 3** Stop. If Y is the LOS for PBTMTP with mixed constraints then it is calculated as:

$$z_{ij} = y_{ij},\ i \in I_1 \text{ and/or } j \in J_1$$
$$z_{ij} = y_{ij},\ \forall(i,j) \in I_2 X J \text{ if } c_{ij} \neq \min_{j \in J_3} c_{ij}$$
$$z_{ij} = y_{ij} + y_{i,n+1}, \forall(i,j) \in (I_2 \bigcup I_3) X\ J_3 \text{ if } c_{ij} = \min_{j \in J_3} c_{ij} \text{ but } c_{ij} \neq \min_{i \in I_3} c_{ij}$$
$$z_{ij} = y_{ij}, \forall(i,j) \in (I_3 X J) \text{ if } c_{ij} \neq \min_{j \in J_3} c_{ij} \text{ and also } c_{ij} \neq \min_{i \in I_3} c_{ij}$$
$$z_{ij} = y_{ij} + y_{m+1,j}, \forall(i,j) \in I_3 X (J_2 \bigcup J_3) \text{ if } c_{ij} = \min_{i \in I_3} c_{ij} \text{ but } c_{ij} \neq \min_{j \in J_3} c_{ij}$$
$$z_{ij} = y_{ij} + y_{i,n+1} + y_{m+1,j}, \forall(i,j) \in I_3 X J_3 \text{ if } c_{ij} = \min_{I_3} c_{ij} = \min_{J_3} c_{ij}$$

For each pair of lexicographic solutions, we find:

$$Min\ f_r = Min\ [\mathrm{T}_r + T_{r'}] = T,\ r = 1,2,\ldots,p$$

Hence, $T = (T_r, T_{r'})$ is the lexicographic optimal solution.

### 13.4.2 Problem with Flow Constraints

**Step 1** For a given PBTMTP with flow constraints, formulate the corresponding balanced transportation problem.

- **Case I:** *If* $P \geq \max(\Sigma a_i, \Sigma b_j)$

$$I^* = I \cup (m+1)$$
$$J^* = J \cup (n+1)$$
$$a_{m+1} = \sum_I b_j + 1$$
$$b_{n+1} = \sum_I a_i + 1$$
$$c_{m+1} = 0 \text{ if } j \in J_1$$
$$c_{m+1,j} = \min_{i \in I_3} c_{ij}, \text{ if } j \in J_2 \bigcup J_3$$
$$c_{i,n+1} = 0, \text{ if } i \in I_1$$
$$c_{i,n+1} = \min_{j \in J_3} c_{ij}, \text{ if } i \in I_2 \bigcup I_3$$
$$c_{m+1,n+1} = 0, \text{ if } i \in I_1$$

- **Case II:** *If* $P \leq \min(\Sigma a_I, \Sigma b_j)$

$$I^* = I \cup (m+1)$$
$$J^* = J \cup (n+1)$$
$$a_{m+1} = \sum_I b_j - P$$
$$b_{n+1} = \sum_I a_i - P$$
$$c_{(i,m+1)} = \min_J(c_{ij}) \text{ or } 0$$

$$c_{(m+1,j)} = \min_I(c_{ij}) \text{ or } 0$$

$$c_{m+1,n+1} = M$$

- **Case III:** *If* $P \leq \min(\Sigma\alpha_i, \Sigma b_j) \leq P \leq \max(\Sigma\alpha_i, \Sigma b_j)$

$$I^* = I \cup (m+1),(m+2)$$

$$a_{m+1} = P - \sum_I a_i$$

$$a_{m+2} = \sum_J b_j - P$$

$$c_{m+1,j} = \min_I c_{ij}, c_{m+2,j} = 0 \ \forall j \in J$$

**Step 2** Generate pairs $T^r$ and find $[T_r, T_{r'}]$, i.e. the time corresponding to primary and secondary destinations respectively using working steps of Kaushal and Arora (2018; see Section 13.4.2). Find $f_r = [T_r + T_{r'}]$.

**Step 3** Stop. If Y is the LOS for PBTMTP with flow constraints is calculated as:

$$if\ P \geq \max(\sum a_i, \sum b_j) \text{ or } P \leq \min(\sum a_i, \sum b_j)$$

$$z_{ij} = y_{ij} \forall (i,j) \in IXJ \text{ if } c_{ij} \neq \min_J c_{ij} \text{ and } c_{ij} \neq \min_I c_{ij}$$

$$z_{ij} = y_{ij} + y_{i,n+1}, \forall (i,j) \in IXJ \text{ if } c_{ij} = \min_J c_{ij} \text{ but } c_{ij} \neq \min_I c_{ij}$$

$$z_{ij} = y_{ij} + y_{m+1,j}, \ \forall (i,j) \in IXJ \text{ if } c_{ij} = \min_J c_{ij} \text{ but } c_{ij} \neq \min_J c_{ij}$$

$$z_{ij} = y_{ij} + y_{m+1,j} + y_{i,n+1}, \forall (i,j) \in IXJ \text{ if } c_{ij} = \min_J c_{ij} \text{ but } c_{ij} = \min_I c_{ij}$$

$$z_{ij} = y_{ij}, \ \forall (i,j) \in IXJ \text{ if } c_{ij} \neq c_{m+1,j}, \ x_{ij} = y_{ij} + y_{m+1,j}, \forall (i,j) \in IXJ \text{ for which } c_{ij} = c_{m+1,j}$$

For each pair of lexicographic solutions, we find

$$\min f_r = \min[T_r + T_{r'}] = \text{T, r=1,2,...,p}$$

Hence, $T = (T_r, T_{r'})$ is the lexicographic optimal solution.

## 13.5 NUMERICAL ILLUSTRATION

### 13.5.1 PBTMTP with Mixed Constraints

Consider a PBTMTP with mixed constraints with three sources and four destinations, as shown in Table 13.1.

Entries in each cell show the time of transportation between source destination pairs. Here, destinations 1 and 3 are the prior destinations whereas 2 and 4 are the secondary destinations. The rightmost entries show the availabilities at each source and bottom entries shows the corresponding demand of each destination.

**TABLE 13.1**
**Entries in Each Cell Shows Transportation Time between Sources and Destinations, the Rightmost Entry Shows the Availability at Each Source, Entries at the Bottom Show the Demand at Each Destination**

| | | | | |
|---|---|---|---|---|
| **12** | 13 | **34** | 40 | ≥10 |
| **7** | 18 | **36** | 7 | =15 |
| **11** | 20 | **30** | 21 | ≥12 |
| ≥ 10 | = 15 | ≤ 13 | ≥ 11 | |

**TABLE 13.2**
**Entries in Each Cell Shows Transportation Time between Sources and Destinations, the Rightmost Entry Shows the Availability at Each Source, Entries at the Bottom Show the Demand at Each Destination**

| | | | | | |
|---|---|---|---|---|---|
| **12** | 13 | **34** | 40 | 0 | 10 |
| **7** | 18 | **36** | 7 | 7 | 15 |
| **11** | 20 | **30** | 21 | 11 | 12 |
| **11** | 20 | **0** | 21 | 0 | 50 |
| 10 | 15 | 13 | 11 | 38 | |

**Step 1** For the given 3 × 4 PBTMTP with mixed constraints, $I = \{1, 2, 3\}$, $J = \{1, 2, 3, 4\}$:

$$a_1 = 10,\ a_2 = 15,\ a_3 = 12,\ b_1 = 10,\ b_2 = 15,\ b_3 = 13,\ b_4 = 11,\ \sum_I a_i = 37,\ \sum_J b_j = 49$$

Now formulate the following balanced transportation problem:

$$I^* = I \cup (3+1) = \{1,2,3,4\}$$
$$J^* = J \cup (4+1) = \{1,2,3,4,5\}$$
$$a_{m+1} = \sum_J b_j + 1 = 50$$
$$b_{n+1} = \sum_I a_i + 1 = 38$$
$$c_{41} = 11,\ c_{42} = 20,\ c_{43} = 0,\ c_{44} = 21,\ c_{45} = 0$$

**Step 2** Generate pairs $T^r$ and find $[T_r, T_{r'}]$, i.e. the time corresponding to primary and secondary destinations respectively using the working steps of Kaushal and Arora (2018; see Section 13.4.2). Find $f_r = [T_r + T_{r'}]$.

- Now partition various time entries given as:

$$t_1(=40) > t_2(=36) > t_3(=34) > t_4(=30) > t_5(=21) > t_6(=18) > t_7(=13)$$
$$> t_8(=12) > t_9(=11) > t_{10}(=7) > t_{11}(=0).$$

Here, $t_q = t_{11}$, so $q = 11$. Let $M_p = \{(i,j): t_{ij} = t^p\}$, $l = \{1,2, \ldots,11\}$, and $\lambda_{q-p+1}$ be the weights attached to the set $M_p$.

$$M_1 = \{(1,4)\}, M_2 = \{(2,3)\}, M_3 = \{(1,3)\}, M_4 = \{(3,3)\}, M_5 = \{(3,4),(4,4)\},$$
$$M_6 = \{(3,2),(4,2)\}, M_7 = \{(2,2)\}, M_8 = \{(1,1)\}, M_9 = \{(3,1),(3,5),(4,1)\}$$
$$M_{10} = \{(2,1),(2,4),(2,5)\}, M_{11} = \{(1,5),(4,3),(4,5)\}$$

- Attach weights, say $\lambda_{q-p+1}$, to each of the above set $M_p$, $p = \{1,2, \ldots,11\}$ s.t $\lambda_{p+1} > \lambda_q$, $\forall p = 1,2, \ldots,q-1$. These weights can be calculated using Mazzola and Neebe (1993).
- While (true)

Find the optimal solution of $\sum_{p=1}^{11} \lambda_p (\sum \sum_{M_p} c_{ij} z_{ij})$ using the UV method.

- Find $T^r = Max(t_{ij}) = 20$.

Here, $T^r$ is M-feasible. For this solution $T^r$ find $[T_r, T_{r'}] = [11, 20]$, i.e. the time corresponding to primary and secondary destinations, respectively.

- Find $f_r = [T_r + T_{r'}] = 31$.
- Set $t_{ij} = M \; \forall t_{ij} > 20$.
- Find the optimal solution of $\sum_{p=1}^{11} \lambda_p (\sum \sum_{M_p} c_{ij} z_{ij})$ using the UV method.
- Find $T^r = Max(t_{ij}) = M$ which is non M-feasible.

**Step 3** Stop. If Y is the LOS for PBTMTP with mixed constraints then it is calculated by the equations mentioned above (see Step 3, Section 13.4.1).

For each pair of lexicographic solution, we find:

$$Min\, f_r = Min\, [T_r + T_{r'}] = 31, r = 1,2,\ldots,11.$$

Hence, $T = (11,20)$ is the lexicographic optimal solution.

### 13.5.2 PBTMTP with Flow Constraints

Consider a PBTMTP with mixed constraints with four sources and four destinations as shown in Table 13.3. Entries in each cell shows the time of transportation between source destination pairs. Here, destinations 1 and 3 are the prior destinations, whereas 2 and 4 are the secondary destinations. The rightmost entries show the availabilities at each source and bottom entries shows the corresponding demand of each destination.

**Step 1** For the given 4 × 4 PBTMTP with flow constraints, $I = \{1,2,3,4\}$, $J = \{1,2,3,4\}$,

$$a_1 = 10,\ a_2 = 15,\ a_3 = 12,\ a_4 = 20,\ b_1 = 9,\ b_2 = 13,\ b_3 = 16,\ b_4 = 19,\ \sum_I a_i = 57,\ \sum_J b_j = 57, \text{P=60}$$

**TABLE 13.3**
**Entries in Each Cell Shows Transportation Time between Sources and Destinations, the Rightmost Entry Shows the Availability at Each Source, Entries at the Bottom Shows the Demand at Each Destination**

| | | | | |
|---|---|---|---|---|
| **12** | 13 | **34** | 40 | ≥10 |
| **7** | 18 | **36** | 7 | ≥15 |
| **11** | 20 | **30** | 21 | ≥12 |
| **9** | 7 | **32** | 16 | ≥20 |
| ≥9 | 13 | ≤16 | ≥19 | |

Now formulate the following balanced transportation problem.

$$I^* = I \cup (5) = \{1,2,3,4,5\}$$
$$a_5 = 3$$
$$b_5 = 3$$
$$c_{51} = 7,\ c_{52} = 7,\ c_{53} = 30,\ c_{54} = 7,\ c_{55} = M$$

**Step 2** Generate pairs $T^r$ and find $[T_r, T_{r'}]$, i.e. the time corresponding to primary and secondary destinations respectively using the working steps of Kaushal and Arora (2018; see Section 13.4.2). Find $f_r = [T_r + T_{r'}]$.

- Now partition various time entries given as

$t_1(=M) > t_2(=40) > t_3(=36) > t_4(=34) > t_5(=32) > t_6(=30) > t_7(=21)$
$> t_8(=20) > t_9(=18) > t_{10}(=16) > t_{11}(=13) > t_{12}(=12) > t_{13}(=11) > t_{14}(=9)$
$> t_{15}(=7)$. Here, $t_q = t_{11}$, so $q = 11$.

Let $M_p = \{(i,j): t_{ij} = t^p\}$, $l = \{1,2,...,15\}$ and $\lambda_{q-p+1}$ be the weights attached to the set $M_p$.

$$M_1 = \{(5,5)\}, M_2 = \{(1,4)\},$$
$$M_3 = \{(2,3)\}, M_4 = \{(1,3)\},$$
$$M_5 = \{(4,3)\}, M_6 = \{(3,3),(5,3)\},$$
$$M_7 = \{(3,4)\}, M_8 = \{(3,2)\},$$
$$M_9 = \{(2,2)\} M_{10} = \{(4,4)\},$$
$$M_{11} = \{(1,2)\}, M_{12} = \{(1,1),(1,5)\},$$
$$M_{13} = \{(3,1),(3,5)\} M_{14} = \{(4,1)\},$$
$$M_{15} = \{(2,1),(2,4),(2,5),(4,2),(4,5),(5,1),(5,2),(5,4)\}$$

- Attach weights, say $\lambda_{q-p+1}$, to each of the above set $M_p$, $p = \{1,2,...,15\}$ s.t. $\lambda_{p+1} > \lambda_q$, $\forall p = 1,2,...,q-1$. These weights can be calculated using Mazzola and Neebe (1993).

- While (true)

  Find the optimal solution of $\sum_{p=1}^{15} \lambda_p(\sum\sum_{M_p} c_{ij}\, z_{ij})$ using the UV method.
- Find $T^r = Max(t_{ij}) = 32$.

  Here, $T^r$ is M-feasible. For this solution $T^r$ find $[T_r, T_{r'}] = [32,16]$, i.e. the time corresponding to primary and secondary destinations respectively.
- Find $f_r = [T_r + T_{r'}] = 48$.
- Set $t_{ij} = M \; \forall t_{ij} > 32$.
- Find the optimal solution of $\sum_{p=1}^{15} \lambda_p(\sum\sum_{M_p} c_{ij}\, z_{ij})$ using the UV method.
- Find $T^r = Max(t_{ij}) = M$ which is non M-feasible.

**Step 3** Stop. If Y is the LOS for PBTMTP with flow constraints then it is calculated by equations mentioned above (see Step 3, Section 13.4.2).

For each pair of lexicographic solution, we find
$Min\, f_r = Min\ [T_r + T_{r'}] = 48, r = 1,2,...,15.$

Hence, $T = (32,16)$ is the lexicographic optimal solution.

## 13.6 COMPUTATIONAL DETAILS

The algorithm was implemented and tested successfully in MATLAB for various random-generated problems of different sizes. Implementation was carried out on Intel Processor i5 with 2.40 gigahertz, 4 gigabytes RAM on a 64-bit Windows operating system. Table 13.4 shows the computational results for some classes of different

**TABLE 13.4**
**Standard Deviation of Run Time (Taken over 100 Instances) of PBTMTP**

| Source | Destination | Mixed Constraints | Flow Constraints |
|---|---|---|---|
| 10 | 10 | 0.229043 | 0.293591 |
| 10 | 20 | 0.173383 | 0.059001 |
| 20 | 20 | 0.085354 | 0.058684 |
| 20 | 30 | 0.097041 | 0.124086 |
| 30 | 30 | 0.194144 | 0.259341 |
| 30 | 40 | 0.247712 | 0.423043 |
| 40 | 40 | 0.125068 | 0.558497 |
| 40 | 50 | 0.381834 | 0.558497 |
| 50 | 50 | 0.597483 | 0.763184 |
| 50 | 60 | 0.829355 | 0.575899 |
| 60 | 60 | 0.557359 | 0.711509 |
| 60 | 70 | 0.450518 | 0.512106 |
| 70 | 70 | 0.608001 | 0.836573 |
| 70 | 80 | 0.444942 | 0.274502 |
| 80 | 80 | 0.085009 | 0.667915 |
| 80/ | 90 | 0.181561 | 0.774244 |
| 90 | 90 | 0.109030 | 0.519990 |

sizes. Standard deviation of run time is reported for various problems of different sizes.

## 13.7 CONCLUDING REMARKS

This chapter presented an extension of PBTMTP. Some mixed and flow constraints were added, and these added constraints can help in handling real-life situations like war, business, etc. Hence, when there is restriction on total requirement transported from given locations or when there are mixed availabilities and requirements, all those situations can be solved easily with the method presented. The algorithm was implemented and tested for various random-generated problems in MATLAB and the standard deviation of the results was also reported.

## REFERENCES

Ahuja, R. K. (1986). Algorithms for the minimax transportation problem. *Naval Research Logistics*, *33*, 725–739.

Appa, G. (1973). The transportation problem and its variants. *Operational Research Quarterly 1970–1977*, *24*(1), 79–99.

Arora, S., & Puri, M. C. (1997). On lexicographic optimal solution in transportation problem. *Optimization*, *39*(4), 383–403.

Arora, S., & Puri, M. C. (1999). On standard time minimization transportation problem. *ASOR Bulletin*, *18*(4), 9–24.

Bansal, S., & Puri, M. C. (1980). A min max problem. *ZOR*, *24*, 191–200.

Bhatia, H. L., Swarup, K., & Puri, M. C. (1977). A procedure for time minimization transportation problem. *Indian Journal of Pure and Applied Mathematics*, *8*(8), 79–99.

Brigden, M. E. B. (1974). A variant of transportation problem in which the constraints are of mixed type. *Operational Research Quarterly*, *25*(3), 437–445.

Burkard, R. E., & Rendl, F. (1991). Lexicographic bottleneck problems. *Operations Research Letters*, *10*(3), 303–308.

Chandra, S., Seth, K., & Saxena, P. K. (1987). Time minimizing transportation problem with impurities. *Asia-Pacific Journal of Operational Research*, *4*(3), 19–27.

Dantzig, G. B. (1963). *Linear programming and extensions*. Santa Monica, CA: RAND Corporation, R-366-PR.

Dantzig, G. B., & Thapa, M. N. (1963). *Linear programming: 2: Theory and extensions*. New York and Berlin: Springer Verlag.

Garfinkel, R. S., & Rao, M. R. (1971). The bottleneck transportation problem. *Navel Research Logistics Quarterly*, *18*, 465–472.

Hammer, P. L. (1969). Time minimizing transportation problem. *Navel Research Logistics Quarterly*, *16*, 345–357.

Hammer, P. L. (1971). Communication on bottleneck transportation problem. *Navel Research Logistics Quarterly*, *18*(4), 487–490.

Hitchcock, F. L. (1941). The distribution of a product from several sources to numerous localities. *Journal of Mathematics and Physics*, *20*(1–4), 224–230.

Issermann, H. (1984). Linear bottleneck transportation problems. *Asia-Pacific Journal of Operational Research*, *1*, 38–52.

Kaur, P., Sharma, A., Verma, V., & Dahiya, K. (2016). A priority based assignment problem. *Applied Mathematical Modelling*, *40*(17–18), 7784–7795.

Kaushal, B., & Arora, S. (2018). Priority based time minimization transportation problem. *Yug/oslav Journal of Operations Research*, *28*(2), 219–235.

Khanna, S., Bakshi, H. C., & Puri, M. C. (1981). On controlling total flow in transportation problem. *Science Management of Trans-System North Holland Publishing Company*, 293–303.

Khanna, S., Bakshi, H. C., & Puri, M. C. (1983). Solving a transportation problem with mixed constraints and a specified transportation flow. *Opsearch*, *20*(1), 16–24.

Klingman, D., & Russel, R. (1974). The transportation problem with mixed constraints. *Operational Research Quarterly*, *25*(3), 447–455.

Mazzola, J. B., & Neebe, A. W. (1993). An algorithm for bottleneck generalised assignment problem. *Computers & Operations Research*, *20*(4), 355–362.

Nikolic, I. (2007). Total time minimizing transportation problem. *Yugoslav Journal of Operations Research*, *17*(1), 125–133.

Orlin, J. B. (1997). A polynomial time primal network simplex algorithm for minimum cost flows. *Mathematical Programming*, *78*(2), 109–129.

Prakash, S. (1982). On minimizing the duration of transportation. *Proceeding of the Indian Academy of Sciences*, *91*(1), 53–57.

Sharma, A., Verma, V., Kaur, P., & Dahiya, K. (2015). An iterative algorithm for two level hierarchical time minimization transportation problem. *EJOR*, *246*(3), 700–707.

Sherali, H. D. (1982). Equivalent weights for lexicographic multi-objective programs: Characterizations and computations. *EJOR*, *11*(1), 367–379.

Sonia, P. M. C. (2004). Two level hierarchical time minimizing transportation problem. *Sociedad de Estadística e Investigacion Operativa Top*, *12*(2), 301–330.

Sonia, P. M. C. (2008). Two stage time minimizing assignment problem. *Omega*, *36*(5), 730–740.

Szwarc, W. (1971). Some remarks on time transportation problem. *Navel Research Logistics Quarterly*, *18*(1), 473–485.

Szwarc, W. (1996). The time transportation problem. *Zastosowania Matematyki*, *8*(1), 231–292.

# 14 ASEAN and India: Exploring the Progress and Prospects in Trade Relationship

*R. K. Sudan*

**CONTENTS**



## 14.1 INTRODUCTION

The Association of Southeast Asian Nations (ASEAN), organized in 1967, is a regional intergovernmental organization comprising 10 Southeast Asian countries that promotes intergovernmental cooperation and facilitates economic, political, security, military, educational, and sociocultural integration among its members and other Asian states. The 10 member countries are Brunei, Cambodia, Indonesia, Laos, Malaysia, Myanmar, Philippines, Singapore, Thailand, and Vietnam. On 1 January 1993, ASEAN officially formed ASEAN Free Trade Area (AFTA), with the goal to cut tariffs on all intrazonal trade to a maximum of 5% by 1 January 2008, with the provision that the weaker ASEAN countries would be allowed to phase in their tariff reductions over a longer period.

At its inception, there were five members. Brunei (1984), Vietnam (1995), Laos, Myanmar (1997), and Cambodia (1999) joined later. Today, there are still 10 members. Work toward further integration continued when ASEAN Plus Three was created in 1997 with China, Japan, and South Korea. This was followed by another expansion by inviting India, Australia, and New Zealand into the East Asia Summit (EAS). The EAS is a pan-Asian forum held annually by the leaders of 18 countries in the East Asian region, with ASEAN in a leadership position. Initially, ASEAN belonged to all member states of ASEAN plus China, Japan, South Korea, India, Australia, and New Zealand, but ASEAN deliberations later were expanded to include the United States and Russia at the Sixth EAS in 2011. China, Japan, South Korea, India, Australia, and New Zealand are special invitees only, not the regular members. The group is now known as ASEAN Plus Six, and stands as a pivotal column of the Asia Pacific's economic, political, security, and sociocultural architecture, as well as playing a role in the global economy.

The year 2017 celebrated 50 years of ASEAN's existence and 25 years of ASEAN–India dialogue. The ASEAN–India breakfast summit was held on November 15, 2018.

## 14.2 AREAS OF COOPERATION BETWEEN ASEAN AND INDIA

As per PMO notification, a number of areas of cooperation have been identified:

**Political and security cooperation:** Regional stability, international security, counterterrorism, cybersecurity.

**Economic cooperation:** ASEAN–India free trade area, conservation and sustainable use of maritime resources, growth of MSME sector, food and energy security, collaboration in space science and satellite imagery, promotion of private-sector engagement and B2B relations.

**Sociocultural cooperation:** Exchange programmes of policymakers, managers, students and academicians; health care and affordable quality medicines; cultural tourism; university student exchange programs; education of women and children; disaster management; biodiversity conservation; climate change.

## 14.3 MISCELLANEOUS

India is committed to enhancing its physical and digital connectivity in line with the MPAC 2025 and the AIM 2020. It is availing itself of the US$1 billion line of credit to promote physical infrastructure and digital connectivity.

India is encouraging the early completion of the India-Myanmar-Thailand Trilateral Highway Project and its extension to Cambodia, Laos, and Vietnam.

India's continues to demonstrate support for ASEAN's efforts in narrowing the development gap within and between ASEAN member states by implementing the IAI Work Plan-3.

## 14.4 POLICY MEASURES AND INITIATIVES

India and ASEAN observed 25 years of dialogue partnership, 15 years of summit-level interaction, and 5 years of strategic partnership throughout 2017 by undertaking over 60 commemorative activities, both in India and through her missions in ASEAN Member States. The commemoration reached its peak with the ASEAN–India Commemorative Summit on the theme "Shared Values, Common Destiny" on 25 January 2018 in New Delhi.

The commemorative activities were highlighted by an ASEAN–India Regional Diaspora event in Singapore, a music festival, an artists' retreat, a youth summit, port calls by Indian naval ships, a connectivity summit, a meeting to reinforce the network of think tanks, a workshop on the blue economy, the Dharma-Dhamma Conference, a hackathon and startup festival, a business and investment meet and expo, a global SME summit, a textiles event, an ICT expo, a business council meeting, a film festival, a Ramayana festival, and the inauguration of an India–ASEAN Friendship Park in New Delhi.

### 14.4.1 Plans of Action

The ASEAN–India Partnership for Peace, Progress and Shared Prosperity, which sets out the roadmap for long-term ASEAN–India engagement, was signed at the Third ASEAN–India Summit in 2004 in Vientiane, as a reflection ASEAN's and India's desire to strengthen their engagement. A plan of action (POA) for the period 2004–2010 was developed to implement the partnership. The third POA (2016–2020) was adopted by the ASEAN–India Foreign Ministers Meeting held in August 2015. Furthermore, ASEAN and India have identified priority areas for the period of 2016–2018 and are already implementing activities under it, which would contribute towards successful implementation of the 2016–2020 POA.

### 14.4.2 Political-Security Cooperation

In light of growing traditional and non-traditional challenges, political-security cooperation is key and an emerging area of importance in India's relationship with ASEAN. The rise in the export of terrorism, increased radicalization through an ideology of hatred, and the spread of extreme violence define the pattern of common security threats to our societies. The Indian partnership with ASEAN attempts to carve out a response that is based on coordination, cooperation, and experience-sharing at various levels.

ASEAN, as a regional grouping based on consensus, has worked diligently over 50 years to help secure peace, progress, and prosperity in the region. India, therefore, views ASEAN at the centre of its Indo-Pacific vision of "Security and Growth for All" in the Region.

The key forum for ASEAN security dialogue is the ASEAN Regional Forum (ARF). India has been participating in annual meetings of this forum since 1996 and has actively joined in its various activities. The ASEAN Defense Ministers' Meeting (ADMM) is the highest defense consultative and cooperative mechanism in ASEAN. The ADMM+ brings together defense ministers from the 10 ASEAN nations plus Australia, China, India, Japan, New Zealand, Republic of Korea, Russia, and the United States on a biannual basis.

### 14.4.3 Economic Cooperation

India-ASEAN trade and investment relations have been developing in a steady-state way, with ASEAN being India's fourth largest trading partner. India's trade with ASEAN stands at US$81.33 billion, which is about 10.6% of India's overall trade. India's export to ASEAN stands at 11.28% of total exports.

Investment flows are also substantial from both sides, with ASEAN accounting for approximately 18.28% of investment flows into India since 2000. Foreign direct investment (FDI) inflows into India from ASEAN between April 2000 to March 2018 was around US$68.91 billion, while FDI outflows from India to ASEAN countries from April 2007 to March 2015, as per info provided by Department of Economic Affairs (DEA), was about US$38.672 billion. The ASEAN–India Free Trade Area has been formalized, entering into force the ASEAN–India Agreements on Trade in Service and Investments on 1 July 2015.

ASEAN and India have been also working on the lines of enhancing the private-sector engagement. The ASEAN India-Business Council (AIBC) was formed in March 2003 in Kuala Lumpur as a forum to induct key private-sector players from India and ASEAN countries on a single platform for business networking and sharing of ideas.

### 14.4.4 Sociocultural Cooperation

India has been organizing a large number of programmes to boost people-to-people interaction with ASEAN, such as inviting ASEAN students to India each year for the student exchange program, providing special training courses for ASEAN diplomats, exchanging parliamentarians, inviting ASEAN students to participate in the National Children's Science Congress, promoting the ASEAN-India network of think tanks, and hosting the ASEAN-India Eminent Persons Lecture Series, etc.

The second edition of the ASEAN–India Workshop on the Blue Economy, jointly hosted with Vietnam, was held on 18 July 2018 in New Delhi.

### 14.4.5 Connectivity

A matter of priority for India is ASEAN–India connectivity. In 2013, India became the third dialogue partner of ASEAN to initiate an ASEAN Connectivity Coordinating Committee–India Meeting. While India has made considerable progress in implementing the India-Myanmar-Thailand Trilateral Highway and the Kaladan Multimodal Project, issues related to increasing the maritime and air connectivity between ASEAN and India and transforming these corridors of connectivity into economic corridors are under consideration. A possible extension of the India-Myanmar-Thailand Trilateral Highway to Cambodia, Laos, and Vietnam is also under discussion. A consensus on finalizing the proposed protocol of the India-Myanmar-Thailand Motor Vehicle Agreement (IMT MVA) has been arrived at. This agreement will have a crucial role in realizing continuous movement of passenger, personal, and cargo vehicles along roads linking India, Myanmar, and Thailand. India announced a line of credit of US$1 billion to promote projects that support physical and digital connectivity between India and ASEAN, and a project development fund with a corpus of INR 500 crore was set up to develop manufacturing hubs in Cambodia, Laos, Myanmar, and Vietnam (CLMV countries) at the 13th ASEAN India Summit held in Malaysia in November 2015.

### 14.4.6 Funds

The ASEAN Multilateral Division offers project-based financial assistance to ASEAN countries. Financial assistance has been provided to ASEAN countries from the following funds.

#### 14.4.7.1 ASEAN–India Cooperation Fund

At the Seventh ASEAN–India Summit in 2009, India announced a contribution of US$50 million to the ASEAN–India Fund to support implementation of the ASEAN–India Plans of Action, which envisage cooperation in a range of sectors as well as capacity-building programmes in the political, economic, and sociocultural areas for deepening and strengthening ASEAN–India cooperation. In order to more these development and capacity-building initiatives forward, India's prime minister has proposed enhancing the ASEAN-India Fund with an additional grant of US$50 million at the 14th ASEAN India Summit in Vientiane in September 2016.

#### 14.4.7.2 ASEAN–India S&T Development Fund (AISTDF)

At the Sixth ASEAN–India Summit in November 2007 in Singapore, India announced the establishment of an ASEAN–India Science & Technology Development Fund with a US$1 million contribution from India to promote joint collaborative R&D research projects in science and technology. This fund become operational in 2009–2010 and expenditures began to be incurred from FY 2010–2011. This fund has been stepped up to US$5 million from 2016–2017.

#### 14.4.7.3 ASEAN–India Green Fund

At the Sixth ASEAN–India Summit on 21 November 2007 in Singapore, India declared the setting up of an ASEAN–India Green Fund with an initial contribution of US$5 million from India to support collaboration activities relating to the environment and climate change. Some of the areas identified for collaboration under the fund are climate change, energy efficiency, clean technologies, renewable energy, biodiversity conservation, and environmental education.

### 14.4.8 ASEAN–India Projects

India has been collaborating with ASEAN by way of implementation of various projects in the fields of agriculture, science and technology, space, environment and climate change, human resource development, capacity building, new and renewable energy, tourism, people-to-people contacts, connectivity, etc.

Some of the more prominent projects, which are either ongoing or in the final stages of approval, include a space project envisaging establishment of a tracking, data reception/data processing station in Ho Chi Minh City, Vietnam; the upgrading of the Telemetry Tracking and Command Station in Biak, Indonesia; the establishment of Centres of Excellence in software development and training in CLMV countries; an e-network for provision of telemedicine and tele-education in CLMV countries; and quick impact projects in CLMV countries.

In addition to these projects, India has been supporting ASEAN, especially CLMV countries, under the Initiatives for ASEAN Integration, which include projects on training of English Language for law enforcement officers in CLMV countries and training of professionals dealing with capital markets in CLMV countries by the National Institute of Securities Management, Mumbai. India also provides scholarships for ASEAN students for higher education at Nalanda University and training of ASEAN civil servants in drought management, disaster risk management, sustainable ground water management, etc.

### 14.4.9 Agriculture

With regard to agriculture, India is cooperating with ASEAN by way of projects such as exchange of farmers, ASEAN–India fellowships for higher agricultural education in India and ASEAN, exchange of agriculture scientists, empowerment of women through cooperatives, training courses on organic certification for fruits and vegetables, etc. These were further strengthened at the Fourth ASEAN–India Ministerial Meeting on Agriculture held in January 2018 in New Delhi, with the endorsement of the Medium Term Plan of Action for ASEAN–India Cooperation in Agriculture and Forestry for 2016–2020.

### 14.4.10 Science and Technology

In the science and technology field, India promotes the ASEAN–India S&T Digital Library, the ASEAN–India Virtual Institute for Intellectual Property, the ASEAN-India Collaborative Project on Science & Technology for Combating Malaria, the ASEAN–India Programme on Quality Systems in Manufacturing, and the ASEAN–India Collaborative R&D Project on Mariculture, Bio-mining and Bioremediation Technologies, etc.

### 14.4.11 Delhi Dialogue

India has had an annual Track 1.5 event, Delhi Dialogue, for discussing political-security and economic issues between ASEAN and India. Since 2009, India has hosted 10 editions of this flagship conference. The 10th edition of Delhi Dialogue was hosted by the Ministry of External Affairs on 19–20 July 2018 in New Delhi, with the theme, "Strengthening India–ASEAN Maritime Advantage".

### 14.4.12 ASEAN–India Centre (AIC)

At the Commemorative Summit held in 2012, the heads of the government recommended establishment of the ASEAN–India Centre (AIC) to undertake policy research, advocacy, and networking activities with organizations and think tanks in India and ASEAN, with the aim of promoting the ASEAN–India Strategic Partnership. Set up in 2013, the AIC has been serving as a resource center for ASEAN member states and India for strengthening ASEAN–India strategic partnerships and promoting India–ASEAN dialogue and cooperation in the areas of mutual interest. AIC has provided inputs to policymakers in India and ASEAN on implementation of ASEAN–India connectivity initiatives by organizing seminars, roundtables, etc. AIC also organizes workshops, seminars, and conferences on various issues relevant to the ASEAN–India strategic partnership. It undertakes regular networking activities with relevant public/private agencies, organizations, and think-tanks in India and ASEAN and EAS countries, with the aim of providing up-to-date information, data resources, and sustained interaction to promote the ASEAN–India strategic partnership.

## 14.5 OVERVIEW OF THE INDIA–ASEAN RELATIONSHIP

A lot much was expected of India since its "Look East" policy was initiated in 1991 along with the economic reforms. It was supposed to be a steady-state path where India and ASEAN countries 25 years down the line are deeply entrenched and integrated through economic, social, cultural, and political associations. The reality today reveals that India is not even close to the trade figures that China has with ASEAN countries. India's latest trade figures amount to close to US$80 billion with ASEAN states, whereas China's trading figures with ASEAN countries has crossed US$450 billion and is expected to touch US$1 trillion by 2020 (www.asiaone.com/business/china-asean-trade-hit-us1-trillion-2020). China's Belt and Road Initiative (BRI) is expected to increase trade between China and ASEAN's six most prominent economies—Malaysia, Indonesia, the Philippines, Singapore, Thailand, and Vietnam—to an investment equivalent to US$2.1 trillion by 2030 (www.thestar.com.my/business/business-news/2017/05/20/chinas-bri-seen-boosting-trade-with-asean-to-us21-trillion/). The huge gap in trade alone is a dampener on India's future with ASEAN.

The terminology was changed from "Look East" to "Act East" in 2014, but the approach has been following the old principles of shared values, a common destiny, shared prosperity, and culture and dwelling on technical jargon of relations becoming "strategic" in nature. While celebrating the 25 years of relations, it is this nature of "strategic partnership" that needs to be given more impetus in ASEAN–India relations for the next 25 years. India needs to effectively engage the Southeast Asian countries to keep the channels of communications open in the South China Sea region amidst overlapping territorial claims with China. In fact, some of the ASEAN countries, aside from Cambodia and Laos, are searching for a balancing power to China in the region and will be keenly observing how India tackles the overlapping conflicts. If the proceedings of the CSCAP (Council for Security Cooperation in Asia-Pacific) meeting in December 2017 are to be weighed-in, then the prevailing sense amongst the ASEAN delegates has been of apprehension towards China's intimidating economic tactics (www.ibtimes.sg/india-needs-harmonise-ties-global-power-centres-22907). Meanwhile, certain foreign policy experts from the Southeast Asian region have also questioned the lack of performance by India, even hinting at a subtle resistance to the concept of "Indo-Pacific" as a replacement to the old "Asia-Pacific". India was also called out on its "Act East"' policy looking more like "At Ease" policy (http://indianexpress.com/article/opinion/columns/asean-cscap-security-look-east-indian-navy-a-strategy-for-the-sea-5013281/).

## 14.6 CHOICE AMONGST OPTIONS OF AREAS OF COOPERATION

Policy makers, planners, and programmers face the following issues:

- Ranking the options of areas and sub-areas based on a logic of need and urgency.
- Present and future cost-benefits of options.
- Internal consistency.
- Balance of long- and short-run perspectives in matters of cost-benefit evaluations of the options.
- Role of non-economic determinants in the decision-making process (e.g. international diplomacy, social, political, legal, and cultural factors).

## 14.7 CONCLUSION

India and ASEAN can contribute to each other in regional economic development. The extent to which this occurs depends on the timing, spacing, and sequencing of policy measures. Cooperation requires mutual faith and trust-building measures. Trade policy must envisage a coordination between various ministries, viz. External Affairs, Commerce, Finance, Corporate Affairs, and Law.

## REFERENCES

*The Economic Times*, November 15, 2018
www.tandfonline.com
www.orfonline.org
www.mea.gov.in
www.pib.nic.in
www.asiaone.com
www.thedialogue.co/revisiting-asean-india-2018

# 15 Impact of Internationalization on Financial Performance: A Study of Family and Non-Family Firms

*Amit Kumar Singh, Amiya Kumar Mohapatra, Varda Sardana, and Shubham Singhania*

**CONTENTS**



## 15.1 INTRODUCTION

The increasing integration of economies all over the world as a result of globalization has paved the way for development of business organizations that connect the developing and developed parts of the globe with each other. A major reason for the economic differences between the developed and developing nations is their degree of internationalization. Internationalization of business calls for expansion of any business firm from its home market into a foreign market. Realizing the importance of foreign trade and investment relations, India passed the New Economic Policy for liberalization of trade norms, privatization of firms, and globalization of the Indian economy in 1991. Nations, including India, have been able to reap the benefits of globalization through foreign collaborations like strategic alliances, formation of joint ventures, and multinational companies, to name a few. The rapid move towards internationalization has made it a popular research theme globally, and amongst them, instances of family firms going global have topped the research charts.

As per Leach and Bogod (1999), a family firm is one wherein a "Single family effectively controls the firm through the ownership of greater than 50 per cent of the voting shares; or a significant portion of the firm's senior management is drawn from the same family". Family business groups have dominated the global as well as Indian business industry. India ranks amongst the top three countries in terms of number of listed family firms. Family businesses that started with the purpose of hustle-free trading and money lending have restructured with the changing dynamics of business. Most of the large trading houses in India are family firms and contribute a major chunk to the trade and investments in the country. This scenario calls for a study on how the family firms going global showcase a performance difference vis-à-vis non-family global firms or domestic firms.

The changing business environment calls for foreign collaborations and other means of expansion as an effective way of coping up with the dynamic market behaviors. The rationale behind business consolidations is "unity is strength". Hence, a collective effort of two groups is more fruitful than their separate moves. The most popular ways that large family businesses have gone global include mergers and acquisitions, joint ventures, and wholly owned subsidiaries. The US and China also have a huge number of publicly listed family firms. The European Union has an extremely high number of family firms, constituting more than 70% of all business firms in the European Union. Studies all over the world have given mixed reviews regarding pre- and post-collaboration performance. While some studies have concluded a favorable impact, others argue that it is non-family firms of smaller sizes that are more likely to go global than large and established family firms that eventually split their operations, just like the case of Reliance Group in India. A number of pre- and post-merger comparative financial studies have exhibited a negative performance post foreign merger.

Considering the mixed reviews, we sought to compare the pre- and post-internationalization financial performance of family and non-family firms.

### 15.1.1 Objectives of Study

The objectives of the study were as follows:

1. To determine the impact of internationalization on financial performance of family firms in India.

2. To determine the impact of internationalization on financial performance of non-family firms in India.
3. To study the difference, if any, between the post-internationalization financial performance of family firms vis-à-vis non-family firms in India.

This chapter has been divided into five sections. The first section consists of the introduction along with the objectives of this study. The second section consists of the review of existing literature in order to determine the gaps. This is followed by our hypothesis for the study. The third section covers the research methodology employed. The fourth section lays down the results, findings, and their interpretations. This is followed by conclusions in the fifth section.

## 15.2 LITERATURE REVIEW

### 15.2.1 Internationalization of Family Firms

Graves and Thomas (2008) laid down three determinants of the internationalization pathways taken by family firms: (1) the level of commitment towards internationalization, (2) the availability of financial resources, and (3) the ability to commit and use those financial resources to develop the required capabilities. Also, the presence of next generation in the family firms increases the propensity and intensity with which these firms go global, as they are more likely to search for opportunities actively in the international or national market to expand their market share to ensure the business sustainability (Knežević & Wach, 2014). On the other hand, in a study conducted by Donckels and Fröhlich (1991) among European countries, it was found that family businesses should be viewed as stable rather than progressive because these firms follow a closed family-related system, and their managers are usually not all-rounders.

In his study of Polish family firms, Wach (2014) conducted a survey among 216 firms, including 88 family businesses, and found that the average time of internationalization is longer in case of family firms than non-family firms. Although Graves and Thomas (2004) found that the extent of internationalization of family firms is less than that of non-family firms, this difference is not persistently significant over time. Also, family firms are less likely to engage in networking with other businesses vis-à-vis non-family firms.

Moreover, family firms are mostly market seekers, unlike non family firms, which are very much capability seekers. However, there is no statistically significant difference between family and non-family firms regarding the preferred mode of internationalization (Daszkiewicz & Wach, 2014). Gallo and Sveen (1991) suggest that a multitude of factors, such as organizational structure, culture, stage of development, firm strategy, and the like, facilitate or restrain the growth of family firms beyond their country's boundaries, and these should be fully taken advantage of.

In another study by Mensching, Calabrò, Eggers, and Kraus (2016) the authors used choice-based conjoint analysis and suggested that the perception of risk and success differs among the types of chief executive officers: family CEOs, non-family CEOs, and CEOs from non-family firms. The difference has been found based on the geographical, psychological, and cultural distance.

### 15.2.2 Positive Impact of Internationalization

Lee et al. (2010) in a major study suggested that breadth (which was measured by the number of countries in which the firm has undertaken direct investments) has positive effects on firm performance (which was measured by Tobin's Q) and depth (measured by the number of foreign investment sites in top two countries divided by total number of foreign investment sites) is negatively correlated with firm performance. Based on a study, family firms, on an average, take more time to go global as compared to non-family firms (Knežević & Wach, 2014).

According to Das and Banik (2015), in the context of Indian firms, the major reasons for going global are resources, technology, efficiency, and strategic assets.

A study undertaken by José Mas-Ruiz, Luis Nicolau-Gonzálbez, and Ruiz-Moreno (2002) in Spain to examine the factors affecting firm performance when the firm expands into foreign markets revealed that, on average, the announcement of a firm's foreign expansion or foreign collaboration has a positive effect on its stock market returns. Rani, Yadav, and Jain (2016) have also found in their Indian study that the market usually reacts positively to merger and acquisition announcements, thereby leading to positive abnormal returns and wealth creation for investors. On the basis of exports, McDougall and Oviatt (1996) observed that firm performance improves after going global.

The profitability, efficiency, liquidity, and market prospect performance variables improve following mergers and collaborations, as per a Jordanian industrial sector study conducted by Al-Hroot (2016). Barutha, Jeong, Gransberg and Touran (2018) found in their study that collaborative projects are always better because as the degree of collaboration increases, the project performance increases in direct correlation. A study of small-scale agricultural businesses in South Africa (Rambe & Agbotame, 2018) demonstrated that the establishment of foreign alliances is positively related to their performance. Similar is the case with a Malaysian study that reported an improvement in long-run operating performance of the sampled firms after acquisitions (Rahman & Limmack, 2004).

Hajela and Akbar (2013) in their study have shown that there is positive relation between the internationalization of the firms and their firm performances, which clearly means that the benefits accruing from going global outweigh the costs.

### 15.2.3 Negative Impact of Internationalization

As per Shrestha et al. (2017), a domestic or foreign merger should not be considered as a definite solution to overcome the challenges in the market, as a firm has to undertake a lot of research about the collaborating partner, resources, and the like before sealing a deal. Their study found that in case of large and stable firms with adequate managerial resources, mergers positively impact the firm performance, but this is not true in the case of many firms that usually don't have the expertise. For such firms, profitability in terms of return on assets (ROA) and return on equity (ROE) is negatively impacted.

On the same lines, Patel (2018) concluded that the post-merger financial performance of firms turn outs to be negative in the Indian scenario with respect to many measures, including return on equity, return on assets, net profit ratio, yield on advance, and yield on investment. Similar results were obtained by Pazarskis, Vogiatzogloy, Christodoulou, and Drogalas (2006), as per whom the profitability of firms entering into collaborations in Greece decreased after such a deal. Moreover, companies that grow through foreign acquisitions yield a relatively lower rate of return than those that grow organically (Dickerson, Gibson, & Tsakalotos, 1997).

### 15.2.4 No Impact of Internationalization

The study conducted by Larasati, Agustina, Istanti, and Wijijayanti (2017) using Indonesian firms' data has shown that various ratios such as debt-to-equity ratio (DER), current ratio (CR), price-earnings ratio (PER), net profit margin (NPM), and total asset turnover (TATO) have no significantly different impact on the company's performance before and after merger and acquisition. A similar result has been obtained by Abbas, Hunjra, Azam, Ijaz and Zahid (2014) in their study of Pakistani banks, as well as by Abdulwahab and Ganguli (2017) in Bahrain. Sharma and Ho (2002), having conducted a study of 36 Australian acquisitions during six years, found no significant improvements in post-acquisition operating performance.

### 15.2.5 Speed of Internationalization

As per Chang and Rhee (2011), more and more firms are expanding rapidly into foreign markets through collaborations, mergers, and acquisitions. Although these trends seem to be contrary to the conventional theory of gradual internationalization and learning from prior experiences, it also decreases the risk of being a late entrant and foregoing the first-mover advantages. This has been proved by their study of Korean firms which shows that undertaking foreign expansions through foreign direct investments (FDIs) at an accelerated speed enhances firm performance in industries where globalization pressures are high and when it is done by firms with superior internal resources and capabilities.

Chang (2007), in his study of 115 Asia-Pacific multinational enterprises (MNEs), found a non-linear relationship between internationalization and firm performance. This shows that as a firm expands internationally, it grows at an increasing pace. The results also indicate that moderate product diversification, foreign expansion speed, and geographic scope can increase the emerging-market MNEs capacities to exploit different market opportunities when they engage in foreign activity. This result was substantiated by another study by Gracia et al. (2015) on Spanish firms which shows the existence of an inverted U-shaped relationship between speed of internationalization and firm value creation. At the same time, putting restrictions on their future growth can lead to a negative impact.

To examine whether a faster foreign entry is always better, Tao and Jiang (2014) undertook a case study of 214 auto FDIs in China. Their results showed an inverted U-shape relationship between expansion frequency and firm performance. In other words, greater firm performance is associated with greater expansion frequency and expansion magnitude.

### 15.2.6 Statement of Hypotheses

$H_{01}$: Internationalization of family firms has no effect on their financial performance post internationalization.
$H_{a1}$: Internationalization of family firms has an effect on their financial performance post internationalization.
$H_{02}$: Internationalization of non-family firms has no effect on their financial performance post internationalization.
$H_{a2}$: Internationalization of non-family firms has an effect on their financial performance post internationalization.
$H_{03}$: There is no difference between the post internationalization financial performance of family firms and non-family firms.
$H_{a3}$: There is a difference between the post internationalization financial performance family firms and non-family firms.

## 15.3 RESEARCH METHODOLOGY

### 15.3.1 Data Sources

We used moneycontrol.com to collect data regarding the various family and non-family firms that went international during the sample period. The various measures of firm performance, pre and post internationalization, were collected from the same source.

### 15.3.2 Time Period

We considered a nine-year-long time period for this study, from 2009–2010 to 2017–2018. Ten family firms and 10 non-family firms that engaged in internationalization during this time period were included in the sample. The main reason for including firms after the year 2009–2010 was to ensure that we were measuring the performance of firms after the global financial crisis, thereby avoiding any structural break.

### 15.3.3 Sample Size

A total of 20 firms were included in the sample, out of which 10 were family firms and 10 were non-family firms.

### 15.3.4 Sampling Technique

Purposive sampling technique was used for the study. A purposive sample, also known as judgmental sample, is a non-probability sample that is selected based on characteristics of a population and the objective of the study. We selected the family firms to be included in the sample if they satisfied the definition of family firms (majority of ownership or control lies with one or two families together) and had undertaken internationalization during the years 2009–2010 to 2017–2018. The same sampling technique was used for selecting non-family firms as well.

### 15.3.5 Variables

The study was undertaken in two phases. For the first phase, wherein we tried to gauge the impact of firms' internationalization on their financial performance, the dependent variable was "Internationalization of firms".

The independent variables consisted of measures for profitability, market prospect, firm market value, and leverage. The following are the variables that were used for measuring the firm performance three years pre-internationalization and three years post-internationalization:

**Return on assets (ROA):** This has been used as a measure of profitability of the firm. ROA is calculated as:

***ROA*** = Net income after taxes/Total asset value

**Earnings per share (EPS):** This has been used as a proxy for a firm's market prospects. A firm's market prospects refer to a firm's forecasted ability to compete in a marketplace. EPS has been calculated as:

**EPS** = Total earnings available to equity shareholders/Total number of outstanding equity shares.

**Tobin's Q**: The Tobin's Q ratio is a measure of firm assets in relation to a firm's market value. It has been calculated using the formula:

$$Q = \left(MVS + D\right)/TA,$$

where MVS = Market Capitalization = Outstanding number of equity shares * Current market price.
D is the debt of the firm.

**Debt-equity ratio:** The debt-equity ratio is a leverage ratio that calculates the weight of total debt and financial liabilities against the total shareholder's equity. It indicates what percentage of total assets of a firm is financed by debt. The D/E ratio is equal to the firm's outstanding debt divided by the firm's total outstanding equity.

### 15.3.6 Statistical Techniques

For the first phase of the study, a paired-sample *t*-test was used for all 10 family firms and 10 non-family firms to check the impact of going international on firms' financial performance. The paired-sample *t*-test or the dependent sample *t*-test compares two means that are from the same individual, object, or related units in order to determine whether the mean difference between two sets of observations is zero. In our study the unit in consideration is the firm, whose financial performance was measured twice—mean performance for three years before internationalization and mean performance for three years after internationalization.

For the second phase of the study, that is, for comparing the performance of family firms vis-à-vis non-family firms after internationalization; we used the independent samples test for proportions. In other words, we compared the proportion of family firms which showed an insignificant difference in their pre- and post-performance, with the non-family firms.

We used two software for our study: IBM SPSS Statistics 20 and Microsoft Excel 2010.

Tables 15.1 and 15.2 provide an overview of the family firms and non-family firms respectively which have been included in the sample for the study.

**TABLE 15.1**

**An Overview of Family Firms Sampled**

| Name of Firm | Controlling Family | Shareholding of Controlling Family | Year of Internationalization |
|---|---|---|---|
| Ashok Leyland Ltd. | Hinduja | 51.12% | 2013 |
| Bharti Airtel Ltd | Mittal | 50.1% | 2012 |
| Hindalco Industries | Birla | 34.66% | 2013 |
| Infosys Ltd. | Narayana Murthy, Nandan Nilekani | 12.75% | 2012 |
| Jet Airways | Naresh Goyal | 51% | 2013 |
| Mahindra And Mahindra Ltd. | Anand Mahindra and Family | 9.14% | 2012 |
| Marico Ltd. | Harsh Mariwala | 59.57% | 2011 |
| Reliance Industries | Ambani | 46.17% | 2010 |
| TCS Ltd. | Ratan Tata and Family | 72.05% | 2014 |
| Wipro Ltd. | Azim Premji | 74.31% | 2009 |

*Source*: moneycontrol.com, trendlyne.com.

**TABLE 15.2**

**An Overview of the Non-Family Firms Sampled**

| Name Of Firm | Year Of Internationalization |
|---|---|
| Arvind Ltd. | 2014 |
| Asian Paints India Private Ltd. | 2010 |
| Aurobindo Pharma Ltd. | 2014 |
| Dabur India Ltd. | 2012 |
| Eicher Motors | 2008 |
| HCL Technologies Ltd. | 2011 |
| ITC Ltd. | 2012 |
| Larsen And Toubro | 2010 |
| Thermax | 2012 |
| Wockhardt Ltd. | 2011 |

*Source*: moneycontrol.com, trendlyne.com.

## 15.4 RESULTS AND FINDINGS

Tables 15.3 and 15.4 show the results of the paired-sample *t*-test for the 10 family firms in our sample. For each of the four variables, that is, ROA, EPS, Tobin's Q, and D/E ratio, the table lays down the pre-internationalization mean, post-internationalization

**TABLE 15.3**
**Results for Family Firms**

| FIRM NAME | ROA | | | EPS | | |
|---|---|---|---|---|---|---|
| | Pre | Post | π-Value | Pre | Post | π-Value |
| ASHOK LEYLAND LTD. | 4.67 | 3.23 | 0.595 | 2.04 | 2.80 | 0.698 |
| BHARTI AIRTEL LTD | 6.61 | 2.29 | 0.352 | 17.09 | 11.26 | 0.442 |
| HINDALCO INDUSTRIES | 2.91 | 0.57 | 0.071 | 15.46 | 2.93 | 0.126 |
| INFOSYS LTD. | 22.05 | 18.42 | 0.004 | 125.00 | 117.92 | 0.895 |
| JET AIRWAYS | −3.86 | 2.12 | 0.572 | −88.25 | 17.98 | 0.53 |
| MAHINDRA AND MAHINDRA LTD. | 5.93 | 3.90 | 0.033 | 48.67 | 61.00 | 0.343 |
| MARICO LTD. | 14.64 | 15.13 | 0.896 | 3.81 | 7.53 | 0.008 |
| RELIANCE INDUSTRIES | 8.84 | 5.67 | 0.166 | 108.39 | 71.46 | 0.175 |
| TCS LTD. | 26.78 | 25.66 | 0.601 | 73.91 | 130.26 | 0.031 |
| WIPRO LTD. | 16.64 | 13.80 | 0.276 | 23.09 | 23.15 | 0.95 |

*Source*: Authors' own calculations.

**TABLE 15.4**
**Results for Family Firms**

| FIRM NAME | Tobin's Q | | | D/E Ratio | | |
|---|---|---|---|---|---|---|
| | Pre | Post | π-Value | Pre | Post | π-Value |
| ASHOK LEYLAND LTD. | 2.49 | 5.21 | 0.021 | 0.94 | 1.73 | 0.059 |
| BHARTI AIRTEL LTD | 2.76 | 2.72 | 0.955 | 0.96 | 1.51 | 0.161 |
| HINDALCO INDUSTRIES | 0.89 | 0.69 | 0.619 | 1.29 | 1.56 | 0.486 |
| INFOSYS LTD. | 6.28 | 4.57 | 0.075 | .000 | .000 | — |
| JET AIRWAYS | −8.66 | −0.96 | 0.386 | −26.11 | −1.44 | 0.362 |
| MAHINDRA AND MAHINDRA LTD. | 2.89 | 2.56 | 0.339 | 1.17 | 1.17 | 0.976 |
| MARICO LTD. | 9.16 | 10.17 | 0.612 | 0.77 | 0.32 | 0.041 |
| RELIANCE INDUSTRIES | 2.78 | 1.33 | 0.138 | 0.60 | 0.56 | 0.767 |
| TCS LTD. | 8.07 | 6.31 | 0.072 | 0.01 | 0.00 | 0.184 |
| WIPRO LTD. | 5.48 | 4.42 | 0.515 | 0.28 | 0.20 | 0.619 |

*Source*: Authors' own calculations.

mean (up to two decimal places), as well as the probability values for the two-tailed *t*-test. We have come across different scenarios of significantly positive, insignificantly positive, and negative impact of internationalization.

Considering ROA as an indicator of financial health, Table 15.3 shows that for almost all family firms the mean ROA after internationalization seems to have fallen (except for Jet Airways and Marico Ltd.), but this difference is insignificant for 8 out of 10 firms, that is, for 80% of the firms (at 0.05 level of significance). The difference is significant only in case of two family firms: Infosys Ltd. and Mahindra & Mahindra Ltd. If we look at absolute mean ROA values, only Jet Airways and Marico Ltd. exhibited a better ROA after going international, although statistically insignificant.

With respect to EPS, we obtained mixed yet insignificant results for the mean values of EPS before and after internationalization for 80% of the firms. Marico Ltd. and Tata Consultancy Ltd., however, have shown a significant improvement in EPS after going international. When it comes to the relation between the firms' market value to book value, as measured by Tobin's Q, only one firm, Ashok Leyland, demonstrates a significant difference in this measure when compared across the periods of internationalization. A good 90% of the firms in the sample have an insignificant *p*-value for Tobin's Q.

Lastly, the D/E ratio for 90% of the sampled family firms is statistically insignificant because the *p*-values are greater than 0.05. Hence, we conclude that there is no difference in the mean values of the D/E ratio before and after going global. Ashok Leyland is the only company with significantly different mean values of the leverage ratios while Infosys has a D/E ratio of zero throughout because of no long-term debt. Although the D/E ratios actually increased in the post-internationalization years for firms like Bharti Airtel Ltd., Jet Airways, and Hindalco Ltd, the results are insignificant.

We also came across a special case of a negative D/E ratio in the case of Jet Airways. The mean D/E ratio was –26.113 in the years before going global but then improved to −1.44, which is still negative. A negative D/E ratio is a result of a negative net worth which typically occurs when a company is unable to raise money to meet its historical losses that accrue and surpass the equity. Another factor that adds to the negative net worth of Jet Airways is a very low book value of –Rs. 628.46 per share.

Overall, we can conclude that out of the sampled firms, the majority of the firms showed an insignificant difference between the pre- and post-internationalization financial performance in terms of profitability, market prospects, firm market value, as well as leverage.

Tables 15.5 and 15.6 show the results of the paired-sample *t*-test for the 10 non-family firms in our sample. For the profitability measure, ROA, the results show that for half of the sampled firms the ROA improved and for the other half the ROA deteriorated after internationalization, but these results are insignificant, except in the case of HCL Technologies. In other words, 9 out of 10 firms show an insignificant difference between their pre- and post- internationalization profitability. Same is the case with EPS, wherein although some firms have witnessed a good mean improvement in their post-foreign collaboration EPS, the results are insignificant. It

**TABLE 15.5**
**Results for Non-Family Firms**

| FIRM NAME | ROA | | | EPS | | |
|---|---|---|---|---|---|---|
| | Pre | Post | π-Value | Pre | Post | π-Value |
| ARVIND LTD | 5.61 | 3.49 | 0.2 | 13.41 | 12.09 | 0.596 |
| ASIAN PAINTS INDIA PVT. LTD. | 17.28 | 16.11 | 0.718 | 57.09 | 77.31 | 0.712 |
| AUROBINDO PHARMA LTD. | 4.80 | 12.79 | 0.226 | 15.36 | 38.45 | 0.178 |
| DABUR INDIA LTD | 18.03 | 17.56 | 0.901 | 4.24 | 6.54 | 0.182 |
| EICHER MOTORS | 2.75 | 7.59 | 0.087 | 9.64 | 126.16 | 0.026 |
| HCL TECHNOLOGIES LTD. | 11.01 | 20.17 | 0.023 | 20.62 | 66.97 | 0.078 |
| ITC LTD. | 19.10 | 20.27 | 0.625 | 8.44 | 10.27 | 0.432 |
| LARSEN AND TOUBRO | 6.71 | 3.48 | 0.063 | 78.23 | 70.94 | 0.704 |
| THERMAX | 7.40 | 4.39 | 0.202 | 26.00 | 21.10 | 0.552 |
| WOCKHARDT LTD. | −5.90 | 14.97 | 0.129 | −31.95 | 85.15 | 0.12 |

*Source*: Authors' own calculations.

**TABLE 15.6**
**Results for Non-Family Firms**

| FIRM NAME | PRICE/BV RATIO (Tobin's Q) | | | D/E RATIO (Leverage) | | |
|---|---|---|---|---|---|---|
| | Pre | Post | π-Value | Pre | Post | π-Value |
| ARVIND LTD | 1.39 | 2.71 | 0.059 | 1.16 | 0.97 | 0.433 |
| ASIAN PAINTS INDIA PVT. LTD. | 9.82 | 12.74 | 0.353 | 0.21 | 0.08 | 0.056 |
| AUROBINDO PHARMA LTD. | 2.36 | 4.33 | 0.358 | 1.12 | 0.44 | 0.061 |
| DABUR INDIA LTD | 12.51 | 12.07 | 0.781 | 0.51 | 0.23 | 0.271 |
| EICHER MOTORS | 1.38 | 4.56 | 0.152 | 0.38 | 0.03 | 0.131 |
| HCL TECHNOLOGIES LTD. | 3.63 | 4.94 | 0.024 | 0.42 | 0.04 | 0.061 |
| ITC LTD. | 8.25 | 8.26 | 0.994 | 0.01 | 0.01 | — |
| LARSEN AND TOUBRO | 5.23 | 2.78 | 0.253 | 1.18 | 1.73 | 0.162 |
| THERMAX | 5.46 | 4.60 | 0.53 | 0.10 | 0.23 | 0.413 |
| WOCKHARDT LTD. | 166.34 | 5.56 | 0.418 | 429.74 | 0.51 | 0.414 |

*Source*: Authors' own calculations.

is just Eicher Motors that has a significantly different value of EPS before and after internationalization, and that too in a positive direction.

Similarly, only 1 out of the sample of 10 non-family firms showed a significant result for Tobin's Q. Wockhardt Ltd. shows a drastic fall in the mean value of Tobin's

**TABLE 15.7**
**Proportion of Family and Non-Family Firms Which Showed No Significant Difference in Their Pre- and Post-Internationalization Financial Performance**

| Type of Firms | ROA | EPS | Tobin's Q | Debt/Equity |
|---|---|---|---|---|
| Family firms | 0.8 | 0.8 | 0.9 | 0.9 |
| Non-family firms | 0.9 | 0.9 | 0.9 | 1 |

*Source*: Authors' own calculations

Q post going global due to the fall in its market value vis-a-vis book value of total assets, but the results are still insignificant.

With respect to the D/E ratio, surprisingly, none of the firms have a significant difference. In other words, of all the non-family firms included in the sample, none show significant difference in their leverage after stepping out of their home countries, although there are certain firms (such as Wockhardt Ltd.) which have witnessed a steep downfall in their leverage.

Overall, it can be said that although the mean values of the variables under consideration differ a lot pre and post internationalization, these differences, for the majority of the non-family firms, are insignificant.

After conducting a paired-sample *t*-test for all the family firms and non-family firms in our sample, the test for difference between proportions was employed manually in order to determine if there was any difference between the proportion of these two types of firms that have insignificant difference in their pre- and post-internationalization financial performance. Table 15.7 provides an overview of these proportions for all the four variables of financial performance in our study.

For the measures of profitability (ROA), market prospects (EPS), and leverage (D/E ratio), the proportion differed significantly among the family versus non-family firms. This means that if both family and non-family firms were to go global, there is a greater chance that family firms would witness a difference in their profitability, market prospects, and leverage after internationalization, instead of non-family firms. The result is opposite in case of the relative measure of firms' market value (Tobin's Q). In case of this variable, there was no significant difference between the proportion, as is very clear by the fact that for Tobin's Q, the proportion of family and non-family firms with no difference in pre- and post-financial performance is exactly the same in our sample.

## 15.5 CONCLUSION

The previous literature has no dearth of studies showing the impact of foreign collaborations, mergers, and acquisitions on the performance of firms—both financial and non-financial. But this study has specially focused on the family firms, and their financial performance post going global vis-à-vis non-family firms. Although there is no single concrete definition of family firms, we have combined various definitions to pick up the gist. In simple words, the term "family firm" refers to a firm in which a majority shareholding or a major controlling stake or a large part of the decision-making (through board of directors or otherwise) lies with a single family or two families together.

This study employed various measures of firm financial performance, in terms of profitability, market prospects, firm book value in relation to market value, and leverage, and compared these dimensions three years before and after a firm opted for stepping out of its home country for commercial purposes. We found that a large proportion of both, family as well as non-family firms, have not shown any significant difference in their financial performance after undergoing internationalization. Moreover, comparing these two types of firms, we couldn't find a significant difference in the proportion of those firms that are insignificantly different in their pre- and post-internationalization market value (as pointed out in the first phase of the study). On the other hand, these proportions were different for the variables of profitability, market prospects, and leverage, and it seems that relatively more number of family firms are likely to witness a difference in their financial performance as a result of internationalization, when compared to family firms.

## 15.6 LIMITATIONS AND FUTURE SCOPE OF RESEARCH

Like all studies, our study also has certain limitations, which form a basis for future research. We only used quantitative variables for measuring financial performance of firms. Some qualitative variables can also be used for supplementing the research. We considered 10 family firms and 10 non-family firms that internationalized during the period 2009–2010 to 2017–2018. A larger sample size could have led to more representative results.

For measuring various dimensions of financial performance such as profitability, leverage, efficiency, etc., more variables such as return on equity, net margin ratio, and the like can be used. We used two-tailed tests in order to see if there is any difference between the performance of firms across the two time periods and across the two type of firms. In future studies, one-tailed tests can be employed to determine if one type of measure or one type of firms have shown an improvement over the other. We have taken three years before and after internationalization to compare the pre- and post-performance. A longer time period can be used for further studies.

## REFERENCES

Abbas, Q., Hunjra, A. I., Azam, R. I., Ijaz, M. S., & Zahid, M. (2014). Financial performance of banks in Pakistan after merger and acquisition. *Journal of Global Entrepreneurship Research, 4*(1), 13.

Abdulwahab, B. A., & Ganguli, S. (2017). The impact of mergers and acquisitions on financial performance of banks in Bahrain during 2004–15. *Information Management and Business Review*, 34–45.

Al-Hroot, Y. A. (2016). The Impact of mergers on financial performance of the Jordanian industrial sector. *International Journal of Management & Business Studies*, 2230–9519.

Barutha, P., Jeong, D., Gransberg, D., & Touran, A. (2018). Impact of collaboration and integration on performance of industrial projects. *Construction Research Congress*, 755–762.

Chang, J. (2007). International expansion path, speed, product diversification and performance among emerging-market MNEs: Evidence from Asia-pacific multinational companies. *Asian Business and Management*, 331–353.

Chang, S-J., & Rhee, J. J. (2011). Rapid FDI expansion and firm performance. *Journal of International Business Studies*, 979–994.

Das, K. C., & Banik, N. (2015). What motivates Indian firms to invest abroad? *International Journal of Commerce and Management, 25*(3), 330–355.

Daszkiewicz, N., & Wach, K. (2014). Motives for going international and entry modes of family firms in Poland. *Journal of Intercultural Management*, 5–18.

Dickerson, A. P., Gibson, H. D., & Tsakalotos, E. (1997). The impact of acquisitions on company performance: Evidence from a large panel of UK firms. *Oxford Economic Papers*, 344–361.

Donckels, R., & Fröhlich, E. (1991). Are family businesses really different? European experiences from STRATOS. *Family Business Review*, 149–160.

Gallo, M. A., & Sveen, J. (1991). Internationalizing the family business: Facilitating And restraining factors. *Family Business Review*, 181–190.

Gracia, R. G., Canal, E. G., & Guillen, M. F. (2015). Accelerated internationalization and performance: A resource-based view framework. *Academy of Management Proceedings.*

Graves, C., & Thomas, J. (2004). Internationalisation of the family business: A longitudinal perspective. *International Journal of Globalisation and Small Business, 1*(1), 7–27.

Graves, C., & Thomas, J. (2008). Determinants of the internationalization pathways of family firms: An examination of family influence. *Family Business Review, 21*(2), 151–167.

Hajela, A., & Akbar, M. (2013). Internationalization of small and medium software firms from India. *International Journal of Technological Learning, Innovation and Development, 26*(1–2), 88–101.

José Mas-Ruiz, F., Luis Nicolau-Gonzálbez, J., & Ruiz-Moreno, F. (2002). Foreign expansion strategy and performance. *International Marketing Review, 19*(4), 348–368.

Knežević, B., & Wach, K. (2014). *International business from the central European perspective.* Zagreb: University of Zagreb, Faculty of Economics and Business.

Larasati, N. D., Agustina, Y., Istanti, L. N., & Wijijayanti, T. (2017). Do merger and acquisition affect on company's financial performance? *Sriwijaya International Journal of Dynamic Economics and Business*, 2581–2912.

Leach, P., & Bogod, T. (1999). *Guide to the family business.* Kogan Page Publishers. Retrieved from https://books.google.co.in/books?hl=en&lr=&id=ZxMk-ylcXLMC&oi=fnd&pg=PR11&dq=family+leach&ots=4tpSctlP1N&sig=avF-d4dJ5sc2p5DRTSawHLaAxeE&redir_esc=y#v=onepage&q=family%20leach&

Lee, T., Chan, K. C., Yeh, J. H., & Chan, H. Y. (2010). The impact of internationalization on firm performance: A quantile regression analysis. *International Review of Accounting, Banking and Finance*, *2*(4), 39–59.

McDougall, P. P., & Oviatt, B. M. (1996). New venture internationalization, strategic change, and performance: A follow-up study. *Journal of Business Venturing, 11*(1), 23–40.

Mensching, H., Calabrò, A., Eggers, F., & Kraus, S. (2016). Internationalization of family and non-family firms: A conjoint experiment among CEOs. *European Journal of International Management, 10*(5), 581–604.

Patel, R. (2018). Pre & post-merger financial performance: An Indian perspective. *Journal of Central Banking Theory and Practice*, 181–200.

Pazarskis, M., Vogiatzogloy, M., Christodoulou, P., & Drogalas, G. (2006). Exploring the improvement of corporate performance after mergers – The case of Greece. *International Research Journal of Finance and Economics*, 184–192.

Rahman, R. A., & Limmack, R. J. (2004). Corporate acquisitions and the operating performance of Malaysian companies. *Journal of Business Finance & Accounting*, 359–400.

Rambe, P., & Agbotame, L. A. (2018). Influence of foreign alliances on the performance of small-scale agricultural businesses in South Africa: A new institutional economics perspective. *South African Journal of Economic and Management Sciences*, 2222–3436.

Rani, N., Yadav, S. S., & Jain, P. K. (2016). Short-term performance of mergers and acquisitions. *Mergers and Acquisitions,* 37–108.

Sharma, D. S., & Ho, J. (2002). The impact of acquisitions on operating performance: Some Australian evidence. *Journal of Business Finance & Accounting*, 155–200.

Shrestha, M., Phuyal, R. K., & Thapa, R. K. (2017). A comparative study of merger effect on financial performance of banking and financial institutions in Nepal. *Journal of Business and Social Sciences Research*, 47–68.

Tao, Q. J., & Jiang, R. H. (2014). Pace of expansion and FDI performance: The case of auto FDIs in China. *American Journal of Industrial and Business Management*, *4*(10), 614.

Wach, K. (2014). Familiness and born globals: Rapid internationalization among polish family firms. *Journal of Intercultural Management*, *6*(3), 177–186.

# 16 The Role of a Responsible Global Citizen (*Gitizen*) in the 21st Century

*Robert Seinfield and Uchit Kapoor*

**CONTENTS**

## 16.1 INTRODUCTION

> "Global Citizenship gives us a profound understanding that we are tied together as citizens of the global community, and that our challenges are interconnected."
>
> —Moon-Ki-Ban, 2008, Ex-UN Secretary-General

Before the outbreak of the First World War, the world must have looked very small to economist John Maynard Keynes, as he aptly writes in his famous 1919 essay, "The Economic Consequences of Peace", "The makers of the world powers could sit in their bed room, sipping morning tea, and order the delivery of the products of future society".

Society is at the stage of enlightenment. We have never been more educated, rich, interconnected, or technology driven. According to the Russian scientist Kardashev (1924), we are at stage 0.2 of the civilization index, reflecting development and utilization of 20% of the planet's energy resources. But if we pay even a marginal amount attention to the state of affairs, we can see that we face global challenges because we lack the solidarity, motivation, and institutional support to solve them.

O'Brien (2000) designates the present period as the "Anthropocene," the first period in geological history where human activities have a significant impact on the Earth system. Trade, production, and consumption are the root cause of all our problems, and they can be eliminated by weaving together the concept of global citizenship.

The concept of global citizenship has evolved over time. In the 1800s, only men and property were included in the concept of citizenship, and that too only national. With the advent of 1900s, women and children also were included. Today, the concept has crossed borders with the liberalization of economies and increased awareness, enactment, and acceptance of political, civil, and social rights and obligations. Current perspectives on citizenship vary in different countries, whether listed or unlisted, and include social rights and obligations and dimensions of national and global citizenship derived from demographic changes, historical regimes, and economic development in various geographic regions. Further, the concept of global citizenship has been enhanced by the establishment of international trade blocks, transnational organizations and corporations, civil society norms and movements, and human rights frameworks. It has to be acknowledged to what extent the concept of global citizenship extends and compliments traditional citizenship (Gies & Wall, 2019).

The aim of this chapter is to spread awareness about the concept of global citizenship, study the challenges that lie ahead, identify methods to eradicate the imminent dangers lurking in the dark, outline the scope of future research, and overcome the limitations to further explore the subject and permanently weave it with the society.

## 16.2 OBJECTIVES

This study has two levels of objectives: national and international.

### 16.2.1 National Objectives

#### 16.2.1.1 To Develop a Sense of Shared Destiny

The first objective of the study is to develop and foster a sense of shared destiny in terms of local, national, sociocultural, political, civil, and political factors amongst various members of national and international associations. Team spirit should remain undeterred in times of prosperity, calamity, or deluge. For achieving this purpose, the process of socialization should be started involving a range of social agents such as the family, peers, and religious groups in an explicit manner. The process of socialization has three stages: arrival, encounter, and metamorphosis which is the last stage after which a concept, individual, idea, or organization is finally accepted in the society.

Formal education is not the only means of spreading socialization; we need a robust public policy at the heart of all social activities. Indeed, beyond mere reiteration of theoretical fundamentals and principles, there is a need for building national and international cohesion by the education bodies (Israel, 2012).

#### 16.2.1.2 To Engage in Civic and Social Action

Active and positive societal participation and transformation for realizing a broader sense of individual and group responsibility towards communities, religions, sects, cultures, regions, nations, and the world as a whole should be articulately embedded in the curriculum, policy, and design of the education system. To set the context, in any given country, this area overlaps many others, such as history, geography, religion, social studies, civics, and political science. Children and adults should be equally aware of the concept to be engrossed in it.

#### 16.2.1.3 To Stratify the Role of Women in Society

It is imperative for a society to create an environment of overall independence with role stratification. It includes increasing and improving the social, economic, political, and legal strength of women—to ensure equal rights to women; to make them confident enough to claim their rights; to freely live with a sense of self-worth, respect, and dignity, both inside and outside of their house and workplace; to make their own choices and decisions; to have equal rights to participate in social, public, and religious activities; to have equal social status in the society; to have equal rights for socioeconomic justice; to determine financial and economic choices; to have equal opportunity for education; to have equal employment opportunity free of any gender bias; to have a safe and comfortable working environment where women have the right to have their voices heard. In the process of stratification of women's role in society, the following objectives with a degree of result orientation have to be kept in mind—employability, talent acquisition, skill enhancement, and replacement in certain job roles that were initially designed only for men.

#### 16.2.1.4 To Imbibe Intercultural Justice

Each nation a country chiefly characterized by identity, cultural diversity, world heritage, arts, languages, world history, indigenous knowledge systems, peace and conflict resolution, learning to live together, education, and intercultural/international understanding. The formulation of values, ethics, and learning outcomes in youth and adults can be differentiated from more modest values like empathy and care to those that are more committed in nature, such as willingness to challenge injustice. It is to be noted that issues of peace and conflict are rooted in the culture. Violent conflicts often and arguably revolve around political, economic, social, cultural, and natural resources. Cultural identities and differences serve as ideological opinions and solutions in context of conflict resolution and imbibing intercultural justice. The poor and downtrodden should be uplifted by the implementation of the above learning outcomes (OXFAM, 2006).

#### 16.2.1.4 To Face All Sorts of Challenges

To face the challenges of a globalized world, the issues of society, economy, and polity have to be understood, studied, and analyzed and solutions have to be crafted. A global village is the best place to live in. Awareness of the wider world and one's own role both as a citizen and consumer with rights and obligations is the foundation stone for the construction of the formidable fortress and realization of the concept of global human community or global citizenship. One of the key deliverables in the concept of global citizenship is the valuation of the cultural diversity comprising languages, arts, religions, and philosophies as components of the common heritage of humanity.

### 16.2.2 International Objectives

#### 16.2.2.1 Commitment to Sustainable Development

The world today is facing a considerable number of interlinked challenges – the global financial crisis, climate change, shrinking biodiversity, declining water and energy resources, threats to food sources, and health risks. The present anthology keeps us revolving knowledge that is connected to power, history, and cultural differences. But, nowadays, mobility and access to new media and academic collaborations have democratized knowledge. Thus, we have to reduce, reuse, and recycle for a sustainable future. Crude oil will totally vanish from the face of the planet by 2050 and the next-generation fuel, ammonium oxide, whose atoms lie tucked up in ice molecules, has been found in sufficient quantity in the polar ice caps, primarily in Alaska and Greenland. Some speculate that there are huge chunks of it buried in Antarctica as well. This next-generation fuel is enough to run the planet for the next 200 years. This is one of the major examples of sustainable development; the concept itself was derived from energy sector.

#### 16.2.2.2 Environmental Safety and Awareness

We live in a world without borders as we knew them. We might be in closer interaction with a fellow academic across the Atlantic Ocean than we are with our own neighbors. Through technological development and means of communication and transportation, we can choose to interact with any part of the world. This means that we might think of ourselves as international rather than national beings (Benhabib 2006). 2nd World Conference on Environmental Safety, Geneva Protocol.

There is an imminent necessity to act as local entrepreneurs, or better as flexible citizens, both nationally and internationally at the same time as the definition of global citizenship puts it and this is possible only through environmental safety and awareness. Our students, youth, and the entire academic staff will have to be deployed to educate people about poverty, climate, human rights, religious dialogue, global health governance and safety, as well as the significance of being a citizen in today's well-connected world. Environmental safety and awareness programmes should be made a part and parcel of routine life. People should be taught how to behave and deal in the event of a natural calamity both at the grassroots level and at the communal level. The safety and awareness concept only accumulates and forms the content of the concept of global citizenship and is also linked personal health and hygiene.

#### 16.2.2.3 Ethical Responsibility and Engagement

Human rights approaches, attitudes and values of caring for others, transformation of personal and social responsibility, societal contribution, skill development through a well-informed, ethical, and peaceful action for exploring own and other's values is the most viable mechanism for sociocultural and political decision-making at the local, regional, national, and global levels.

#### 16.2.2.4 Upheaval of Global Human Values

Human values can be changed only when informed and critically literate people from all corners of society come together and create an atmosphere of fostering such value systems, completely understanding the underlying assumptions and power dynamics of interwoven cultures.

A "spiral-societal" approach is needed which can be elaborated by the inclusion of global citizenship education at all levels of education: primary, secondary. and higher. Teachers should be well versed with the concept, its requirements and scope of implementation, and further study and research. Levels of education and student age groups vary between countries; these groups are merely meant to be indicative. Users at all levels should feel free to select, adapt, and organize their own learning objectives in ways that are suited to their country, context, and preparedness.

#### 16.2.2.5 Social Connectedness and Respect towards Diversity

People differ in their levels of identity and thoughts, the factor which entwines them or separates them is the diversity and that entails a certain degree of respect which is the chief characteristic of this planet. Diversity is characterized by differences in psychographic parameters such as style, appearance, knowledge, loyalty, etc., which results in differences of opinion. Other kinds of diversities are political diversity, biodiversity, cultural diversity, economic diversity, zoological and botanical diversity, and above all historical and evolutionary diversity. Now, it depends on us whether we want to harness this aspect or close our windows. It is this diversity that makes us beautiful. The only way to garner respect for diversity and spur positive change is global citizenship.

## 16.3 LITERATURE REVIEW

The concept of global citizenship starts with organizational citizenship behavior, that is relationships with peers, subordinates, supervisors, and the entire surrounding environment with immediate and surround effect and impact on society's health as aptly quoted by Mossai (1997). According to T. D. Allan (2007), post the era of spiral-societal approach, the concept of global citizenship became a public deliberation unbounded to the political, jurisdictional, and territorial boundary of a geographical state resulting in creation and protection of global rights and responsibilities. Sentiment and awareness about global citizenship has increased as concern for society, culture, policy, and academia forms the essence of the concept of global citizenship. Davies (2009) mentioned that democratic decision-making and community service build resources, time, and energy of the country. The concept of global citizenship has also been mentioned in a speech by the US president. He said that we have drifted apart and have forgotten our shared destiny but that the burdens of global citizenship continue to bind us together. Our candidates must be equipped with social, ethical, and civic competences, with initiative and with entrepreneurial spirit. We must provide a learning environment and an atmosphere that are conducive to self-development. The Nobel Laureate Eugene Stieglitz (2009) elucidates the concept in the most intricate manner by saying that global citizenship can be seen as a modern version of "bill-dung" style of society, stimulating autonomy and critical thinking, and fostering an ability to gauge the interdependence that characterizes the world of today. Not everybody respects diversity, shoulders the responsibility, shares resources and information, empathizes, connects openly, competes healthily and dies graciously, which is the essence of global citizenship and emulates the importance of living together.

## 16.4 RESEARCH TOOLS AND RESULTS

### 16.4.1 KMO-Bartlett's Test

Here from Table 16.1 we can see that the significance level remains same for every change in the sample. We have used SPSS to do the analysis. Accuracy level changes but remains within a specified range, owing to the non-shift in the geopolitical climate. Chi-square value changes signify co-relation between different values and preferences of sub-groups of people. A group of 150 people was chosen from Wisconsin, Washington, and California.

**TABLE 16.1**
**KMO-Bartlett's Test Results**

| Generic Preferences | KMO Measure Of Sampling Accuracy | Bartlett's Test of Sphericity Approximation | |
|---|---|---|---|
| | | Chi-Square | Significance Level |
| **Linguistic preferences** | 0.826 | 338.780 | 00.00 |
| **Logical/mathematical preferences** | 0.868 | 707.817 | 00.00 |
| **Visual/spatial preferences** | 0.820 | 677.971 | 00.00 |
| **Musical preferences** | 0.848 | 980.437 | 00.00 |
| **Kinesics preferences** | 0.858 | 386.436 | 00.00 |
| **Natural preferences** | 0.825 | 400.003 | 00.00 |
| **Interpersonal preferences** | 0.867 | 608.780 | 00.00 |
| **Intra-personal preferences** | 0.892 | 142.096 | 00.00 |

**TABLE 16.2**
**Results of Paired Sampling with Measures of Central Tendency**

| Compared Study Variables (N = 49) | Mean Difference | Standard Deviation | *t*-Test | *p*-Value |
|---|---|---|---|---|
| **Average of pre-adoption period** | .48 | 1.28 | 2.62 | .01* |
| **Average of 1 year post adoption** | .31 | .80 | 2.76 | .01* |
| **Average of 2 year post adoption** | .64 | .165 | 2.73 | .00* |
| **Average of 3 year post adoption** | .63 | 1.33 | 3.34 | .00* |

* Significant($p$ = .05, two-tailed).

### 16.4.2 Paired Sampling with Measure of Central Tendency of Pre- and Post-Adoption Periods of Liberal Government Policies

From Table 16.2 we observe the following. The trend shows $p$-value greater than .05. There is a significant difference between averages of pre- and post-adoption periods. The pre-ESOP window is also under comparison. There is more subtle difference between 2 and 3 year difference averages 3 year difference window is the widest.

## 16.5 CHALLENGES

Global citizenship has two categories of challenges. Some of the main challenges have arisen as consequences of globalization, a term used to describe the relationship between states and communities created by geographical spread of ideas and norms due to the increased speed of travel. The necessity of trade between different regions resulted from a richness in natural resources and increasing population levels. It gave rise to new trade routes, such as the famous Silk Road, the Suez Canal, etc. Unsustainable levels of carbon dioxide emissions in atmosphere have made life miserable for all forms inhabiting the planet. This is the most pressing challenge. Many species are going extinct before they are even discovered. E. O. Wilson refers this phenomenon as the Eremocene, the Age of Loneliness. Next is the inter-connection of financial crises and ecological aberrations that make the world suffer pain agony when due to crunch and unemployment, the society faces a brutal consequence and collapse of machinery and all constituent units. The ultimate safeguard against this is prediction and prevention (Lioudis, 2019). The inability to restrict the spreading of epidemics following natural disasters such as the devastating tsunami of 2004, the spread of Ebola in the fall of 2014, the unprecedented health risks, perils, and hazards presented as imminent dangers, particularly to the undernourished, portend future challenges. In the 19th century, people, goods, and services used to freely move across borders, but due to historical interventions and excavations, the concept of global citizenship has given rise to a new danger—global terrorism—which has its roots in commerce and trade regimes of today. The freedom and ability to cross borders can be both a blessing and a curse, resulting in far-ranging consequences. We as global citizens have to free ourselves from prejudices of actions and words. The collapse of judicial and constitutional machinery of at least 10 countries in the fall of 1914 was a concrete example of the dire and bizarre results of an over-globalized world. Our world system is not prepared to enact and implement global commercial, political, and legal regimes due to religious, cultural, economic, and philosophical differences between various sects of society. There is an absence of acceptance of global norms, tolerance, and expertise.

Secondary challenges are a result of modern or proto-globalization leading towards increasing trade links and cultural exchanges across the world during early 19th century. Mostly, this was due to European empires that had for more than a century controlled a large portion of land area itself and had in turn impacted and created modern, quick, populist, and rather complex systems which were produced in medieval Europe after the sinking of the Spanish Armada and the start of the Renaissance (1535–1875). Some notable secondary challenges stand as non-coherent states on grounds of religion. Religion is product of history and some out-worldly incidents which have shaped our entire social structure in due course of time. It is so close to people's sentiments that the membrane is difficult to crack and penetrate into settling new embodiments there. There is also a level of difficulty in bounding nations due to differences in language, art, culture, customs, and habits which are more of individual in nature than communal. The difference in language is a major bottleneck and barrier in communication between the states and people. Diversity in geo-politics is another striking impediment in the way of global citizenship. Geographical movements and migrations have always been there since prehistoric times. Some regions, whether naturally or due to manmade reasons, are more accessible than others. This is the major reason for development of some regions more than others. An innovation in transport technology gives a huge impetus to cross-border trade. Those states that have superior transport infrastructure have better access to trade and commerce, filling bankers' pockets as compared to those states that are still evolving. Last but not the least, the military also creates regions that are almost inaccessible to people or difficult to cross (e.g. Nevada Areas 51 and 52, various military installations in the US). These regions have been created for national security, prowess, and military muscularity. This strategy is adopted mostly by those countries which have a long border, vicious neighbor, or vast economic assets to protect and are used for deployment of tactical weapons, weapons of mass destruction, and surveillance.

From Figure 16.1, it is clearly observed that during the 13th century most tribes were nomadic and inhabited Central Asia, Europe, China, and Mongolia. Later, due to migratory movements around 14th and 15th centuries, the population base shifted to the Middle East, North America, Africa, and Indian sub-continent, from where today's population trends began. This happened mainly due to change of geo-climatic zones, modernization, and increased transportation and infrastructure.

It is concluded from Figure 16.2 that fidelity is considered to be a crucial factor in deciding the behavior of society in different countries for the obvious reason that acceptable and unacceptable behaviors form the structural edifice of the concept of global citizenship. A scale of 0 to 100 is chosen for studying the proximity of acceptable/unacceptable behaviors of different countries which can also be related to the establishment of tolerance zones. From the

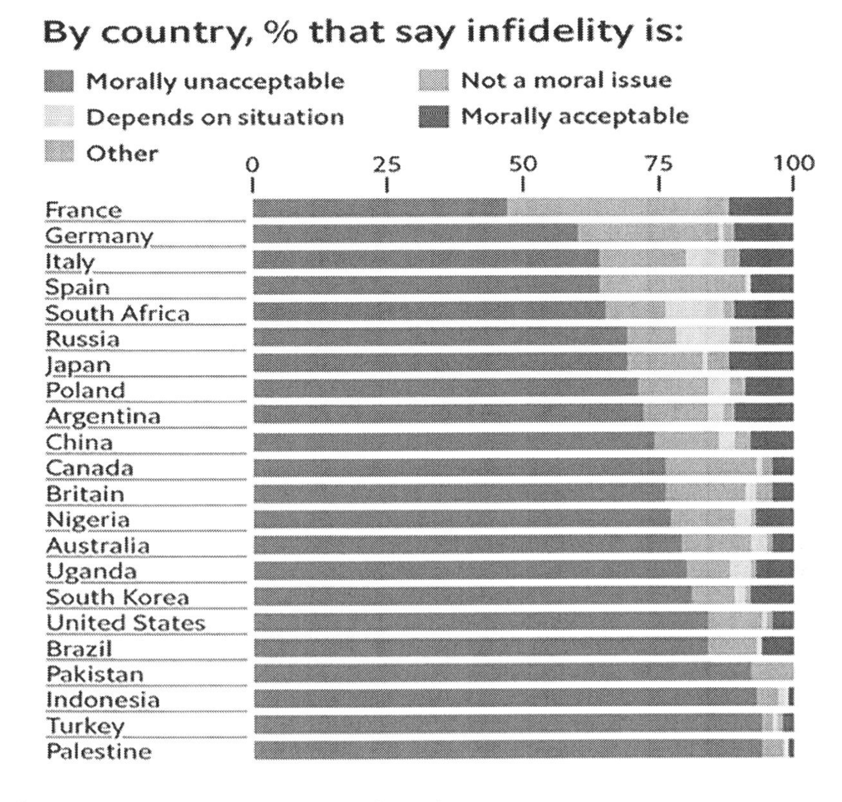


**FIGURE 16.1** Fidelity Rates in Major Countries.

*Source*: *Journal of Education* (2011).

densely populated and globally related zones across of the world, a sample of 22 countries was chosen to study the trend of global citizenship on the basis of acceptable and unacceptable behaviors of social structures operating in them which in turn depend upon psychographic parameters like lifestyle, education, personality, awareness, and loyalty. Based upon response generated by public and analysis of trend, it is finally concluded that France has 47.5% morally unacceptable behavior, 40% not a moral issue and 12.5% morally acceptable behavior. Germany has 54% morally unacceptable behavior, 37.5% not a moral issue, 1.5% depends upon situation, and 11% morally acceptable behavior. Italy has 56% morally unacceptable behavior, 22% not a moral issue, 10% depends upon situation, and 12% morally acceptable. Spain has 56% morally unacceptable behavior, 33% not a moral issue, 1% depends upon situation, and 10% morally acceptable behavior. South Africa has 57% morally unacceptable behavior, 19% not a moral issue, 10% depends upon situation, 1.5% other reasons, and 12.5% morally acceptable behavior. Russia has 64% morally unacceptable behavior, 13% not a moral issue, 10% depends upon situation, 3% other reasons, and 10% morally acceptable. Japan has 64% morally unacceptable, 17.5% not a moral issue, 1.5% depends upon situation, 4.5% other reasons, and 12.5%

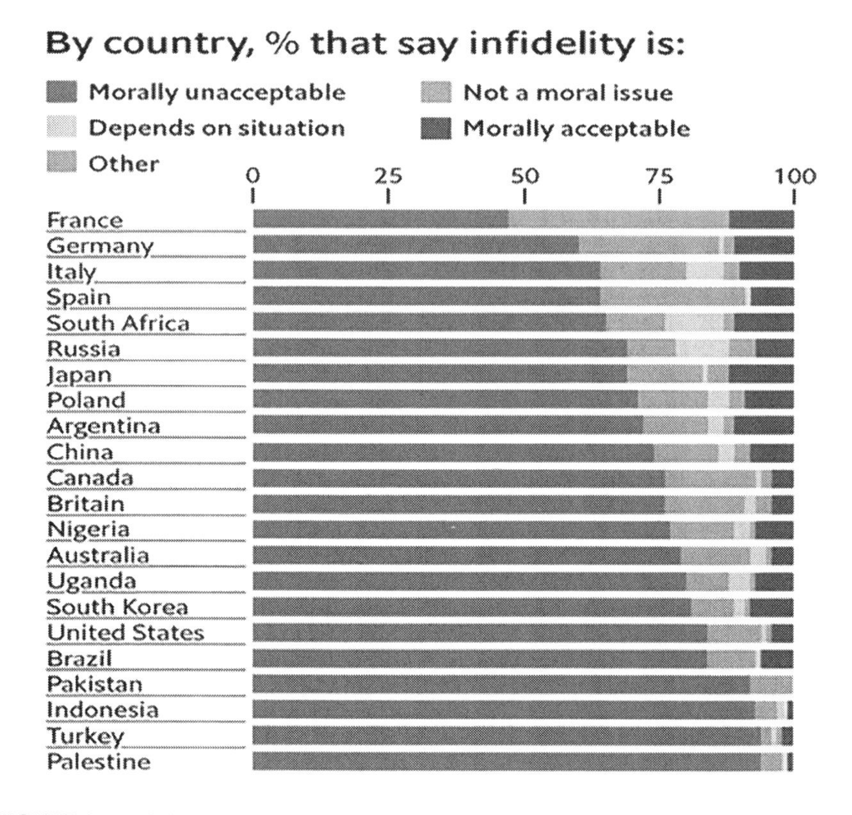


**FIGURE 16.2** Migratory Trends and Inhabited Regions in the 13th Century A.D.

*Source*: *Journal of Global Optimization* (1997).

morally acceptable. Poland has 66% morally unacceptable behavior, 11.5% not a moral issue, 3% depends upon situation, 2% other reasons, and 7.5% morally acceptable. Argentina has 67% morally unacceptable behavior, 17% not a moral issue, 3% depends upon situation, 2% other reasons, and 11% morally acceptable behavior. China has 68% morally unacceptable, 19% not a moral issue, 2% depends upon situation, 2% other reasons, and 9% morally acceptable behavior. Canada has 69% morally unacceptable behavior, 27.5% not a moral issue, 0.5% depends upon situation, 0.75% other reasons and 4% morally acceptable behavior. Britain has 69% morally unacceptable behavior, 23.5% not a moral issue, 1.5% depends upon situation, 2% other reasons, and 4% morally acceptable behavior. Nigeria has 69.5% morally unacceptable behavior, 29.75% not a moral issue, 2.5% depends upon situation, 0.25% other reasons, and 8% morally acceptable behavior. Australia has 70.5% morally unacceptable behavior, 22.5% not a moral issue, 2.75% depends upon situation, 0.25% other reasons, and 4% morally acceptable behavior. Uganda has 71% morally unacceptable behavior,

20.25% not a moral issue, 1.8% depends upon situation, 0.2% other reasons, and 6.75% morally acceptable behavior. South Korea has 71.5% morally unacceptable behavior, 20.75% not a moral issue, 1% depends upon situation, 0.25% other reasons, and 7% morally acceptable behavior. The US has 73% morally unacceptable behavior, 21.5% not a moral issue, 0.75% depends upon situation, 0.75% other reasons, and 4% morally acceptable behavior. Brazil has 73% morally unacceptable behavior, 20.25% not a moral issue, 0.25% depends upon situation, and 6.5% morally acceptable behavior. Pakistan has 89% morally unacceptable behavior and 11% not a moral issue. Indonesia has 89.5% morally unacceptable behavior, 9% not a moral issue, 1% depends upon situation, and 0.5% morally acceptable behavior. Turkey has 92% morally unacceptable behavior, 3% not a moral issue, 2% depends upon situation, and 3% morally acceptable behavior. Palestine has 92% morally unacceptable behavior, 6% not a moral issue, 1% depends upon situation, and 1% morally acceptable behavior. Turkey and Palestine have highest morally unacceptable behavior, i.e. 92%, and Japan as the highest morally acceptable behavior, i.e. 12.5%. Pakistan has a bizarre state of affairs with 89% morally unacceptable behavior and rest 11% not a moral issue at all without the other categories.

## 16.6 FUTURE

This subject needs further education and exploration by academicians to identify the global concerns of the global citizens. Each country of a specific trade or regional block has its own advantages, historical baggage, and demerits on account of population, such as in most Asian countries and occupation. Occupation often depends upon geography and abundance of natural resources in a state, which, in turn, decide the macroeconomic parameters of industry. Actionable solutions have to be implemented at the international level to guard against controversies and resentments. Major issues that need to be highlighted include global poverty, communicable diseases, education, migration, climate change, access to water, international trade, hunger, financial stability, and regional conflicts. These areas need extensive research to be brought to the notice on a global scale. Lead education bodies should conduct joint studies and government should provide sufficient avenues for grants boosting the entire process.

## 16.6 LIMITATIONS

Poverty and inequality limit research in this area. Regional data due to political influence is often distorted, and accurate figures cannot be arrived at most of the times, which hampers the final solution to be implemented or suggested. It is also difficult to access government research labs. Trade reforms, malnutrition, and human development create impediments in the way of research on the topic due to migratory trends exhibited across the world.

## REFERENCES

Allan, T. D. (1998). Global citizenship: A political cemetery. *Journal of Civilization*, *12*(3), 45–55.

Benhabib, S. (2006). Proceedings from 2nd WCES'06. *The Second International Conference on World Environment and Safety*. 2 October. Retrieved from https://www.brown.edu

Butenko, S. (1997). The migratory trends and consequences. *Journal of Global Optimization*, *12*(3), 35.

Davies, L. (2009). *Humans and the Homo Sapiens*. A Documentary. 5 July.

Freud, S. (1905). Cognitive Dissonance: The various types and reasons so far. *Journal of Cognitive Learning, 23*(1), 45.

Gies, W. T., & Wall, C. (2019). The Eighteenth Centuries: Global networks of enlightenment. *Journal of Global Intellectual History, 10*(2), 3.

Israel, R. C. (2012). What does it take to be a Global Citizen? *Kosmos Journal of Global Transformation, 45*(1), 22.

Kardashev, N. (1984). On the possibility of inevitability and possible structures of civilizations. *Sizing Up the Civilizations*, *15*(3), 497–507.

Keynes, J. M. (1919). The economic consequences of peace. *The Library of Economics and Liberty*. 5 February. Retrieved from https://www.econlib.org/library/Enc/bios/Keynes.html

Lioudis, K. N. (2019). The collapse of Lehman Brothers: A case study. *Journal of Investment Banking, 11*(1), 34–36.

Moon, K. Ban. (2008). Proceedings from UNOAM'08: The Sixty Fourth United Nations Conference on Achievement of Millennium Development Goals. 25 September. Retrieved from https://www.un.org/millenniumgoals/2008highlevel/index.shtml.

Mossai, D. (1997). Global Citizenship: A global phenomenon. *Harvard Business Review*, *44*(3), 45.

O'Brien, K. (2000). Double exposure: Assessing the impacts of climate change within the context of economic globalization. *Journal of Global Environmental Change, 7*(1), 221–232, https://reseacrhgate.net/publication/323368867

OXFAM. (2006). https://www.oxfam.org.uk/education/resources/education-for-global-citizenship-a-guide-B – Schools.

Pablo, N. (2011). The morals of fidelity and nationalism. *Journal of Education Sciences*, *9*(2), 56.

Stieglitz, J. (2009). Famous Statement. 4 October. Retrieved from https://www.econlib.org › library › Enc › bios › Stiglitz

Taylor, F. M. (1997). The theory of global citizenship: Timeframe and history. *Journal of International Trade and Policy, 9*(2), 55–60.

# 17 Small and Medium-Sized Enterprise Networks and Their Contribution to Territorial Development

*Amina Omrane*

**CONTENTS**



## 17.1 INTRODUCTION

The community of researchers who are interested in the small and medium-sized enterprises, henceforth SMEs, have not reached consensus on one definition (Vilette, 2010). This difficulty in finding a common perception seems to be explained by the great diversity and heterogeneity encompassed by the term SME. According to this perspective, SME is synonymous with diversity, so that a good number of researchers were harnessed with the task to identify the criteria of differentiation and similarity making it possible to categorize the SMEs in more or less homogeneous typologies (Allali, 2007) in order to reconsider the conceptualization of SMEs.

In the past, family SMEs often had a negative image, since they were thought to be characterized by inertia, paternalism, and rigidity with few competitive advantages. In more recent years, this view has changed, and the family company has become rather a sign of strength, dynamism, good quality, and an indicator of economic development. From this same point of view, Basly (2007, p. 2) advanced that "the awareness of its (the family SME) economic importance in terms of participation in the national production of the countries and in terms of employment and economic and financial performance constitutes an imperative reason explaining the renewed interest for the questions which are relative for him". More particularly, innovative networks of SMEs seem to play a significant role in local, regional, and territorial development (Leducq & Lusso, 2011; Torre & Traversac, 2011; Campagne & Pecqueur, 2014).

Despite their vulnerability (due to their small size, limited resources and funds, informal human resource management function, intuitive and implicit strategy, informal communication, infrastructure barriers, close links between all the functions, etc.) (Torrès & Julien), family and non-family SMEs often have clarifying and supporting networks which seem to play an important role in developing economies (Joyal, 1997; Torrès, 2002; De Oliveira, 2016).

As an example, industrial districts or regional clusters have gained in productivity in a globalized economy. This gain is explained by their contribution to the revival and the territorial development of certain areas or cities so much "sunk without trace" or completely lacking in commercial activity (Piore & Sable, 1984; Cooke et al., 1997; Porter, 2003; Colletis-Wahl & Pecqueur, 2001; Pecqueur, 2008; Carré & Levratto, 2011).

In this perspective, certain disadvantaged geographical areas have been able to find the way to competitiveness thanks to networks of family SMEs that constitute them. Indeed, such family innovative networks are characterized by certain distinctive particularities and patterns based on their roots in the area, flexible specialization, and family socio-territorial "linking" capital (Anderson, Mansi, & Reeb, 2003; Habbershon, Williams, & MacMillan, 2003; Angeon et al., 2006; Allouche, Amann, Jaussaud, & Kurashina, 2008; etc.).

The current study addresses the following main research questions:

1. What are the main characteristics of SMEs (family and non-family ones) installed in specific territories?
2. To what extent do these SMEs work together and collaborate in order to compound innovative networks, enabling them to leap into new markets, ahead of their competitors?
3. How do such innovative SMEs networks contribute to the regional and economic development of their territories (technopoles, clusters, local production systems, etc.), while growing internationally?

## 17.2 SME VULNERABILITIES

Despite of the inherent weaknesses in SMEs, many scholars have pointed out that SMEs benefit from simple organizational structures that allow them to provide a high level of flexibility (Wolff & Pett, 2006). Indeed, thanks to their simple structure and few resources, SMEs can better adapt to environmental changes than big/large firms with complex structures, time-consuming activities, and costly processes (Tidd, 2001; Man, Lau, & Chan, 2002; Qian & Li, 2003; Cited by Pierre & Fernandez, 2018).

Schumacher (2001) underlined in his book entitled '*Small Is Beautiful*' (1973) that SMEs could drain multiple advantages whenever they are created to spread a policy that restrains the territory of intervention of their actions emphasizing the central role of the human being in developing small projects which create synergies between the territory capabilities and its human resources. These SMEs, which could grow via microcredits and small cooperatives, could supplement strategies and actions of big firms or "white elephants" which require heavy (and sometimes unproductive) investments. Such SMEs could also contribute to endogenous and integrated regional and national development, mainly in developing countries.

According to the group of researchers on economy and management of SMEs- called 'GREPME', based in Québec, the SME is a centralized organization that is weakly specialized and which possesses a simple internal and external information system as well as intuitive and weakly informal strategies (Julien, 2000; Marchesnay, 2014).

SMEs generally have some specific characteristics which don't enable it to evolve at the same rhythm as big firms. However, they can develop strategies in order to survive with them in a fast-changing, turbulent market.

Among the SMEs' characteristics, the following ones should be retained: the butterfly effect; the scarcity of resources; proximity management; informal, simple, and intuitive information systems; straightforward strategies; and flexible Human Resources Management (henceforth HRM) practices.

The *butterfly effect* means that the flapping of the wings of a butterfly in China could ultimately result in a hurricane in Florida. In other words, a small event that has no effect on big firms could have a great impact on a small firm. This effect is broken down into the microcosm effect, the proportion effect, and the egotrophy effect (Paradas, 2007).

The *microcosm effect* signifies that the manager focuses his attention on the short-term and immediate actions and concentrates on the geographical space that surrounds him and that is psychologically behind him.

The *egotrophy effect* implies that the manager of a SME is usually a self-absorbed person and that his way of thinking is focused only on his closest collaborators. In other words, it means that the enterprise is centralized and focused on its values, as well as the manager's personality and his own aspirations.

The *proportion effect* translates into the fact that as long as the workforce is relatively small, each actor owns a high place in the company. For example, when the entrepreneur hires a new employee, if this latter is not "the good man at the good place", that recruitment will engender many problems. It means that the impact of each movement is inversely proportional to the number of the organizational actors

(Torrès, 2003). This effect seems to derive from another one: the number effect which indicates that as long as the workforce is relatively small, the manager knows each employee well.

### 17.2.1 SME Strategies

SME strategies are intuitive, implicit, flexible, and informal (Busenitz & Barney, 1997; Mazzarol & Reboud, 2009). The strategy of an SME emanates from a small group of dominating managers/entrepreneurs who may focus on reaction, as well as a proximity management system which is based on a shorter-term strategic perspective (Bos-Brouwers, 2009; Cited by Staniewsk, Nowacki, & Awruk, 2016).

Strategic decisions which are mostly taken by the owner-manager are made in emergency situations and in an iterative process by which events could change the vision and the SME's actions). The strategic process thus relies on the activities, experience, judgement, and the way of thinking of the owner-manager (Wang, Walker, & Redmond, 2007; Lima & Filion, 2011).

The strategies that are deployed by SME managers are mainly specialization, collaboration, networking, and internationalization (Mbengue & Ouakouak, 2012). Indeed, managers often choose either strategic niches in order to evolve in a non-competitive market (Lima, 2003) or networking and collaborative strategies based on cooperation which enable them to adapt to a constantly shifting environment (Hanna & Walsh, 2002). Such strategies allow SMEs to maintain their competitive advantages, to internationalize their activities, and to supplement resources and skills they are lacking by inserting them into visionary and clarifying networks.

### 17.2.2 SME Information Systems

It is argued that the SME information systems are simple, flexible, undersized, direct, and informal (Torrès, 2002, 2004a,b; Julien, 2000). The simplicity of the internal information system is explained by the high physical proximity between the manager (who has a low interest in the strategic value of the information) and the main actors of the SME (Marchesnay, 2014). However, the simplicity of the external information system is due to the spatial proximity, that is, a nearby market which allows physically and psychologically the manager to establish direct ties with suppliers, customers, investors, bankers, etc.

Because SMEs have less bureaucracy, their organizational structure is flat. They have fewer formal systems and procedures and fewer planified activities (Wang et al., 2007). Moreover, their internal communication systems are usually direct, simple, oral, and verbal, built on trust and informal relationships (Staniewsk, Nowacki, & Awruk, 2016). This lack of formalization could be explained by SMEs' attempts to minimize the costs of operations.

### 17.2.3 SME Human Resources Management (HRM)

SMEs are simple organizations with little hierarchy; they cannot afford the same advantages as the big firms. This is due to their limited financial resources, the few social advantages they can benefit from, as well as the restrained internal mobility they can propose to their employees. Indeed, SMEs don't have the required skills in order to develop and implement the concrete and formalized tools of HRM. More precisely, HRM is linked to the manager and an omnipresent centralized power. This latter often has poor managerial skills and refuses to delegate responsibilities to his subordinates, so that the level of resistance to change is generally high in the SME (Forbes & Milliken, 1999; Mazzarol & Reboud, 2014). This issue could result from a lack of appropriate qualifications of such managers who are not willing to tolerate risk in order to adopt changes and implement new innovative solutions (Horibe, 2003; Cited by Staniewsk et al., 2016). In line with this assumption, Robert-Huot and Cloutier (2014) reported that HRM practices and procedures in SMEs are informal due to the lack of expertise in HRM methods and techniques.

As a result, the three levels of the HRM are confused and juxtaposed: the administrative level (rules and regulations, etc.), the strategic level (organizational and management choices, etc.), and the staff politics (skills, management, training, remuneration, etc.) (Bootz et al., 2011).

Hence, SME employees often suffer from a lack of creativity and reactivity in response to fast-changing needs and market expectations (Staniewsk et al., 2016).

### 17.2.4 SME Proximity Management Systems

For most SMEs, the management system is a mix of proximity: the functional, the spatial, the temporal, and the hierarchical proximities form a coherent framework which creates the conditions necessary for action and reflection (Torrès, 2004a, 2004b).

The hierarchical proximity is reflected in the absence of the middle management, the centralization of decisions, as well as the lack of authority and management delegation (Julien, 2000; Courrent & Torrès, 2005; Marchesnay, 2014).

The functional proximity is characterized by the versatility of the owner-manager who will concentrate and take on many functions (such as commercial, marketing, logistics, etc.) without a division of labor (Torrès, 2004b).

The proximity coordination implies that the methods and modes of coordination are mainly the direct supervision, the verbal and the oral communication via the usage of speech (Torrès, 2004a, 2004b, 2006).

The temporal proximity is explained by the preference for the short-term, the role of intuition in the strategy elaboration process, the near-absence of planning, as well as the quick reactivity, flexibility, and versatility of the SME.

### 17.2.5 SME Resources

Resources represent a set of available factors owned by a company, such as time, financing, material, patent, software, human skills, etc. However, capabilities of the firm represent its ability to deploy and exploit those resources according to its processes, routines, and all its activities (Amit & Schoemaker, 1993; Forsman, 2011).

Because of the scarcity of resources in terms of time, qualified staff, financing (and access to credits and loans), as well as energy, SMEs do not encourage the development of a learning culture, and often do not use a formalized and detailed strategic analysis. They have many supply difficulties nationally and internationally. Moreover, SMEs lacks the financial and temporal resources that allow them to do everything on their own internally (Rogers, 2004). Therefore, it appears that innovativeness plays an important role for SMEs with limited resources (Rhee, Park, & Lee, 2010).

## 17.3 SME NETWORKS AND THEIR ROLE IN TERRITORIAL AND LOCAL DEVELOPMENT

Despite their vulnerabilities, SMEs compensate for their lack of resources through cooperation and networking, enabling them to foster their innovative activities and processes, improving, in turn, regional development.

The openness to globalization should be done by building on the deep roots and the identity of enterprises and territorial actors while also cultivating their differences and by themselves in the big countries of the world (Godet, 1997).

In this perspective, it should be taken into consideration that the notion of "area" shouldn't be analyzed via a micro-economic perspective that considers it by the elements that compose it. It is, nevertheless, recommended to propel a meso-economic vision that emphasizes the networks, the relations, as well as the interdependences that might influence the economic evolution of the environment. In line with this reflection, it has been argued that SME networks can play an important role in the territorial development of an area by boosting the local actors to mobilize its territorial resources (Bros-Clergue, 2006).

These SME networks are composed of social actors who can interact and play a central role in the construction of a collective group and a system of governance that is well structured by social capital which grows every day. This social capital helps develop the ties of the networks with external stakeholders, especially with local/national public authorities who enable these networks to be financed and to gain in reputation and legitimacy. It also supports the guidance of information actions in order to anchor the networks in their territory and consolidate their own values and their contribution to the territory's dynamics (Bories-Azeau & Loubès, 2013).

These SME networks could also contribute to developing resources, structuring relations based on confidence, improving efficiency (information transmission, reduction of transaction costs, and opportunism-related to the network members' relations, etc.), and developing dynamic innovation capabilities as well as technological and relational learning (Joyal & Deshaies, 2000; Ferrary & Pesqueux, 2004).

SME networks are constituted by the public authorities as well as the territorial collectivities and enterprises which have the prerequisite skills and resources to construct a place of governance by elaborating diagnoses/strategies, analyzing institutional/public changes, and coordinating actions of economic/local development.

### 17.3.1 Local and Territorial Development: Genesis, Origins, and Definitions

The territory is not only a local system, a geographical and a natural space which is well organized and developed, but also a social structure and a living space that encompasses human relations between multiple actors. These actors adhere to local projects, innovate, and mobilize resources and skills thanks to their sense of belonging and to their appropriation (Maillat & Lecoq, 2006; Truda, 2007; Pacquot, 2011; Moine, 2014). According to Le loup, Moyart, and Pecqueur (2004), the territory is an open, dynamic, complex system which is socially constructed by the intersection of networks (physical, human, formal, and informal ones), strategies, as well as by the interdependence between partners. It's then a place of production, negotiation, and sharing (cited by Redondo-Toronjo, 2007).

Note that that a territory is not limited to a rural space or to the suburbs of cities or to urban areas. It is not constrained to specific environments. What is important for a territory is the local development that it implies as well as the growing awareness of its actors (even those working in organizations, Non Governmental Organizations (henceforth NGOs)., administrative structures, local collectivities, etc.), who should advocate for job creation, entertainment, citizens' security and health, elimination of geographic and socioeconomic disparities between areas, etc.

Local and territorial development is considered as a response to globalization, as well as to the socioeconomic exclusion and marginalization of some zones and capitalist politics.

This concept first emerged with regards to underdeveloped and developing countries of the Third World where decades of development initiated by UN (United Nations) organizations have failed. These failures prompted a reexamination of the role NGOs so that they work for a restrained scale of economic development, taking into consideration not only the local concerned culture, but also the resources (human, financial, etc.) made available so that energies and skills are not dispersed.

According to André Joyal (1994), local development is a socioeconomic intervention strategy by which local and national representatives of the public, the private, or the social fields work together in order to value and upgrade the financial, technical, and human resources of a collectivity as well as to combat the devitalization (or the decline) of the area. For this purpose, these representatives are called to join their forces and to associate their efforts in a sectoral or an inter-sectorial private or public work structure, guided by a central objective of local economic development. Joyal (1994) has also advanced that the regional and the territorial development holds four dimensions:

1. *Cultural dimension*: Local leadership, the level of implication of the actors-members of the area, the motivation and the willingness to work in order to enhance the development of the collectivity.
2. *Socioeconomic dimension*: Consolidating a qualitative social change (the quality of housing, access to health care and treatment, training, etc.) in order to promote a good process of local development.

3. *Environmental dimension*: Socioeconomic development of the territory should not jeopardize the quality of life in the ecosystem surrounding it.
4. *Spatial dimension*: Qualifying an endogenous development taking into consideration the geographic limit of the perimeter where actors operate while consolidating their territorial identity.

In this same logic, two aspects should be taken into consideration in order to enhance a local development:

- The participation of citizens and the cohesion between them.
- The role played by the public powers and the local collectivities and their mobilization and solidarity to face the crises as well as to preserve the territory.

In other words, certain immaterial factors, such as the social capital, the social representatives, the social capabilities of development, and socio-local governance, differ from one territory to another.

Nevertheless, in order to improve the local development of a territory, territorial dynamics should be improved and maintained by relying on three dimensions of an area (Maillat, 1994):

1. Micro-analytic dimension: Constitutes a market structure that facilitates the transaction costs limitation.
2. Cognitive dimension: Encompasses the local know-how.
3. Organizational dimension: Highlights the interdependency between the local actors.

### 17.3.2 SME Network Types and Their Contributions to Promote the Innovation of the Territory and Its Development

Carluer (2006) identified six types of networks: clusters, the technopoles, service-based spaces, innovative areas, industrial districts, and learning areas.

A *learning area* is a system dominated by immaterial components and characterized by its capacity to attract the high-demand skills in this age of knowledge-intensive capitalism. It is known as a repository and a collector of knowledge and ideas, a source of innovation and economic growth, as well as a vehicle for globalization (Florida, 1995, p. 527). Silicon Valley (in the US) is the most commonly cited example that demonstrates how entrepreneurs, skilled high-tech engineers, and computer scientists from around the world are gathered in a one common village, affording a series of related infrastructures (with common restaurants, restrooms, sport rooms, cafes, etc.). They can easily collaborate and share knowledge continuously to invent new technologies, software, applications, etc. with personalized information.

However, an *innovative/creative milieu* is a territorialized space where socially embedded interactions between economic agents and collaborative relationships between firms are growing thanks to common resources management and project-related joint activities. This concept has been developed mainly by a group of European researchers on an innovative milieu called "GREMI". Since the mid-1980s, this group of 25 researchers have put substantial efforts into theorizing and verifying milieu characteristics and effects with respect to various types of regions, investigating several European examples (Maillat, 2006 Fromhold-Eisebitch, 2004). They have pointed out how universities, research laboratories, public institutions, as well as firms, can coordinate within a limited geographical area, leading to a large number of innovating enterprises. The example cited by Fromhold-Eisebitch is related to the German region of Aachen dominated by new technology–driven firms (such as Aixtron, Parsytec, and Head Acoustics), taking the place of old ones such as manufacturing, coal, steel, and textiles. Public research centers, startups, university technology hubs, etc., contribute to the shift of this milieu.

In this same perspective, it is argued that a *service-based space* is a center of knowledge creation constituted by a set of enterprises associated to research and training centers and diverse other public or private entities whose initiatives are supported by local collectivities (learning by commuting). It is assimilated into an informal technopole, based on information and communication technologies (Cooke, 2002).

The *industrial district* is a socio-territorial entity which is marked by the presence of a community of persons and enterprises making the same product or gravitating around a certain type of production (Porter, 2000; Karlsson, Johansson, & Stough, 2005). The district of Montpellier (in France) is, for example, a local collectivity which aims to develop as good actors-relays and company incubators (such as Cap Alpha and Cap Oméga) working for the promotion and creation of innovative SMEs and strengthening connections between local actors with public and parapublic networks (like ANVAR-The national agency for research valorization, and CRITT-The regional center of innovation and technology transfer) in order to facilitate the transfer of information (Torrès, 2002).

Piore and Sabel (1989) have, for example, advanced that the networks of companies localized in the north of Italy have played an important role in the industrial success of the whole country, as a consequence of the passage from an accumulation mode to another one, via the paradigm of flexible specialization.

The *technopole/science park* is generally composed of heterogeneous actors/institutions/companies participating collectively in the conception, production, and diffusion of production processes, products, services, etc. It brings together in one location the components necessary for innovation. Globally, almost 20 technopoles compete with Silicon Valley. We can cite: Silicon Hills of Texas (US), 22@Barcelona (Spain), the Raheje IT Park of Bangalore (India), the Cambridge Science Park (England), or Dresden (Germany), etc. More precisely, in France, for example, the technopole of Montpellier, composed essentially of small companies without an industrial history, is located next to a huge university pole and some institutions of research.

In this same perspective, Torre (2006) pointed out that *localized productive systems* (LPS; a concept developed by Courlet and Pecqueur, 1996, 2013, in Grenoble and comparable to that of industrial districts) constitute networks of productive SMEs specialized around a product or a profession. Such SME networks are capable of building and developing some relationships of complementarity, collaboration, and cooperation, similar to local competitive communication nodes and links evolving in a restrained open and competitive geographic space. This view is similar to that of Porter (1990) who identified four constituents of regional competitive advantage called the diamond: (1) the production resources, (2) a local market with good quality and quantity, (3) the socioeconomic and legal environment, and (4) the local fabric rich of suppliers and related industries.

In compliance with previous assumptions, we can deduce that several clusters of enterprises are made up of networks of SMEs linked or not to a basic enterprise (and generally located near the universities conducting to the development of spin-offs). Those clusters allow enterprises to take advantage of innovative capacity development, the acceleration of the new venture creation, as well as the promotion of the productivity.

In Morocco, and especially in regions like Aïn Chock, Al Fida Derb Sultan, and Maârif, some informal family SMEs, characterized by their ethnic and familial solidarity, are active in the local market. However, they have evolved in order to become export-oriented firms. In the long run, such SMEs work for their relocation to industrial zones of Casablanca or Ben Msik.

At this level, it is essential to note that, in emergent contexts like Tunisia, competitive clusters are partly constituted by innovative networks of family SMEs which struggle to promote the regional and territorial development. The case of handicraft companies located at Sfax, the competitive clusters based in Monastir, in Borj Cedria Science Park, or in the technopoles of Sousse demonstrate how collaboration, exchange, and networking between public institutions, companies, associations, and actors can lead to regional development and economic growth.

In other words, SME networks can contribute to territorial development and innovation by (Moulaert & Sekia, 2003; Moulaert, Martinelli, Gonzalez, & Swyngedouw, 2007):

- Supporting the competitiveness of enterprises and looking for innovative and technologic partnerships with enterprises, big firms, and research centers.
- Bringing a bigger visibility to the territory in order to captivate new skills and ventures.
- Consolidating their economic fabric and reinforcing their identity.
- Animating the territory (and its competitiveness) which federate enterprises around simple and common thematics via good relays.
- Knitting a mesh of relations between firms, universities, and research centers via structured projects (with good technologic platforms).
- Participating in an innovative dynamics of the territory by promoting its attractiveness.
- Diffusing an innovative culture into the territorial structures and between the territorial actors in order to benefit from the national and international notoriety throughout the visibility of the competitive clusters.
- Accessing to a prospective vision concerning the future challenges and implications of the field.

## 17.4 INTERNATIONALIZING OF THE SME MILIEU: IMPORTANCE OF GLOCALIZATION AND LOCAL GOVERNANCE IN TERRITORIAL DEVELOPMENT AND INNOVATION

In an internationalizing area characterized by fast-changing events occurring from a turbulent environment, glocalization allows SMEs to think globally and to act locally by adapting to this unpredictability and increasing the development of their territory. Moreover, local governance contributes to the local development via the actions engaged by the territorial actors, NGOs, and other entities in order to improve the attractiveness of the territory and to consolidate its image.

### 17.4.1 Role of Glocalization in Territorial Development

Given that glocalization constitutes a keystone of such innovative networks that work together in order to expand into international markets (Torrès, 2002), family SMEs are fighting towards a "measured" opening which would allow them to "think globally and act locally" taking into consideration the specificities and vulnerability of their territories. Such an arrangement appears so delicate that it requires a certain balance between regional particularisms and international trends. Efforts must then be maintained and sustained in order to ensure that deal. At this level, is important to mention that the concept of glocalization was popularized by Porter (1990, p. 78) to design "a geographic concentration of enterprises related to each other, of specialized suppliers, of service providers, of related industries' firms, and associated institutions in a particular field, which confront each other and cooperate". The theory of glocalization entails the proximity principle which plays an important role in the global insertion of SMEs and the conciliation between the globalization constraint and the proximity logic related to their management style.

### 17.4.2 Role of Territorial Governance in an Internationalizing SME Environment

First of all, it should be mentioned that local and short-distance structures, as well as small institutions and NGOs with an international scope, could largely impulse the attractiveness and the international development of the region. Indeed, such structures could provide the interface among the community, at the local and the global level, and infrastructures of communication which ensure an homogeneous state vision (such as proximity of airports, highways, of partner zones as well as big, prestigious, and multinational firms, etc.).

Moreover, i n a territory with an internationalizing innovative environment, territorial actors such as public or private entities (like chambers of commerce) can act to support foreign trade and to facilitate the internationalization of SMEs as well as the local entrepreneurial fabric (Torrès, 2002). This is a space where relationships of long-lasting collaboration and exchange are established in order to enhance learning, innovation, and creativity of actors and their access to the global market. In this perspective, Torrès (2002) cited four characteristics of an internationalizing milieu (4 Ds): diversity, density, dynamics, and directionality.

Three decades ago central government was considered to be the principal actor of self-regulation between fields and in every sector; however, this view changed rapidly in the late 1980s and early 1990s. The literature has emphasized the role of local and regional governments in cluster development and in the creation of an innovative internationalizing milieu. Territorial governance has entered the scene of economic development, and the territories must benefit from a local governance (Helmsing, 2001).

Indeed, when public action becomes a prerequisite for countries that have weak socioeconomic development, governance is of increased importance, especially in international forums. In line with this view, Gaudin (2002)reported that this new form of governing is cooperative and different from the previous hierarchical model (by which the state made a sovereign control on groups and citizens). This form of governance applies the values of the enterprise (such as the new techniques of management), as well as the social values (democracy, environment, human rights, etc.) in accordance with the business ethics and a long-term economic and profit perspective.

The World Bank added that governance represents "the manner in which power is exercised for the management of social and economic resources of the country to reach a development objective". Its action is restricted to four essential domains: (1) the capacity and the efficiency in the public services management, (2) admissibility and prevision, (3) the legal framework of development, and (4) information and transparency.

The Institute of Governance of Private and Public Organizations (2008) has proposed a definition of the SME governance by stipulating that:

> the SME's governance is defined as a set of reports between an owner-manager and a group of persons gathered into an advisory council or a board of directors. These persons who are predominantly independent of the direction, and of the stakeholders of control, accepted to support this owner-manager with their expertise, and their experience in order to improve decisions and to ensure the sustainable growth of the enterprise.

The concept of territorial governance is emerging to ensure the relative autonomy of local development processes, to consolidate democracy and the role of civic society in decision-making, to grant another image to the public action, to encourage the participation of citizens and the civic society, and to develop partnerships between different actors of specific politics (Wilson, 2000; Truda, 2007; Torre & Traversac, 2011).

Territorial governance is thus emanating not only from local citizens and actors, but also from national organizations and other local collectivities which play a major role in network activation and meshing between institutions and enterprises, enhancing transnational network synergies among them (Fourcade, 1993). These synergies enable the territory to be internationalized, connected with the rest of the world and not enclosed in solitary confinement. Indeed, the sustainability of a territory is widely based on the national extra-territorial and international relationships that local actors establish with other partners to ensure its durability throughout a networking dynamics.

Local and bridging actors, such as universities, research and technical centers, as well as technopoles, could also contribute to the emergence and development of innovative SMEs and the dissemination of innovation. Collectivities must also advocate for the companies' access to resources and skills distributed around the world and their communication with other firms located in foreign and international areas, as part of a global network (taking into consideration the proximity principle regulating the functioning of SMEs).

All these local organizations and actors advocate for a local governance which avoids the simple reproduction of imported development models which are most often inappropriate.

According to Jean (2000, p. 224), the territorial governance is "a specific mode of regulation of power and decision taking in a given community. It designs more than the government and the politic governance". It is reflected by the collective ability of learning and management that certain groups of individuals or human societies possess and develop in order to anticipate situations, to recognize a clarifying operating process leading to sustainable improvements.

Local governance has four dimensions (Lusthaus, Adrien, & Pestinger, 1999):

1. *Organizational dimension*: In the organization, the components or the elements of the collective action could be localized, such as the skills, the proactive systems, the leadership, etc.
2. *Systemic dimension:* The society is a set of multilevel and correlated systems: the public, the private, and the philanthropic systems are interrelated. In this perspective, three levels of capabilities can be determined: the individual, the sectorial, and the environmental.
3. *Participative dimension*: The two essential values of governance are change and learning. At this level, participative processes could influence the management approaches and the mode of government of public affairs to fight against political distortions and to ensure market effectiveness. Public interventions could contribute to favorable economic development.
4. *Institutional dimension*: At this level, the focus is placed on laws, rules, cultural values and beliefs, politic attitudes, etc. in order to stimulate the knowledge creation and the access to the actors' formal and informal games.

Thus, because competition remains global and because the competitiveness of SMEs requires a high territorial anchorage, the proximity dimension is well taken into consideration at the managerial level, and the proximity territory may facilitate the access of the SMEs and their orientation towards an international opening. Competitive advantages could also be gained by the global insertion of SMEs and their integration into the territory or the region thanks to the proximity principle development. This integration enables the SMEs to access to resources and skills throughout the world (financing, partnerships, technology, etc.), as well as to identify and to be embedded into the best networks that create prosperity, regulation, and distribution. Emphasis is then made on making of social, political, and economic local governance aiming at ensuring the territorial cohesion and development (Torrès, 2002; Swyngedouw, 2003).

In this perspective, the *territorial intelligence* seems to play an essential role for the local/economic development of the territory (Herbaux & Masselot, 2007; Truda, 2007). It should identify the external changes (related to technological, demographic, sociologic, and political trends), causes, and engines over the next 5 to 10 years, their opportunities and their threats, as well as the strengths and weaknesses of the territory's SMEs. The territorial intelligence should also determine the weightings of these changes, the plausible deadlines of their impacts on the regional economy, as well as the way by which the regional policy and the entrepreneurial dynamics of the region are coherent with such engines (Girardot, 2008). Its purpose is to analyze territories and territorial information shared subsequently with the help of information and communication technologies (Saccheri, 2008).

Widmer (2014) recommends, for example, a territorial information system which allows the modeling of the territory, to benchmark it to other territories and to make an update of it via its cartography and its spatial representations. The model can incorporate knowledge of the territory's actors, their skills, and their interests. A culture based on a customer relationship management (CRM) must be focused on sharing, dialogue, and negotiation to develop territorial marketing by good mastery of the territory users'/ clients' aspirations and needs and by fostering competition between territories.

## 17.5 CONCLUSION

Despite its vulnerability and limits, a territory—which is mainly constituted by a set of SMEs (suffering from some weaknesses related to their scarcity of resources and skills) – could be considered as an internationalizing area. It may easily adapt to a complex and changing environment thanks to different factors. As shown in Figure 17.1, the following are especially concerning: (1) innovative SME networks acting in the territory and (2) the territorial strategies orbiting around innovative

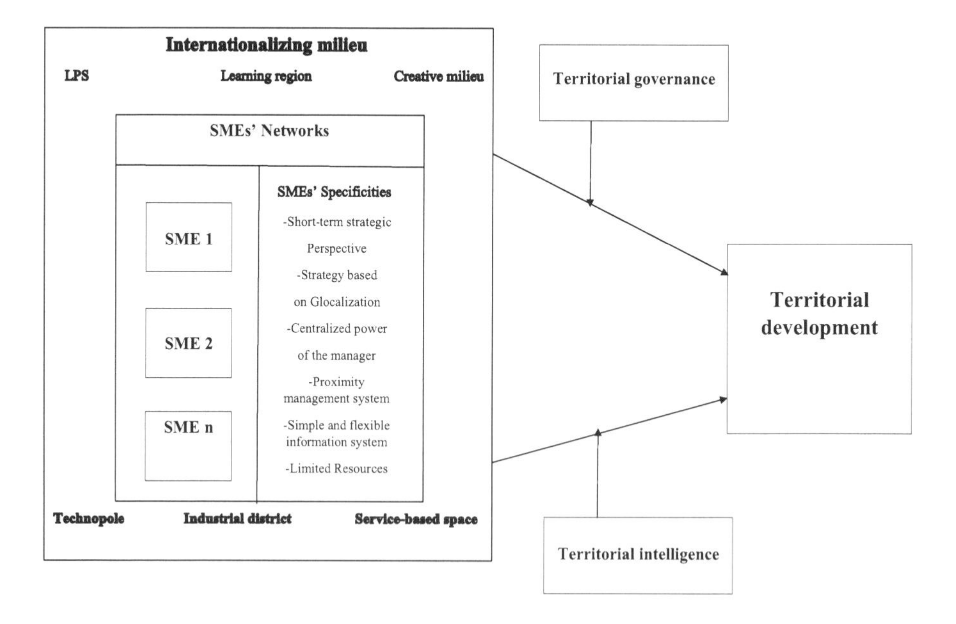


FIGURE 17.1 The theoretical proposed model.

glocalization, territorial intelligence, and local governance. In this perspective, local actors, public authorities, as well as some specific local entities should play an important role on the development of the territory in question.

## REFERENCES

Allali, I. (2007). For clarifying networks of SMEs: A visionary approach (Trad. Pour des réseaux de PME clarifiants: une approche visionnaire), HEC Montréal. *Revue Gestion, 32*(2), 81–90. Retrieved from www.cairn.info/revue-gestion-2007-2-page-81.html.

Allouche, J., Amann, B., Jaussaud, J., & Kurashina, T. (2008). The impact of family control on the performance and financial characteristics of family versus non-family businesses in Japan: A matched pair investigation. *Family Business Review, 21*(4), 315–319.

Amit, R., & Schoemaker, P. J. (1993). Strategic assets and organizational rent. *Strategic Management Journal, 14*(1), 33–46.

Anderson, R., Mansi, S. A., & Reeb, D. (2003). Founding family ownership and the agency cost of debt. *Journal of Financial Economics, 68*(2), 263–285.

Angeon, V., Caron, P., & Lardon, S. (2006). From social ties to the construction of sustainable territorial development? (Trad. Des liens sociaux à la construction d'un développement territorial durable: quel rôle de la proximité dans ce processus ?). *Développement durable et territoire,* File N°7. Proximity and Environment. (Trad. Proximité et environnement), Retrieved from http://developpementdurable.revues.org.

Basly, S. (2007). The internationalization of family SME: An organisational learning and knowledge development perspective. *Baltic Journal of Management, 2*(2), 154–180.

Bootz, J-P. (2015). How to reconcile self-organization and control in driven communities of practice?: a scoping review (Trad. Comment concilier auto-organisation et contrôle au sein des communautés de pratique pilotées?: une scoping review). *Management International, 19*(3), 15–30.

Bories-Azeau, I., & Loubès, A. (2013). The evaluation of workforce and competency forecasting (GPEC) mechanisms at the international level: toward a practices' renewal? (Trad. L'évaluation des dispositifs de GPEC à l'échelle territoriale: vers un renouvellement des pratiques ?). *Management & Avenir, 59*(1), 157–175.

Bos-Brouwers, H. E. J. (2009). Corporate sustainability and innovation in SMEs: evidence of themes and activities in practice. *Business Strategy and the Environment, 19*(7), 417–435.

Bros-Clergue, M. (2006). Differentiating territories: Which tools of management? (Tra. Différencier les territoires: quels outils de management?), RECEMAP Proceedings, 1–12, Retrievedfromhttp://www.unice.fr/recemap/contenurevue/Articles/Revue_Recemap7_Bros.pd

Busenitz, L. W., & Barney, J. B. (1997). Differences between Entrepreneurs and Managers in Large Organizations: Biases and Heuristics in Strategic Decision-Making. *Journal of Business Venturing, 12*(1), 9–30.

Campagne, P., & Pecqueur, B. (2014). Territorial development. An emerging reply to Globalization. (Trad . *Le développement territorial. Une réponse émergente à la mondialisation).* Paris, Charles Léopold Mayer Editions (ECLM), 268 p.

Carluer, F. (2006). Enterprises' networks and territorial dynamics: A strategic analysis (Trad. Réseaux d'entreprises et dynamiques territoriales: une analyse stratégique). *Geography Economy Society, 8*(2), 193–214.

Carré, D., & Levratto, N. (2011). Territorial dynamics, Urban district and location of firms. *Innovations, 35*(2), 183–206.

Colletis-Wahl, K., & Pecqueur, B. (2001). Territories, development and specific resources: What analytical framework? *Regional Studies, 35*(5), 449–459.

Cooke, P. (2002). *Knowledge economies – Clusters, learning and cooperative advantage.* Routledge Editions, London.
Cooke, P., Uranga, M. G., & Etxebarria, G. (1997). Regional Innovation Systems: Institutional and organisational dimensions. *Research Policy, 26*(4), 475–491.
Courlet, C., & Pecqueur, B. (1992). Industrial systems localised in France: A new model of development (Trad, Les systèmes industriels localisés en France: un nouveau modèle de développement, in Benko, G. & Et Lipietz, A. (Éds.), *Regions that succeed: Districts and networks: New paradigms of economic geography* (Trad, Les régions qui gagnent. Districts et réseaux: Les nouveaux paradigmes de la géographie économique) (pp. 81–102). PUF Editions, Paris.
Courlet, C., & Pecqueur, B. (1996). Industrial districts, localized production systems, and development. (Trad. Districts industriels, systèmes productifs localisés et développement), in Abdelmalki, L. L. & Courlet, C. (Eds). New logics of development. (Trad. Les nouvelles logiques du développement), (pp. 91–101). L'Harmattan Editions, Paris,
Courlet, C., & Pecqueur, B. (2013). The territorial economy. (Trad. L'économie territorialre). Presses universitaires de Grenoble Editions, 144p.
Courrent, J. M., & Torrès, O. (2005), "A proxemic approach of small business: the case of business ethics", *50th International Council of Small Business (ICSB),* Washington, United States.
De Oliveira, J. A. P. (2016). *Upgrading clusters and small enterprises in developing countries: Environmental, Labor, Innovation, and Social Issues,* Routledge Editions, 192p.
Fernandez, A. S., Le Roy, F., & Gnyawali, D. R. (2014). Sources and management of tension in co-opetition case evidence from telecommunications satellites manufacturing in Europe. *Industrial Marketing Management, 43*(2), 222–235.
Ferrary, M., & Pecqueux, Y. (2004). Organization in network: myths and realities. (Trad. L'organisation en réseau: Mythes et réalités), Presses universitaires de Frances Editions, 296 p.
Florida, R. (1995). Toward the learning region. *Futures, 27*(5), 527–536.
Forbes, D. P., & Milliken, F. (1999). Cognition and corporate governance: Understanding board of directors as strategic decision: Making groups, *Academy of Management Review, 24*(3), 489–505.
Forsman, H. (2011). Innovation capacity and innovation development in small enterprises: A comparison between the manufacturing and service sectors. *Research Policy, 40,* 730–750.
Fourcade, C. (1993). Territorial government and industrial district: The example of Montpellier. (Trad. Gouvernement territorial et district industriel: l'exemple de Montpellier). *Revue Internationale PME,* 6(1), 101–121.
Fromhold-Eisebith, M. (2004). Innovative milieu and social capital: Complementary or redundant concepts of collaboration-based regional development. *European Planning Studies, 12*(6), 745–767.
Gaudin, J. P. (2002). Why governance? (Trad. Pourquoi la gouvernance?). *Presses de Sciences Po.* La bibliothèque du citoyen. 179–180.
Girardot, J. J. (2008). *Evolution of the concept of territorial intelligence within the coordination action of the European network of territorial intelligence.* ReS-Ricerca e Sviluppoper le politiche sociali, Numero 1–2, C. E. I. M. Editrice, Mercato SanSeverino (SA).
Godet, M. (1997). Handbook of strategic prospective. (Trad. Manuel de prospective stratégique). Volume 1: An intellectual indiscipline (Trad. Une indiscipline intellectuelle) (293p). Volume 2: The art of method. (Trad. L'art et la méthode), (441p), Dunod Editions.
Habbershon, T. G., Williams, M. L., & MacMillan, I. C. (2003). A unified systems perspective of family firm performance. *Journal of Business Venturing, 18(4),* 451–466.
Hanna, V. & Walsh, K. (2002). Small firm networks: A Successful approach to innovation?. *R&D Management,* 32(3), 201–207.

Helmsing, A. H. J. (2001). Externalities, learning and governance: New perspectives on local economic development. *Development and Change*, *32*(2), 277–308.

Herbaux, P., & Masselot, C. (2007, October). Territorial intelligence and governance. In *International conference of territorial intelligence* (pp. 509–521). Huelva (CAENTI), France.

Horibe, F. (2003). Innovation, Creativity and Improvement. Working the right level to prosperity. *The Canadian Manager*, 28(1), 20–30.

Institute of Governance of Private and Public Organizations (Institut sur la gouvernance d'organisations privées et publiques (IGOPP)) (2008). Rapport du Groupe de travail sur la gouvernance des PME au Québec. Pour développer des entreprises championnes. InIGOPP. Accueil, L'IGOPP, Groupes de travail. Retrieved from http://igopp.org/ligopp/groupe-de-travail/ (page consultée le 21 avril 2015).

Jean, B. (2000). A new territorial governance for supporting rural restructuration: The put in perspective of European, American, and Canadian approaches; (Trad. Une nouvelle gouvernance territoriale pour accompagner la restructuration rurale: la mise en perspective des approches européenne américaine et canadienne). In: Carrier, M. & Côté, S. (Dirs) *Gouvernance et territoires ruraux. Éléments d'un débat sur la responsabilité du développement. Presses de l'Université du Québec*, Québec.

Joyal, A. (1994). Community economic development: The Montreal examples. In *Community economic development* (Vol. II). Canada.

Joyal, A. (1997). SME and territorial development. (Trad. PME et développement territorial). In 'SME: Balance Sheet and perspectives' (Trad. "PME: Bilan et perspectives"), under the supervision of Pierre-André Julien, 2nd Edition, (pp. 67–96), Laval Inter-university Presses, Québec.

Joyal, M., & Deshaies, L. (2000). Information networks of SMEs in non metropolitan milieu (Trad. Réseaux d'information des PME en milieu non métropolitain). *Cahiers de géographie du Québec*, *44*(122), 189–207.

Julien, P-A. (2000). *The state of the art in small business and entrepreneurship* (2nd ed.). Brookfield: Ashgate.

Karlsson, C., Johansson, B., & Stough, R. R. (2005). *Industrial clusters and inter-firm networks*. Hardback. Cheltenham: Elgar.

Leducq, D., & Lusso, B. (2011). Innovative cluster: conceptualization and spatial implementation (Trad. Le cluster innovant: conceptualisation et application territoriale), *Cybergeo: European Journal of Geography* [online], Espace, Société, Territoire, Document N°521, Retrieved from http://journals.openedition.org/cybergeo/23513.

Leloup, F., Moyart, L., & Pecqueur, B. (2004). *Territorial governance as new mode of territorial coordination ?* (Trad. *La gouvernance territoriale comme nouveau mode de coordination territoriale?*.4th proximity days.

Lima, S. (2013). Multi-Sited Territories and Migratory Circulation. *L'Espace géographique, 42*(4), 340–353. Retrieved from https://www.cairn.info/revue-espace-geographique-2013-4-page-340.htm.

Lima, E., & Filion, L. (2011). Organizational Learning in SMEs' Strategic Management: A Descriptive and Systemic Approach, Working paper, ICSB World Conference, Stockholm.

Lusthaus, C., Adrien, M-H., & Pestinger, M. (1999). *Capacity development, definitions, issues, and implications for planning, monitoring, and evaluation*. Universalia Occasional Paper, N. 35, p. 3.

Maillat, D. (1994). Spatial behaviors and innovative environments. (Trad. Comportements spatiaux et milieux innovateurs», in Aury, J-P., Bailly, A, Derycke, P-H, Huriot, J-M (dir.), Encyclopedia of spatial economy: Concepts, behaviors-Organization. (Trad. Encyclopédie d'économie spatiale: Concepts- Comportements- Organisation), Economica Editions, Paris, 255–262.

Maillat, D. (2006). Territorial dynamic, innovative milieus and regional policy. *Entrepreneurship & Regional Development*, 7(2), 157–165.
Maillat, D., & Lecoq, B. (2006). New technologies and transformation of regional structures in Europe: The role of the milieu. *Entrepreneurship and Regional Development*, *4*(1), 1–20.
Man, T., Lau, T., & Chan, K. (2002). The competitiveness of small and medium enterprises. *Journal of Business Venturing*, *17*(2), 123–142.
Marchesnay, M. (2014). Strategic scanning of small entrepreneurs: A pragmatic view. *Journal of Innovation Economics & Management*, *14*(2), 105–120.
Mazzarol, T., & Reboud, S. (2009). *The strategy of small firms, strategic management and innovation in the small firm*. Cheltenham, United Kingdom, Edward Elgar Publishing Editions.
Mazzarol, T., & Reboud, S. (2014). *Key problems facing SME owner-managers in strategy and inno-vation: Evidence from a diagnostic survey*. 28th ANZAM Conference. Sydney, Australia, 3–5 December.
Mbengue, A., & Ouakouak, M. L. (2012). Rational strategic planning and enterprise performance: an international study. (Trad. Planification stratégique rationnelle et performance de l'entreprise: une étude internationale). *Management international / International Management,* 16 (4), 117–127.
Moine, A. (2014). Territory as a cross-disciplinary tool for shared diagnostics. (Trad. Le territoire comme un outil de trans-disciplinarité vers des diagnostics partagés). CIST2014. Fronts and frontiers of territory sciences. (Trad. Fronts et frontières des sciences du territoire), Collège international des sciences du territoire (CIST), March, Paris, France. 284–290. Retrieved from https://hal.archives-ouvertes.fr/hal-01353474/document.
Moulaert, F., Martinelli, F., Gonzalez, S., & Swyngedouw, E. (2007). Social innovation and governance in European cities: Urban development between path dependency and radical innovation. *European Urban and Regional Studies*, *14*(3), 195–209.
Moulaert, F., & Sekia, F. (2003). Territorial innovation models: A critical survey. *Regional Studies*, *37*(3), 289–302.
Pacquot, T. (2011). What is a territory? (Trad. Qu'est-ce qu'un territoire?). *Vie sociale, 2*(2), 23–32.
Paradas, A. (2007). Mutualising training and hiring in SMEs. A variety of responses. (Trad. Mutualiser la formation et le recrutement dans les PME. Une variété de réponses). *La Revue des Sciences de Gestion*, *226–227*(4), 147–155.
Pecqueur, B. (2008). Territorial dynamics: Towards a new model of development facing globalization: Chapter book. In M. Joes, A. Querejeta, C. I. Landart, & J. Wilson (Eds.), *Networks, governance, and economic development: Bridging disciplinary barriers* (pp. 30–39). Edward Edgar Publishing Editions, Cheltenham/Massachusetts.
Pierre, A., & Fernandez, A-S. (2018). Going deeper into SMEs' innovation capacity: An empirical exploration of innovation capacity factors. *Journal of Innovation*, *1*(25), 139–181.
Piore, M. J., & Sabel, C. F. (1984). *The Second Industrial Divide: Possibilities for Prosperity,* (354 p), Basic Books, New York.
Piore, M. J., & Sabel, C. (1989). *The roads of prosperity: From large-scale production to adaptable specialization* (Trad. Les chemins de la prospérité: de la production de masse à la spécialisation souple), Hachette Editions. Paris: France.
Porter, M. E. (1990, March-April). The competitive advantage of nations. *Harvard Business Review*. New Yok.
Porter, M. E. (2000). Locations, clusters and company strategy. In *Oxford Handbook of Economic Geography,* G. L. Clark, M. P. Feldman, & M. S. Gertler. Editions. (pp. 253–274). Oxford: Oxford University Press.
Porter, M. E. (2003). The economic performance of regions. *Regional Studies, 37*(6-7), 549–578.

Qian, G., & Li, L. (2003). Profitability of small and medium sized enterprises in high tech industry: The case of the biotechnological industry. *Strategic Management Journal, 24*(9), 881–887.

Redondo-Toronjo, D. (2007). Territory, governance and territorial intelligence. (Trad. Territoire, gouvernance et intelligence territorial). *Bulletin de la Société Géographique de Liège-BSGLg*, 49(1). Retrieved from https://popups.uliege.be:443/0770-7576/index.php?id=1648.

Rhee, J., Park, T., & Lee, D. H. (2010). Drivers of innovativeness and performance for innovative SMEs in South Korea: Mediation of learning orientation. *Technovation, 30*(1), 65–75.

Robert-Huot, G., & Cloutier, J. (2014). Human Resources management in SMEs and new enterprises: a statement (Trad. La gestion des ressources humaines dans les PME et les nouvelles entreprises: un bilan). 12Th CIFEPME International Francophone Congress in Entrepreneurship and SMEs, Agadir, Morocco. Retrieved from https://www.airepme.org/images/File/AGADIR2014/Robert-Huot_Cloutier.pdf

Rogers, M. (2004). Networks, firm size and innovation. *Small Business Economics, 22*(2), 141–153.

Saccheri, T. (2008). *Territorial intelligence and participation*. ReS-Ricerca e Sviluppo per le Politiche Sociali, Numero 1–2, C. E. I. M. Editrice, Mercato San Severino (SA).

Schumacher, E. F. (2001). *Small is beautiful: A study of economics as if people mattered* (p. 286). Vancouver, BC, Canada: Hartley and Marks Publishers.

Staniewsk, M. W. I., Nowacki, R., & Awruk, K. (2016, September). Entrepreneurship and innovativeness of small and medium-sized construction enterprises. *International Entrepreneurship and Management Journal, 12*(3), Springer, 861–877.

Swyngedouw, E. (2003). *Globalisation or glocalisation? Networks, territories, and re-scaling* (European Forum, 46). Retrieved from www.europaforum.or.at/site/Homepageifhp2003.

Tidd, J. (2001). Innovation management in context: Environment, organization and performance. *International Journal of Management Review, 3*(3), 169–183.

Torre, A. (2006). Collective action, governance structure and organizational trust in localized systems of production: The case of the AOC organization of small producers. *Entrepreneurship & Regional Development, 18*(1), January, 55–72.

Torre, A., & Traversac, J. B. (Eds.). (2011). *Territorial governance: Local development, rural areas and Agrofood systems* (p. 29). New York: Springer Verlag and Heidelberg.

Torrès, O. (2002). *Small firm, glocalization strategy and proximity*. ECSB – Research in Entrepreneurship and Small Business – 16th Conference Barcelona, 21–22 November, Best Paper Award, p. 12.

Torrès, O. (2003). Essay of proximity conceptualization of enterprises' smallness of enterprises and proximity effect magnification. (Trad. Petitesse des entreprises et grossissement des effets de proximité), *Revue française de gestion, 144*(3), 119–138.)

Torrès, O. (2004a). *The proximity law of small business management: Between closeness and closure*. 49th International Council of Small Business (ICSB). Johannesburg, South Africa.

Torrès, O. (2004b). The SME concept of Pierre-André Julien: An analysis in terms of proximity. *Piccola Impresa, Small Business, 3*(2), 51–61.

Torrès, O. (2006). *The proximity law of small business finance*. 51st International Council of Small Business. Australia: Melbourne, 18–21 June.

Torrès, O., & Julien, P-A. (2005). Specificity and denaturing of small business. *International Small Business Journal, 23*(4), 355–377.

Truda, G. (2007, October). *The dynamics of a territory: The main actors of sustainable development in the Irno valley*. International Conference of Territorial Intelligence, Huelva 2007, Huelva, Spain, pp. 337–351, 2008 <halshs-00523759>

Vilette, M.-A. (2010, September). *Shared-time work: between special shape of work transformation and a tool of introduction of HRM in SMEs.* (Trad. Le Travail à temps partagé, entre forme particulière de transformation du travail et outil d'introduction de la GRH dans les PME), PHD Dissertation, Auvergne University, Retrieved from https://tel.archives-ouvertes.fr/tel-00719589.

Wang, C., Walker, E. A., & Redmond, J. L. (2007). Explaining the lack of strategic planning in SMEs: The importance of owner motivation. *International Journal of Organizational Behaviour, 12*(1), 1–16.

Widmer, S. (2008). Which territorial intelligence for sustainable development of urbanized territoires? The case of countries of Montbéliard (Trad. Quelle intelligence territoriale pour le développement durable des territoires urbanisés ? Le cas du Pays de Montbéliard), 6th International Conference of Territorial Intelligence"Tools and methods of Territorial Intelligence", Besançon, Franc, Retrieved from https://halshs.archives-ouvertes.fr/halshs-00985842/document.

Wilson, R-H. (2000, April–June). Understanding local governance: An international perspective. *RAE- Revista de Administração de Empresas*, São Paulo, *40*(2), 51–63. Administração Pública.

Wolff, J. A., & Pett, T. L. (2006). Small-Firm Performance: Modeling the Role of Product and Process Improvements. *Journal of Small Business Management*, *44*(2), 268–284.

# 18 Influence of Branding of Financial Instruments on Investment Decisions: Mediating Role of Behavioral Biases

*P. C. Sharma*

## CONTENTS



## 18.1 INTRODUCTION

Stock investing is not only influenced by financial factors like ROI, EPS, or P/E ratio. The universe of stock investing includes various marketing factors, advertising factors, and also various emotional factors causing psychological biases. Considering the present scenario where individual investors are turning out to be highly vigilant in financial matters and yet contributing a substantial amount of stakes along with emotions in the market (Nofsinger & Richard, 2002), financial planners face the challenge of understanding the investment behavior of their clients. In this context, brand reputation and brand trust play a pivotal role in affecting the final decision of an investor, as a brand includes the consumer's complete experience with both the product and the company. Psychological biases also attract equal attention due to their extremely effective background influence. Same marketing ideology about branding is applicable for studying brand reputation of companies in stock markets worldwide as it channelizes the influential role of reputation in pulling modern investors for investment. Despite the highly standardized and regulated nature of financial investment industry, firms are striving for greater engagement through personalization and emotional connection, since they operate in a segment where trust is paramount. Investors may form relationships with brands on the basis of several characteristics of these brands, like brand image. According to Cova (1999) and Sheth and Parvatiyar (1995), the image or brand does not represent the product; it's the product that represents the image. Larsen and Buss(2002) claims that companies manage their reputations mainly for financial reasons. Following the recommendations of Brown (1952) and Tucker (1964), in this study, we conceptualized brand trust as the outcome of brand reputation. Banks (1968) found a fairly strong relationship between trust and actual investment decision. The concept of branding has been majorly studied in the context of FMCG products, primarily in the marketing domain. This chapter is a novel effort to study this phenomenon in the context of finance and eventually in the investment management domain as branding has a very strong psychosomatic impact on investor's memory which has a prolonged effect because there are various behavioral biases which repeatedly affect the investment decision of an investor and at times are so influential that the investor ends up making erroneous decisions. So linking the effects of branding in context of brand reputation and brand trust with the investment decision and analyzing the effect of behavioral biases as a mediator amongst the two would be interesting. Regarding the current situation, a novel model that takes into consideration traditional finance, behavioral finance, and the effect of branding to provide superior information about individual investor decision-making processes under the influence of behavioral biases is indispensable. This model is a natural progression from previous branding models as it also includes behavioral bias. The new model highlights the importance of customer insight, analytics, brand, and customer experience and proposes to analyze the branding factors which have an impact on the investment decisions of individual investors. The endeavor is to validate whether investors are aware of the effect of branding practices in the presence of behavioral biases in their choices. This model covers a big literature gap of unavailability of an inter-disciplinary behavioral finance model for measuring market investment behavior as a result of effects of branding under the influence of behavioral biases. A new perspective of irrational thinking and resultant decision-making of individual investors has been presented. This is also the novel contribution of this chapter. The combination of precursors of brand reputation, brand trust, and behavioral biases which has been considered for model framing has not been used in any previous model in the similar arrangement. Various behavioral biases such as conservatism, disposition effect, anchoring behavior of investors, and herding behavior play a great role in financial markets in effecting investment decisions. Data for this research work were collected through a self-structured questionnaire from investors in the regions of Delhi, Mumbai, Chennai, and Kolkata. Analysis of this research work predicts that the effect of branding on the individual investor's decision-making process is credible and once an investor develops trust in a brand behavioral biases are less effectual. This study will help firms go beyond product and investments. This model also emphasizes the need to understand the customer journey across multiple channels, going beyond last touch attribution to improve investor engagement and finally his investment decision. Studying the investment decision under the influence of brand reputation and brand trust will help practitioners, financial service providers, sponsors, financial institutions, and corporate giants to preplan their public offerings according to the expected investor reaction due to their inherent behavioral bias and can offer them better schemes for fund allocation and can also take reasonable steps in improvising their brand impact for following some out-of-the-box practices. Specifically, the key objectives of the study are: (1) to examine the relationship between brand reputation and brand trust, (2) to examine the extent to which brand reputation and brand trust are influenced by behavioral biases, (3) to develop a model based on investment decision conceptualization and its antecedents, and (4) to evaluate reliability and validity of the model. In this research work, theory of reasoned action, customer-based brand equity theory, brand position theory, and prospect theory are the backdrop theories to take the research work further. All these theories have been interpreted from the perspective of an individual investor.

The rest of the chapter has been arranged as follows. The next section reviews the existing literature on various variables affecting brand reputation and brand trust, how brand reputation and brand trust are affected by behavior biases, and how all these factors together affect the investment decision. Section 18.3 describes the research methodology followed for data collection, sample profiling, etc. The data used to test the hypotheses along with structural model are described in Section 18.4. In Section 18.5, the effects of exogenous variables, mediators, and control variables on endogenous variables have been analyzed. The last section concludes the study through findings, discussion, implications, and conclusion.

## 18.2 LITERATURE REVIEW

According to Kotler (1999), a brand is essentially a seller's promise to consistently deliver a specific set of features, benefits, and services to the buyers (Aaker, 1996). This is an attractive idea in financial markets, where it is difficult to differentiate products practically. Financial investments are tangible only up to the extent of the amount of investment, as various hidden charges, opportunity costs, inflation interest rate, etc. remain undercover and therefore difficult to evaluate prior to investment or even after investment. Fund selection behavior of investors persuades marketing decisions of investment management companies, and thus it has attracted the interest of modern investment researchers. The academic investment literature did not clearly recognize the role of brands as relationship builders, as it has been argued that brands are primarily transaction facilitators (Coviello & Brondie, 2001; Coviello, Brondie, Danaher, & Johnston, 2002). But according to current marketing literature, buyers develop relationships with the product (Saren & Tzokas, 1998; Lye, 2002), and their knowledge and feelings about the brand influence their evaluation of the products carrying this brand (Aaker & Keller, 1990; Dacin & Smith, 1994; Brown & Dacin, 1997). It has also been suggested that even children develop relationships with brands and the connections with brands develop strong links between childhood and adolescence (Chaplin & John, 2005) and those childhood memories influence the manner in which they relate to brands for life.

Forward-looking marketers and research agencies acknowledge the importance of this approach and incorporate relationship-based ideas such as trust with brand management (Esch, Langner, Schmitt, & Geus, 2006). This importance to the concept of branding is overdue. Along with branding, behavioral biases also demands required spotlight. Behavioral biases are systematic patterns of deviation from rationality in decision-making (Kahneman & Tversky, 1979). The research work of Gupta (1991) argues that framing a portfolio for a client requires understanding of his psyche, as it is much more than simply selecting securities for investment. Capon and Fitzsimons (1994) in a study mentioned that there are various indications that support the fact that, apart from risk and return, other factors also effect investment decisions such as demographic background, past experience, reputation of the company, etc. Chaudhary (2013) found that emotional and cognitive biases have a strong influence on investors' decision-making processes. An emerging body of literature has validated the role of brand reputation (Bromley, 2002; Shapiro, 1983) and brand trust (Boon & Holmes, 1991; Deutsch, 1960) in influencing decision-making. In this study all the exogenous variables are hypothesized to have significant relationship with the endogenous variable. So, the following main hypothesis has been proposed. Along with these, two sets of 14 sub-hypothesis each have been mentioned and defended in the sections follows.

$H_1$: All the exogenous and endogenous variables have a significant relationship.

### 18.2.1 Brand Reputation

A growing body of literature has led to an abundance of alternative definitions of reputation. Gotsi and Wilson (2001) conclude that reputation should be viewed as "a stakeholder's overall evaluation of a company over time." Reputation is a socially shared impression, a consensus about how a firm will behave in any given situation (Bromley, 2002). Work of Fombrun and Rindova (2000) and Mazzola, Gabbionta, and Ravasi (2004) have emphasized on the plurality of perceptions and representations about a company, but it has mainly focused on the customers', employees', and the general publics' perspectives. In this context, reputational perceptions by individual investors are a novel field of research. Both academicians and practitioners believe that brand reputation is becoming increasingly important. To be successful and hence profitable, brands should have a positive reputation (Herbig & Milewicz, 1995). Shapiro (1983) concluded that the value of a firm's overall reputation is easily seen in its relationship to a firm's revenues: as a firm's reputation increases, so do its sales. To know about the association amongst a brand's reputation and the related investment decision, the following hypotheses have been framed:

$H_{1a}$: An investor's perception that a brand has a good reputation is positively related to the investment decision.

$H_{1b}$: An investor's perception that a brand has a good reputation is positively related to the investor's trust in that brand.

### 18.2.2 Brand Trust

Trust is defined as the expectation of the parties in a transaction and the risks associated with assuming and acting on such expectations (Deutsch, 1958). Lewis and Weigert (1985) argue that trust is not mere predictability but confidence in the face of risk. This line of argument is also followed by other researchers (Deutsch, 1960; Helm, 2007; Schlenker, Helm, & Tedeschi, 1973; Boon & Holmes, 1991). Brand trust is defined as "the willingness of an average consumer to rely on the ability of the brand to perform its stated function" (Chaudhuri & Holbrook, 2001). Conceptualizations of trust in the investment literature, however, have generally been lacking. In the investor market, there are too many anonymous investors, making it unlikely that the financial institution could develop personal relationships with each investor. Thus, investment marketers may have to rely on a symbol—the brand—to build the relationship. The brand becomes a substitute for human contact between the financial institution and its investors, and trust may be developed with it:

$H_{1c}$: An investor's perception that a brand is trustworthy is positively related to the investment decision.

### 18.2.3 Brand Characteristics as Antecedents to Brand Reputation and Brand Trust

The characteristics of the company behind a brand also influence its reputation and the degree to which investors trust the brand. A consumer's knowledge about the company behind a brand is likely to affect his assessment of the brand. The characteristics of the company proposed to affect a brand's reputation and eventually investor's trust in a brand are brand competence (Bendapudi & Berry, 1997; Zucker, 1986; Andaleep & Anwar, 1996), brand experience (Churchill & Surprenant, 1982; Bendapudi & Berry, 1997), brand satisfaction (Martensen, Gronholdt, & Kristensen, 2000; Bendapudi & Berry, 1997), and family branding Aaker and Keller (1990).

#### 18.2.3.1 Brand Competence

A competent brand is one that has the ability to fulfill a customer's need (Butler & Cantrell, 1984; Butler, 1991). Deutsch (1960), Cook and Wall (1980), and Sitkin and Roth (1993) all considered a brand's competence as an essential element influencing its reputation and eventually trust. An investor may find out about an investment avenue's brand's competence through direct investment or word-of mouth communication. Once convinced that a brand is able to serve its requirements, an investor may be willing to rely and invest in that brand. So to verify the same, we propose:

$H_{1d}$: An investor's perception that a brand is competent is positively related to the reputation of that brand.

$H_{1e}$: An investor's perception that a brand is competent is positively related to his trust in the brand.

#### 18.2.3.2 Brand Experience

Brand experience refers to a consumer's past encounters with the brand. Zucker (1986) suggested that, in the development of process-based trust, reciprocity is the key. In organizations, repeated contacts through time suggest a long-term commitment (Arrow, 1984; Powell, 1990) and in investment domain same commitment is visible through repeated investments in different schemes of a particular company. Security and stability of such recurring reciprocal exchanges is the outcome of brand reputation (Powell, 1990). Similarly, as an investor gains more experience with a brand, the investor understands the brand better and grows to trust it more. To summarize:

$H_{1f}$: An investor's experience with a brand is positively related to his perception about the reputation of that brand.

$H_{1g}$: An investor's experience with a brand is positively related to his trust in that brand.

#### 18.2.3.3 Brand Satisfaction

Brand satisfaction can be defined as the outcome of the subjective evaluation that the chosen alternative brand meets or exceeds expectations (Bloemer & Kasper, 1995). This is in line with the disconfirmation paradigm of consumer satisfaction, where the comparison between customer expectations and actual performance features strongly in the definition of satisfaction. Research in equity and social exchange theory suggests that the equity of outcomes affects behavior in subsequent periods (Adams, 1965; Kelly & Thibaut, 1978) and this ultimately leads to reputational brand building. Research work of Ganesan (1994) mentioned that in a continuing relationship, satisfaction with past outcomes indicates equity in the exchange. This eventually develops trust in brand which has long-term effect on future investment plans and referral processes to kins. Therefore we propose:

$H_{1h}$: An investor's satisfaction positively influences his perceived brand reputation.

$H_{1i}$: An investor's satisfaction with a brand is positively related to investor's trust in that brand.

#### 18.2.3.4 Family Branding

Family branding is a standard business practice for products with good attributes. The reputation of a brand depends upon the brand's history, its quality, and its associations. With respect to the general mechanism, Aaker and Keller (1990) find experimental evidence that the perceived quality of one product affects the expected quality of another one. Family branding, that is a company placing the same brand name on all products in a product line, enjoys the distinct advantage of instant recognition, benefiting from the "halo effect" of the brand's established reputation. A new entry using the family brand name gains instant credibility and visibility from the brand's established reputation and brand trust. Also an old marketing ideology says that customers buy products but choose brands. The same ideology of family branding is applicable in investment industry also. As investment is more of a psychological game (Kahneman & Twersky, 1979) and it is more influenced by past experiences and word-of-mouth information by peer groups so it leaves a greater impact on investment decisions:

$H_{1j}$: An investor's perception about a family brand is positively related to the reputation of the brand.
$H_{1k}$: An investor's perception about a family brand is positively related to his trust in the brand.

### 18.2.4 Behavioral Biases

The micro theory of behavioral finance focuses on the behavior of individual investor. According to micro theory, investors' decisions are subject to either emotional or behavioral biases or cognitive errors. Biases come from the feelings, intuition, or impulsive thinking, whereas cognitive errors occur due to misunderstanding of data, faulty reasoning, or memory errors. Both types of biases can lead to poor investment decisions (Muhammad, 2009). When looking at different empirical studies done in relation to the behavioral biases, most of the top researchers in the field have focused on behavioral biases such as disposition effect, anchoring, herd behavior, and conservatism. Herding in financial markets can be described as reciprocated replication resulting in to convergence of action (Hirshleifer & Teoh, 2003). This is the most frequent blunder which investors make by chasing the investment decisions taken by the majority (Kallinterakis, Munir, & Radovic-Markovic, 2010). It initiates a "snowball-effect" that is very complicated to discontinue (Welch, 2000). Tversky and Kahneman (1974) quoted about anchoring that in numerous conditions, people make estimations on the basis of their past experiences that is attuned to capitulate their future expectations in the form of returns. These different initial values results in different estimates. According to the research by Singh (2012), uneducated or less aware investors are most likely to have anchoring tendencies. Disposition effect, which is an extension of Kahneman and Tversky's (1979) prospect theory, was primarily defined by Shefrin and Statman (1985). They revealed that, on average, investors' trades show evidences of disposition effect in which they under react to market information and hold their losing funds for a long time. Numerous researchers have used the prospect theory to explain the phenomenon of disposition effect (Barberis & Huang, 2001; Henderson, 2012). Conservatism is the failure to adequately incorporate new information into one's views. Though incremental information may warrant a change to one's initial view, an investor may conserve his prior view, finding it very difficult to shift their stance. Conservatism badly affects both individual and professional investors alike. Research findings of the work of Bakar and Yi (2016) suggest that the majority of the investors have a tendency to invest in familiar shares of well-known companies. It is supported by the assurance that the returns will be superior if they invest in familiar shares. These results are also supported by the study of Kengatharan and Kengatharan (2014). Considering the crucial role played by behavioral biases with linkage to investment decision-making, it is imperative to discover how they influence brand reputation and brand trust and ultimately investment decisions:

$H_{1l}$: Behavioral biases have a negative relationship with the investment decision.
$H_{1m}$: Behavioral biases have a significant relationship with brand reputation.
$H_{1n}$: Behavioral biases have a significant relationship with brand trust.

Considering the backdrop theories according to customer-based brand equity theory and theory of reasoned action, experience a consumer has had before and after purchasing a service or product determines his subsequent purchase of the service from the same organization. In investment decision context an investor will invest in a particular investment avenue only if he has earned yields similar to his expectations or more. If the past history of investment behavior is integrated to predict and measure brand reputation and brand trust, their prediction and measurement will be more stable over time and accurate. Brand position is defined as an activity of creating a brand in such a manner that it occupies a distinctive place and value in the target customer's mind. It also refers to target consumer's reason to buy a particular brand in preference to others (Kapferer, 1997). In investment decision context a financial product has to be placed in the minds of investors in such a manner that for that particular category of investment investors should consider only that particular brand as that brand has been positioned as offering all the characteristics desired by the investor. Nobel Laureate Kahneman and Tversky (1979) initially explained prospect theory and revealed that utility or satisfaction is not simply a function of risk and return, but it also focuses on purchase price as a reference point. For example, investors feel regret when they are losing money on an investment and feel joy when they are making money relative to their initial investment. According to prospect theory, losses for the typical investor hurt roughly twice as much as gains feel good.

Drawing from the discussion on market investment decision, reputation, and trust in this section, we define trust in a brand as a consumer's willingness to rely on the brand as an outcome of its reputation. We also propose that reputation as well as trust in a brand leads to the final investment decision; reputation and trust have been considered as perceptual phenomenon, for the purpose of this research.

## 18.3 RESEARCH METHODOLOGY

Due to the fact that the research had to be carried out on a large number of dispersed investors in various locations, the questionnaire survey was considered the most suitable method for this research (Taylor, Sinha, & Ghoshal, 2006). The questionnaire included 56 statements out of which 48 were meant to acquire information about the investor behavior, whereas the rest were drafted to get demographic information. These 48 statements were constructed on the Likert scale, which is a symmetric unidimensional scale where all the items measure the same thing but at various levels of approval or disapproval. A five-point Likert scale has been used in the questionnaire (Taylor et al., 2006). The sample profile was framed on the basis of two decision criteria: age of the respondent (above 30 years) and number of years of investment experience in the stock market (at least 5 years). The total number of responses collected by the questionnaire survey was 1118. In order to keep the sample profile complete, 42 partially filled responses, where answers to more than 5 statements were omitted, were filtered and eliminated from the experienced investor sub-sample to attain the final sample profile of 1076. This yielded a response rate of 96.24%. The general profile of respondents was comparable with reference to gender, age, education, and income as there was an almost equal proportion of male (50.6%) and female (49.4%) respondents. Mean age of respondents was 43.47 years and median monthly income was between 50,000 and 1,00,000 INR. The study included control variables like two demographic factors (age and gender) and one sociological factor (education level).

## 18.4 ANALYSIS

During data screening in a pilot study some tricky and strenuous questions were reframed according to respondents temper and understanding. Being Likert-scale data, the possibility of the presence of outliers was rejected. While checking the normality of the data, on the basis of using the value of 7, given by West, Finch, and Curran (1995) as the benchmark for the assessment of the kurtosis values, along with Mardia's (1970, 1974) normalized estimate of multivariate kurtosis, it was identified that sample was non-normal. According to Hair, Anderson, Tatham, and Black (1998), Likert-scale data usually are non-normal. Multivariate data analysis was performed after the measurement model. Linearity in this research study was performed by curve estimation regression for all direct effects in the model. The results demonstrate that the relationships amid the variables are adequately linear except between brand competence and trust, family branding and trust, and trust and behavioral biases; though, no curve estimation was significant either. Consequently, we left the relationship in our model conditional to trimming during subsequent analyses. Multicollinearity in this research work has been checked by variable inflation factor (VIF) for all exogenous variables simultaneously by running a multivariate regression. Values of VIF in all the cases were less than 2, which is within the acceptance range. This shows that all the exogenous variables in the study are able to explain an exclusive discrepancy in the concerned endogenous variable. Reliability statistics of all the factors are shown in Table 18.1.

The hypothesized relationships in the model were examined concurrently via structural equation modeling. We have removed unreliable, not very sure, negative comments, more effective and enjoyable experience items owing to their poor factor loadings. Modification indices were referred to identify if there was any possibility to improve the model. Accordingly, we covaried the error terms amid comparatively better performance and customer orientation, past encounters and invest in brands, and prediction and regret. Results of confirmatory factor analysis specify that the goodness of fit for our measurement model is sufficient. This suggests that the hypothesized model fits the data.

To check the convergent validity AVE was calculated. AVE for all the factors was higher than 0.50. To analyze discriminant validity square root of the AVE were compared with all inter-factor correlations, as shown on the diagonal in Table 18.2. All factors exhibited sufficient discriminant validity since the diagonal values are higher

**TABLE 18.1**
**Reliability Statistics**

| Factor | Reliability Statistics |
|---|---|
| Brand Reputation | 0.82 |
| Brand Trust | 0.88 |
| Brand Competence | 0.73 |
| Brand Experience | 0.79 |
| Brand Satisfaction | 0.81 |
| Family Branding | 0.71 |
| Behavioral Biases | 0.85 |
| Investment Decision | 0.71 |

**TABLE 18.2**
**Validity Measures**

| | CR | BrComp[a] | BrExp[b] | BrSat[c] | FBrand[d] | BrTrust[e] | BrReput[f] | BBias[g] | InvDec[h] |
|---|---|---|---|---|---|---|---|---|---|
| BrComp[a] | 0.856 | **0.836** | | | | | | | |
| BrExp[b] | 0.901 | 0.597 | **0.857** | | | | | | |
| BrSat[c] | 0.717 | 0.573 | 0.807 | **0.853** | | | | | |
| FBrand[d] | 0.795 | 0.680 | 0.762 | 0.701 | **0.804** | | | | |
| BrTrust[e] | 0.857 | 0.241 | 0.391 | 0.805 | 0.616 | **0.842** | | | |
| BrReput[f] | 0.903 | 0.737 | 0.684 | 0.382 | 0.725 | 0.791 | **0.845** | | |
| BBias[g] | 0.786 | −0.536 | −0.516 | −0.586 | −0.575 | −0.437 | −0.625 | **0.782** | |
| InvDec[h] | 0.861 | 0.564 | 0.675 | 0.687 | 0.611 | 0.713 | 0.652 | −0.777 | **0.861** |

*Note*: * Based on (Fornell & Larcker, 1981): AVE in the diagonal and inter construct correlation off-diagonal. BrComp[a] = Brand Competence, BrExp[b] = Brand Experience, BrSat[c] = Brand Satisfaction, FBrand[d] = Family Branding, BrTrust[e] = Brand Trust, BrReput[f] = Brand Reputation, BBias[g] = Behavioral Bias, InvDec[h] = Investment Decision.

than the correlations. Along with these, composite reliability for each factor was also calculated. In all cases the composite reliability was greater than the minimum threshold of 0.70, indicative of reliability in factors.

The data for independent as well as dependent variables were collected using a single instrument, so a common method bias test was carried out to check whether a method bias had influenced the results of the measurement model. The unmeasured latent factor method suggested by Jarvis, Mackenzie, and Podsakoff (2003) was used in this study. A comparison of standardized regression weights before and after adding the common latent factor verified that none of the regression weights were considerably influenced by the common latent factor as the reported changes were less than 0.200 and the CR and AVE for each construct still met the minimum desired standards. However, to be on the conventional side, the common latent factor was retained for structural model by imputing composites in AMOS while the CLF was present, and thus we have CMB-adjusted values. Table 18.2 illustrates that the highest correlation (r = 0.807) was amongst brand experience and brand satisfaction. This correlation verifies the age-old notion of better experience and high level of satisfaction. The correlation (r = −0.777) between the factors behavioral biases and investment decision also is considerable. The noteworthy thing is that they both share a significantly negative relationship which depicts that strong influence of behavioral biases leads to wrong investment decision-making. Next correlation (r = 0.762) was between the factors brand experience and family branding which shows that investors retain their past experiences with a particular brand and in follow-up decisions the impact of those experiences is truly reflected in the form of investment decisions. Additionally, the lowest correlation (r = 0.241) was between brand competence and brand trust. In reality, strong brand competence creates brand reputation which in turn develops investors' trust in the brand. So these two factors share a significantly low correlation. As the correlations are considerably lesser than 0.80 in total values, there is unlikely to be any statistical concern of multicollinearity in the concerned data (Hair et al., 1998).

### 18.4.1 Hypothesis Testing

The available literature has accentuated that financial investment decision-making is majorly affected by demographic and sociological factors. For hypothesis testing we set up two pairs of hypotheses: $H_1$ for testing the association amongst exogenous and endogenous variables, and $H_2$ for testing the association amongst exogenous, mediator, and endogenous variables. Mediation tests were conducted in detail with bootstrapping. This route was essential to facilitate sufficient power to test each set of hypotheses, and with the aim of maintaining theoretical clarity and parsimony of the model.

### 18.4.2 Structural Model

The fitted structural model demonstrates adequate fit. The control factor age shows a considerable influence on dependent variable, as theorized and which is in sync with the available literature as age grows investors becomes more brand loyal and they want to invest in only well-reputed brands. Gender has a marginal significant effect on the dependent variable because, as depicted in the available literature, females are more risk averse and they are more brand loyal. Both of these results are in corroboration to the findings of the work done by Maheshwari, Lodorfos, and Jacobsen (2014). But this philosophy is changing as both males and females are becoming more cautious about the brands in which they are investing. Education level came as a surprise and did not show any significant impact on the dependent variable because data demographics have shown that most of graduates are into investing since more than 10 years whereas post graduates or professionals are into this since last 5 years only. Considering the control variables, five different models in two different groups were framed where each model considered the effect of an individual demographic or sociological factor. The first group comprised different regression models. In the regression models, the first model comprised all the exogenous variables, which showed an $r^2$ of 54.3%. The remaining three models showed a negligible changed $r^2$ on the sequential inclusion of exogenous variables with control variables as models 2, 3, 4, and 5 showed an incremental $r^2$ of 2.6%, 2%, 1.6%, and 1.3% respectively. Considering the second group and taking *p*-values into account, the *p*-values of age and gender were significant, education was insignificant in relation to trust and investment decision, and gender in context of reputation was marginally significant.

Parameter estimates presented in Table 18.3 show improvised values as compared to CFA statistics. This depicts that the structural model connects all the exogenous variables and endogenous variables along with the mediating variables in a more parsimonious manner.

Mediation was tested using 5000 bias corrected bootstrapping re-samples in AMOS. A two-step process was used for assessing the direct and indirect associations amongst the study variables. The structural model with all the significant and insignificant paths is demonstrated in Figure 18.1, and statistics are provided in Table 18.4. The mediating effects of the constructs of brand reputation, brand trust, and behavioral biases on brand competence, brand experience, brand satisfaction, and family branding were systematically tested. Different levels of mediation have been reported by the statistics on the basis of the association amongst variables. The outcomes of the hypothesis tests are recapitulated in Table 18.4.

Path analysis was performed to test the hypotheses generated. Figure 18.1 and Table 18.4 show the results. The standardized parameter estimates value is 0.760, suggesting that variance to the extent of 76% in investment decision can be explained by brand trust, brand reputation and behavioral biases collaboratively. Table 18.4 shows that the 14 relationships have been tested in group one, out of which 8 have been accepted. According to existing literature, our data advocates that brand reputation is the outcome of brand competence (Bendapudi & Berry, 1997; Zucker, 1986; Andaleep & Anwar, 1996), family branding (Aaker and Keller, 1990), and brand experience (Churchill & Surprenant, 1982; Bendapudi & Berry, 1997). Accepted hypotheses validate the findings of Adams (1965), Kelly and Thibaut (1978), and Ganesan (1994) that brand trust is the outcome of brand reputation. As brand trust

**TABLE 18.3**
**Structural Equation Modeling Results**

| Metric | Observed Value | Recommended Value |
|---|---|---|
| **Cmin/df** | **1.591** | **Between 1 and 3** |
| CFI | 0.939 | >0.950 |
| GFI | 0.901 | >0.900 |
| RMSEA | 0.018 | <0.060 |
| PCLOSE | 0.089 | >0.050 |
| SRMR | 0.043 | <0.090 |

**TABLE 18.4**
**Hypothesis Summary Table**

| Relationship | Evidence | Supported/ Not Supported |
|---|---|---|
| $H_1$: All the exogenous and endogenous variables have a significant relationship. | | |
| $H_{1a}$: An investor's perception that a brand has a good reputation is positively related to the investment decision. | .025[a] (.406)[b] (8.832)[c] | No |
| $H_{1b}$: An investor's perception that a brand has a good reputation is positively related to the investor's trust in that brand. | .528[a] (.031)[b] (41.624)[c] | Yes |
| $H_{1c}$: An investor's perception that a brand is trustworthy is positively related to the investment decision. | .690[a] (.001)[b] (53.188)[c] | Yes |
| $H_{1d}$: An investor's perception that a brand is competent is positively related to the reputation of that brand. | .679[a](***)[b] (50.953)[c] | Yes |
| $H_{1e}$: An investor's perception that a brand is competent is positively related to his trust in the Brand. | .031[a] (.392)[b] (9.310)[c] | No |
| $H_{1f}$: An investor's experience with a brand is positively related to his perception about the reputation of that brand. | .789[a] (***)[b] (61.299)[c] | Yes |
| $H_{1g}$: An investor's experience with a brand is positively related to consumer's trust in that brand. | .387[a] (.228)[b] (9.121) [c] | No |
| $H_{1h}$: Individual investor's satisfaction positively influences its investor perceived reputation. | .317[a] (.382)[b] (9.820) [c] | No |
| $H_{1i}$: An investor's satisfaction with a brand is positively related to investor's trust in that brand. | .758[a] (.007)[b] (61.412)[c] | Yes |
| $H_{1j}$: An investor's perception about a family brand is positively related to the reputation of that brand. | .607[a] (.026)[b] (43.404)[c] | Yes |
| $H_{1k}$: An investor's perception about a family brand is positively related to his trust in the brand. | .421[a] (.423)[b] (10.121)[c] | No |
| $H_{1l}$: Behavioral biases have a negative relationship with the investment decision. | .842[a] (.003)[b] (72.329)[c] | Yes |
| $H_{1m}$: Behavioral biases have a significant relationship with brand reputation. | .731[a] (.002)[b] (66.431)[c] | Yes |
| $H_{1n}$: Behavioral biases have a significant relationship with brand trust. | .457[a] (.481)[b] (12.329)[c] | No |

(*Continued*)

**TABLE 18.4**
**(Continued)**

| Relationship | Evidence | Supported/ Not Supported |
|---|---|---|
| $H_2$: All the exogenous and endogenous variables in the presence of mediator have a significant relationship. | | |
| $H_{2a}$: Brand Reputation positively mediates the relationship between Brand Competence and Brand Trust. | Direct w/o Med: 0.489***<br>Direct w/ Med: 0.634(0.138)<br>Indirect: 0.552*** | Yes: Full mediation |
| $H_{2b}$: Brand Reputation positively mediates the relationship between Brand Experience and Brand Trust. | Direct w/o Med: 0.389***<br>Direct w/Med: 0.441(0.178)<br>Indirect: 0.571*** | Yes: Full mediation |
| $H_{2c}$: Brand Reputation positively mediates the relationship between Family Branding and Brand Trust. | Direct w/o Med: 0.657(0.004)<br>Direct w/ Med: 0.524***<br>Indirect: 0.398 (.0021) | Yes: Partial mediation |
| $H_{2d}$: Brand Reputation positively mediates the relationship between Brand Satisfaction and Brand Trust. | Direct w/o Med: 0.517(0.010)<br>Direct w/Med: 0.562(0.202)<br>Indirect: 0.356 (0.281) | No: No mediation |
| $H_{2e}$: Brand Reputation positively mediates the relationship between Brand Competence and Investment Decision. | Direct w/o Med: 0.521***<br>Direct w/Med: 0.624(0.187)<br>Indirect: 0.589*** | Yes: Full mediation |
| $H_{2f}$: Brand Reputation positively mediates the relationship between Brand Experience and Investment Decision. | Direct w/o Med: 0.598(0.002)<br>Direct w/Med: 0.512***<br>Indirect: 0.280(.0011) | Yes: Partial mediation |
| $H_{2g}$: Brand Reputation positively mediates the relationship between Family Branding and Investment Decision. | Direct w/o Med: 0.624***<br>Direct w/Med: 0.696(0.193)<br>Indirect: 0.602*** | Yes: Full mediation |
| $H_{2h}$: Brand Reputation positively mediates the relationship between Brand Satisfaction and Investment Decision. | Direct w/o Med: 0.608(0.010)<br>Direct w/Med: 0.673(0.226)<br>Indirect: 0.310 (0.173) | No: No mediation |
| $H_{2i}$: Brand Trust positively mediates the relationship between Brand Competence and Investment Decision. | Direct w/o Med: 0.523(0.010)<br>Direct w/Med: 0.577(0.216)<br>Indirect: 0.341 (0.192) | No: No mediation |
| $H_{2j}$: Brand Trust positively mediates the relationship between Brand Experience and Investment Decision. | Direct w/o Med: 0.698***<br>Direct w/Med: 0.620***<br>Indirect: 0.318 (.0009) | Yes: Partial mediation |
| $H_{2k}$: Brand Trust positively mediates the relationship between Family Branding and Investment Decision. | Direct w/o Med: 0.492(0.010)<br>Direct w/Med: 0.517(0.304)<br>Indirect: 0.362 (0.283) | No: No mediation |
| $H_{2l}$: Brand Trust positively mediates the relationship between Brand Satisfaction and Investment Decision. | Direct w/o Med: 0.627(0.001)<br>Direct w/Med: 0.696***<br>Indirect: 0.689(0.003) | Yes: Full mediation |
| $H_{2m}$: Behavioral Biases positively mediates the relationship between brand reputation and investment decision. | Direct w/o Med: 0.446***<br>Direct w/ Med: 0.582(0.214)<br>Indirect: 0.561*** | Yes: Full mediation |
| $H_{2n}$: Behavioral Biases negatively mediates the relationship between brand trust and investment decision. | Direct w/o Med: 0.521(0.013)<br>Direct w/Med: 0.568(0.217)<br>Indirect: 0.224 (0.160) | No: No mediation |

*Note*: [a] = Standardized regression weights, [b] = $p$-values, [c] = CR values.

is the outcome of brand reputation, so predictors of brand reputation primarily collaborate for reputation building of the brand not for developing trust in the brand as development of trust is the following stage in the process. Factors contributing significantly and primarily in developing brand trust are brand experience and brand satisfaction. Here, it's significant to state that brand experience is the only common factor which significantly affects both brand reputation and brand trust because a good experience with a newbie brand in the stock market also leads to the development of trust in that brand even if presently brand has not created any kind of reputation in the minds of investors as yet.

To test the mediation effect, indirect effects have been measured by performing bootstrapping. It demonstrates whether a mediator variable considerably transmits the influence of an independent variable to a dependent variable. In this context, the second group tested 14 kinds of associations among the exogenous and endogenous variables in the presence of mediators. Hypotheses testing the mediating role of brand reputation between brand competence, brand experience, brand satisfaction, family branding, and brand trust show that all the mediating effects were significant except for brand satisfaction. These findings related to brand satisfaction and brand trust are in sync with the research of Anton (2015) and Ercis, Unal, Candan, and Tildirim (2012). Behavioral biases, namely herding, conservatism, anchoring, disposition effect, etc. affect brand reputation in a robust manner, and the final investment decision is the outcome of these effects. Brand reputation has a significant direct relationship with the investment decision but it is comparatively less significant, being 65%, whereas under the influence of behavioral biases it is 71% which is strongly significant. But once brand trust is developed as a result of brand reputation it is not much affected by the impact of behavioral biases, as trust is a very robust phenomenon and is not easily prone to any wobble effect due to biases. In order to reveal the extent to which identified exogenous variables affect investment decision the total effect has been considered. On the basis of total effect statistics, family branding has the most significant impact on brand reputation and behavioral biases have the stronger impact on brand reputation. Taking into consideration the effect of brand reputation and behavioral biases on investment decision, behavioral biases have been found to be statistically more impactful and comparatively leave long-lasting effects and implications on investment decision.

## 18.5 DISCUSSION AND CONCLUSION

There are various traditional models like efficient market hypothesis and arbitrage pricing theory which work in specific circumstances. They may even work over the long run. But as said by Keynes (1923), "in the long run, we are all dead" is rightly applicable to today's investors due to their impatience and emphasis on more practical time horizons. The long-term models just cited require one to remain fully invested and endure losses for extended periods of time to avoid the inevitable pitfalls of market timing. Considering the existing literature gaps, the first research question of the study was to find whether investment decisions are influenced by an investment avenue's brand reputation, and it was revealed by analysis that yes a brand reputation (like BNP Paribus, HDFC, etc.) plays an important role in investment decisions.

Second was to find out how investors take their decisions as a result of their trust in a particular investment avenue due to it being an old or newly established brand, and in this context it was revealed that trust in brand affects decisions more than reputation and are more stable. This was followed by the inclusion of the behavioral biases and how the ultimate investment decision is effected by their presence in investor psychology. It is concluded that branding does play a relevant role in framing investors' perceptions about a particular investment avenue and also on investment decisions taken as a result of brand reputation or brand trust. It is also influenced by the presence of behavioral biases in an investor's psychology. This research has also revealed that few respondents were aware of the innate behavioral mechanisms that influences their decision-making process. Insights which this study proposes are that financial investment branding procedures should strive to create long-term customer relationships by seeking new ways to engage with investors. Contradicting the traditional finance theories, which assume rationality of investor, behavioral biases have a momentous effect on investors' decision-making processes and brand reputation is greatly influenced by behavioral biases in comparison to brand trust. The practical implication of the study is to facilitate market intermediaries to work well for investors. In this context financial brands can build differentiation through smart investor segmentation. Here, investment journey mapping can be very useful, as brands can disrupt the category norm by providing a stellar experience at every touch point, whether online, on the phone, or in the branch. Innovative service design can win loyalty from present as well as potential investors who are frustrated with lackluster banking experiences, especially in the case of public banks. The study also framed a theoretical model which illustrates the association amid brand reputation, brand trust, and investment decision. The theoretical model has been framed using different branding factors. Although various researchers have examined investors' behavioral traits, this is the first study that has emphasized the role of branding through brand reputation and brand trust, effects of behavioral traits, and their ultimate impact on investment decision. The combination of precursors of brand reputation and brand trust along with behavioral biases which has been embraced for model framing has not been used in any previous model in the similar arrangement. It is essential to highlight that these results are over and above any significant influence of age, gender, and education. By controlling for these variables, we present a predominantly conservative test of the role of brand reputation and brand trust through behavioral biases on investment decisions.

In concurrence to this research contribution, future scope of this study is that this model can be used to investigate the effect of earnings, dividends, split announcements on investment decision with linkage to behavioral biases, and ultimate investment decision. This will facilitate the related research and financial advisors also. Another budding aspect of research is the topic of neuro economics. The latest medical imaging technology has now made it possible to observe the brain activities during the decision-making process. This assists us in comprehending the character and rationale for certain behavioral biases. Other imaging studies have also authenticated that rational parts of our brain are linked with the emotional or limbic sections of our brain. This line of enquiry recommends the existence of a big opportunity of understanding and improving decision-making. The research in area of behavioral finance in collaboration with behavioral economists, medical practitioners, and marketers has attracted many researchers and scholars considering it an emerging area with huge scope of study and findings. It has been found that there is huge scope for further research in region and country specific as most of the researchers are concentrated in the US, China, Spain, Germany, Israel, and Australia. Thus, this provides scope for research in India and other South East Asian countries. It is suggested for further studies to consider the effects of other behavioral finance factors which have not been embraced in this study and also to take larger sample to validate the findings of this research. Besides this, it is also suggested to take other economic factors like foreign exchange, inflation rate, exchange rate, etc. that can also affect the decision-making of investors besides the behavioral finance factors.

Concluding the research work it has been found that individual investors have agreed that brand competence and family branding affect their perception related to a brand's reputation which ultimately develops their trust in that brand, but satisfaction from a brand directly develops trust in the brand, as it was very apparent that investors firmly believed that once they had good experience with a brand then any kind of momentum volatility in market or in brand's reputation will not be able to shift their brand loyalty and they will be firm on their investment decision. Here brand experience has shown a dual relationship as an experience with a brand simultaneously builds reputation and develops trust in the brand. Data also revealed that decision-making with the perception of only brand reputation is highly influenced by behavioral biases but for the investment instruments when brand reputation has been extended in the form of brand trust behavioral biases do not have any mediating effect which represents that for a trusted brand investors are not affected by behavioral biases as they have prolonged associated loyalty with the brand and accordingly they take decisions. Analyzing the motivations of investor to engage in relationships with brands, several empirical studies found that investors consider brand reputation and trust as the best risk-reducing strategy for certain financial instruments (Sheth & Parvatiyar, 1995). Under this framework we theorized the role of brand reputation and brand trust in the final investment decision. The contribution of this study is twofold. Firstly, we demonstrate that there is a strong and negative relationship between behavioral biases and investment decision. Confirming Chaudhuri and Holbrook's (2001) findings, the results of the study suggest that brand reputation and brand trust are separate constructs, although brand trust is the outcome of brand reputation and they collectively influence investment decisions. Secondly, we focused on identifying the most important antecedent of brand reputation in influencing individual's investment decision and that is family branding. Also in case of brand trust and brand reputation, brand trust is the most influential affecting factor. These are interesting findings for theory and practice as well. Hence, financial marketers can increase the quality spread of family brands. These research insights are very crucial as primarily they fill the existing literature gap and secondly they open up numerous research opportunities in the domains of marketing, investment, and behavioral finance due to inclusion of a multidisciplinary approach heading towards a common goal of identifying and studying investor's investment decisions as a result of the identified non-quantitative factors. Also individual innovation is huge in the automotive and fashion categories with due consideration of behavioral biases, but financial brands have been slow to consider major changes. Findings of this study can help the financial brands to understand investor behavior and perception and place their brand in investor's mind accordingly.

## REFERENCES


Aaker, D. A. (1996). Measuring brand equity across products and markets. *California Management Review*, *38*(3), 102–120.

Aaker, D. A., & Keller, K. L. (1990). Consumer evaluations of brand extensions. *Journal of Marketing*, *54*, 27–41.

Adams, S. J. (1965). Inequity in social exchange. In L. Berkowitz (Ed.), *Advances in experimental social psychology* (Vol. 2, pp. 267–299). New York: Academic Press.

Andaleep, S. S., & Anwar, S. F. (1996). Factors influencing customer trust in salespersons in a developing country. *Journal of International Marketing*, *4*(4), 35–52.

Anton, S. (2015). Brand trust and brand loyalty- An empirical study on Indonesia consumers. *British Journal of Marketing Studies*, *4*, 37–47.

Arrow, K. (1984). *The limits of organization*. New York: Norton.

Bakar, M., & Chui, Y. A. (2016). The impact of psychological factors on investors' decision making in Malaysian stock market: A case of Klang Valley and Pahang. *Procedia Economics and Finance*, *35*(12), 319–328.

Banks, S. (1968). The relationships of brand preference to brand purchase. In H. H. Kassarjian & T. S. Robertson (Eds.), *Perspectives in consumer behavior* (pp. 131–144). Glenview, IL: Scott, Foresman and Company.

Barberis, N., Ming, H., & Tano, S. (2001). Prospect theory and asset prices. *Quarterly Journal of Economics*, *116*, 1–53.

Bendapudi, N., & Berry, L. L. (1997). Customers' motivations for maintaining relationships with service providers. *Journal of Retailing*, *73*(1), 15–37.

Bloemer, J. M. M., & Kasper, H. D. P. (1995). The complex relationship between consumer satisfaction and brand loyalty. *Journal of Economic Psychology*, *16*, 311–329.

Boon, S. D., & Holmes, J. G. (1991). The dynamics of interpersonal trust: Resolving uncertainty in the face of risk. In R. A. Hinde & J. Groebel (Eds.), *Cooperation and prosocial behavior* (pp. 190–211). Cambridge: Cambridge University Press.

Bromley, D. B. (2002). Comparing corporate reputations: League tables, quotients, benchmarks, or case studies? *Corporate Reputation Review*, *5*, 35–50.

Brown, G. H. (1952). Brand loyalty – Fact or fiction. *Advertising Age*, *23*(9), 52–55.

Brown, T., & Dacin, P. (1997). The company and the product: Corporate associations and consumer product responses. *Journal of Marketing*, *61*, 68–84.

Butler, J. K. (1991). Toward understanding and measuring conditions of trust: Evolution of a condition of trust inventory. *Journal of Management*, *17*, 643–663.

Butler, J. K., & Cantrell, S. R. (1984). A behavior decision theory approach to modeling dyadic trust in superiors and subordinates. *Psychological Reports*, *55*, 19–28.

Capon, N., & Fitzsimons, T. G. (1994). Affluent investors and mutual fund purchases. *International Journal of Bank Marketing*, *12*(3), 17–25.

Chaplin, L. N., & John, D. R. (2005). The development of self-brand connections in children and adolescents. *Journal of Consumer Research*, *32*, 119–129.

Chaudhary, A. K. (2013). Impact of behavioral finance in investment decision and strategies-a fresh approach. *International Journal of Management Research and Business Strategy*, *2*(2), 85–92.

Chaudhuri, A., & Holbrook, M. (2001). The chain of effects from brand trust and brand affect to brand performance: The role of brand loyalty. *Journal of Marketing*, *65*(2), 81–93.

Churchill, G. A., & Surprenant, C. (1982). An investigation into the determinants of customer satisfaction. *Journal of Marketing Research, 19*, 491–504.

Cook, J., & Wall, T. (1980). New work attitude measures of trust, organizational commitment, and personal need non-fulfillment. *Journal of Occupational Psychology, 53*, 39–52.

Cova, B. (1999). From marketing to society: When the link is more important than the thing. In D. Brownlie, M. Saren, R. Wensley, & R. Whittinton (Eds.), *Rethinking marketing* (pp. 64–83). London: Sage.

Coviello, N., & Brondie, R. (2001). Contemporary marketing practices of consumer and business-to business firms: How different are they. *Journal of Business and Industrial Marketing, 16*, 382–400.

Coviello, N., Brondie, R., Danaher, P., & Johnston, W. (2002). How firms relate to their markets: An empirical investigation of contemporary marketing practices. *Journal of Marketing, 66*, 33–46.

Dacin, P., & Smith, D. (1994). The effect of brand portfolio characteristics on consumer evaluations of brand extensions. *Journal of Marketing Research, 31*, 229–242.

Deutsch, M. (1958). Trust and suspicion. *Journal of Conflict Resolution, 2*, 265–279.

Deutsch, M. (1960). Trust, trustworthiness and the F-scale. *Journal of Abnormal and Social Psychology, 61*, 138–140.

Ercis, A., Unal, S., Candan, F., & Tildirim, H. (2012). *The effect of brand satisfaction, trust and brand commitment on loyalty and repurchase intentions*. 8th International Strategic Management Conference Procedia – Social and Behavioural Sciences.

Esch, F. R., Langner, T., Schmitt, B., & Geus, P. (2006). Are brands forever? How knowledge and relationships affect current and future purchases. *Journal of Product Brand Management, 15*(2), 98–105.

Fombrun, C. J., & Rindova, V. (2000). The road to transparency: Reputation. In M. Schultz, M. J. Hatch, & M. H. Larsen (Eds.), *The expressive organization* (Vol. 7, pp. 7–96). Oxford: Oxford University Press.

Fornell, C., & Larcker, D. F. (1981). Evaluating structural equation models with unobservable variables and measurement error. *Journal of Marketing Research, 18*, 39–50.

Ganesan, S. (1994). Determinants of long-term orientation in buyer-seller relationships. *Journal of Marketing, 58*, 1–19.

Gotsi, M., & Wilson, A, M. (2001). Corporate reputation: Seeking a definition. *Corporate Communications, 6*, 24–30.

Gupta, L. C. (1991). *Share Holders survey: Geographic distribution*. New Delhi: Manas Publications.

Hair, J. F., Anderson, R. E., Tatham, R. L., & Black, W. C. (1998). *Multivariate data analysis* (5th ed.). Englewood Cliffs: Prentice Hall.

Helm, S. (2007). The role of corporate reputation in determining investor satisfaction and loyalty. *Corporate Reputation Review, 10*(1), 22–37.

Henderson, V. (2012). Prospect theory, liquidation and the disposition effect. *Journal of Management Sciences, 58*(2), 445–460.

Herbig, P., & Milewicz, J. (1995). The relationship of reputation and credibility to brand success. *Journal of Consumer Marketing, 12*(4), 4–10.

Hirshleifer, D., & Teoh, S. (2003). Herd behavior and cascading in capital markets: A review and synthesis. *Journal of European Financial Management, 9*(1), 25–66.

Jarvis, C., Mackenzie, S., & Podsakoff, P. (2003). A critical review of construct Indicators and measurement model specification in marketing and consumer research. *Journal of Consumer Research, 30*(2), 199–218.

Kahneman, D., & Amos, T. (1979). Prospect theory: An analysis of decision making under risk. *Econometrica, 47*(2), 263–291.

Kallinterakis, V., Munir, N., & Radovic-Markovic, M. (2010). Herd behaviour, illiquidity and extreme market states: Evidence from Banja Luka. *Journal of Emerging Market Finance*, *9*(3), 305–324.

Kapferer, J. N. (1997). *Strategic brand management, creating and sustaining brand equity long term* (2nd ed.). London: Kogan Page.

Kelly, H. H., & Thibaut, J. W. (1978). *Interpersonal relations: A theory of interdependence.* New York: John Wiley & Sons.

Kengatharan, L., & Kengatharan, N. (2014). The influence of behavioral factors in making investment decisions and performance: Study on investors of Colombo stock exchange, Sri Lanka. *Asian Journal of Finance & Accounting*, *6*(1), 1–23.

Keynes, M. J. (1923). *A tract on monetary reform* (8th ed.). London: Palgrave Macmillan & Co. Ltd.

Kotler, P. (1999). *Marketing management analysis, planning, implementation and control* (9th ed.). Englewood Cliffs, NJ: Prentice Hall College Inc.

Larsen, R. J., & Buss, D. M. (2002). *Personality psychology domains of knowledge about human nature* (4th ed.). Boston: McGraw-Hill.

Lewis, J. D., & Weigert, A. (1985). Trust as a social reality. *Social Forces*, *63*(4), 967–985.

Lye, A. (2002). *Don't throw the baby out with the bath water: Integrating relational and the 4P's concepts of marketing.* The 10th International Colloquium in Relationship Marketing, pp. 223–233.

Maheshwari, V., Lodorfos, G., & Jacobsen, S. (2014). Determinants of brand loyalty: A study of the experience-commitment-loyalty constructs. *International Journal of Business Administration*, *5*(6).

Mardia, K. (1970). Measures of multivariate skewness and kurtosis with applications. *Biometrika*, *57*, 519–530.

Mardia, K. (1974). Applications of some measures of multivariate skewness and kurtosis in testing normality and robustness studies. *Sankhya, B*, *36*(2), 115–128.

Martensen, A., Gronholdt, L., & Kristensen, K. (2000). The drivers of customer satisfaction and loyalty: Cross-industry findings from Denmark. *Total Quality Management*, *11*, 544–553.

Mazzola, P., Gabbionta, C., & Ravasi, D. (2004). *Exploring reputation on financial markets: A preliminary study.* 8th International Conference on Corporate Reputation, Identity and Competitiveness, Fort Lauderdale.

Muhammad, N. (2009). Behavioral finance vs. traditional finance. *Advance Management Journal*, *2*(6), 17–28.

Nofsinger, R., & Richard, A. (2002). *Individual investments behavior.* New York: McGraw-Hill.

Powell, W. W. (1990). Neither market nor hierarchy: Network forms of organization. In B. M. Staw & L. L. Cummings (Eds.), *Research in organizational behavior* (Vol. 12, pp. 295–336). Greenwich, CT: JAI Press.

Saren, M., & Tzokas, N. (1998). The nature of the product in market relationships: A plury-signified product concept. *Journal of Marketing Management*, *14*, 445–464.

Schlenker, B. R., Helm, B., & Tedeschi, J. T. (1973). The effects of personality and situational variables on behavioral trust. *Journal of Personality and Social Psychology*, *25*, 419–427.

Shapiro, C. (1983). Premiums for high quality products as returns to reputations. *Quarterly Journal of Economics*, *98*, 659–679.

Shefrin, H., & Statman, M. (1985). The disposition to sell winners too early and ride losers too long: Theory and evidence. *Journal of Finance*, *40*(3), 777–790.

Sheth, J., & Parvatiyar, A. (1995). The evolution of relationship marketing. *International Business Review*, *4*(4), 397–417.

Singh, B. (2012). A study on investors' attitude towards mutual funds as an investment option. *International Journal of Research in Management*, *2*(2), 61–70.

Sitkin, S. B., & Roth, N. L. (1993). Explaining the effectiveness of legalistic 'remedies' for trust/distrust focused issue: The legalistic organization. *Organizational Science*, *4*(3), 367–392.

Taylor, B., Sinha, G., & Ghoshal, T. (2006). *Research methodology: A guide for researchers in management and social sciences*. New Delhi: Prentice Hall of India Private Limited.

Tucker, W. T. (1964). The development of brand loyalty. *Journal of Marketing Research*, *1*, 32–35.

Tversky, A., & Kahneman, D. (1974). Judgment under uncertainty: Heuristics and biases. *Science*, *185*(4157), 1124–1131.

Welch, I. (2000). Herding among security analysts. *Journal of Financial Economics*, *58*(3), 369–396.

West, S. G., Finch, J. F., & Curran, P. J. (1995). Structural equation models with non-normal variables: Problems and remedies. In R. H. Hoyle (Ed.), *Structural equation modeling: Concepts, issues and applications* (pp. 56–75). Newbery Park, CA: Sage.

Zucker, L. G. (1986). Production of trust: Institutional sources of economic structure, 1840–1920. In B. M. Staw & L. L. Cummings (Eds.), *Research in organizational behavior* (Vol. 8, pp. 53–111). Greenwich, CT.

# 19 Proposed Model of Evaluating Entrepreneurial University Ecosystems from a Talent Development Perspective

*Belma Rizvanović, Aneesh Zutshi, Tahereh Nodehi and Antonio Grilo*

CONTENTS



## 19.1 INTRODUCTION

Rapid changing social and business environments are visible in every aspect of the market. On the industry level, technology is creating diverse and novel jobs, and future generations will have to be equipped with different skills and talents in order to efficiently adapt to the new age. Various skills adapting to constant changes and diverse talents can be considered as one of the key pillars for the future of work from an industrial perspective as well as from an entrepreneurial perspective. As entrepreneurship can be identified as proactive movement in the modern economic theory and practice of today's industry, by including entrepreneurship courses and practices as part of the curriculum, business educators have the opportunity to better prepare students for a changing environment (Smith, 2003).

In order to seek out the best possible ways to solve a problem, collaborate within different groups, and be in a position where entrepreneurship with startups is seen as a solution, talent must be recognized and developed with a specific set of required skills, motivation, and awareness. Considering universities are a direct connection between new generations entering the work market and industry, followed by the possibility of founding a startup, the need for an active entrepreneurial ecosystem is vital.

Having all these facts in mind, development of talent should be strategically amplified and integrated into the curriculum of educational institutions, in order to shape and empower innovative, empathic, ethical, creative problem-solvers. The students within the active university entrepreneurial ecosystems should be able to confront complex global problems with the skills and entrepreneurial mindset required to adapt to the current and future digital age.

The proposed Evaluation model of Entrepreneurial University Ecosystem from Talent perspective (the Evaluation Model) has the goals of identifying the points of development in university entrepreneurial ecosystems and to serve as tool for making concrete actions towards talent awareness and advancement. The Evaluation Model has four components: motivation towards entrepreneurship, awareness of entrepreneurial activities, skill set, and usage of external learning resources.

### 19.1.1 Talent Development Necessity within University Entrepreneurial Ecosystem

The development of sophisticated technology trends has greatly influenced today's market. From a business perspective, new companies—startups—are created rapidly with practically no boundaries, solving different challenges on the market. From the labor perspective, workers are expected the have a diverse set of skills, knowledge, and abilities to adapt to uncertainty and different scenarios in the company and the market. Looking from the perspective of workers/graduates who are just entering the market, being agile, adapting to new circumstances, and having an entrepreneurial mindset are expected characteristics. This fact puts universities in a position to constantly focus on identifying current industry trends and creating activities on how to best prepare the future workforce, resulting in having a more entrepreneurial ecosystem.

An entrepreneurial university can be defined as a university that reveals new ways of addressing the pressures and challenges which derive from an uncertain and unpredictable environment and global movements (Hannon, 2013). In order to have an empowering environment to overcome these challenges, universities create an active entrepreneurial ecosystem and networks, connecting different stakeholders from industries, and going beyond of just being an educational institution. An entrepreneurial university is a natural incubator that is coordinating strategies across critical activities (e.g. teaching, research and entrepreneurship) and finding productive ways to provide a learning atmosphere in which the academic and broader community can explore, test, and exploit ideas (Kirby, Guerrero, & Urbano, 2011). While defining entrepreneurship ecosystem, (Isenberg, 2010) points out a set of individual elements entrepreneurial ecosystem consists of—such as leadership, culture, capital markets, and open-minded customers—that combine in complex, unorthodox ways. In this way, the university (including all the academic community) is an active collaborator and co-creator of business activities in the new market reality. Within this setting, the challenge of university courses often not providing students the context and the link between their actions and real-life outcomes (Rombach et al., 2008) is minimized.

It is the connecting knowledge, infrastructure, and real, hands-on entrepreneurial experience that ensures university environments have the necessary results. With the opportunity of real entrepreneurial experience, the university environment becomes even more important because it is the factor which gives the needed edge for the future possibilities of developing a specific set of desired skills for the market. Not all students will found a startup, but designing an environment where students (and stakeholders) will have novel, structured ways of gaining skills and real hands-on experience will definitely be an advantage to each student for any future activities. This research was focused on the university environment from the perspective of students. Following this notion, a literature review was conducted to establish which elements, forms, and connections are formed in an active entrepreneurial university and identified elements were later used as a framework for the Evaluation Model.

To make an active entrepreneurial ecosystem, the OECD developed *A Guiding Framework for Entrepreneurial Universities*, meant to serve as a guide for universities to strategically incorporate entrepreneurship in the curriculums and make connected ecosystems (OECD, 2012). The *Guidelines* note that in order to develop an entrepreneurial culture, the entrepreneurial activities should be part of the overall strategy. The form of entrepreneurial universities is followed by development of formal and informal external factors and internal factors relating to learning capabilities and resources (Kirby et al., 2011). As the authors propose, the formal factors include the entrepreneurial organizational and governance structure (courses, incubators, science parks, technology transfer, and startup support) and the formation of strategic alliances with external stakeholders (links with industry) while establishing flexible organizational structure (Kirby et al., 2011).

Connecting more to possibility of impact, Guerrero, Cunningham, and Urbano (2015) propose a framework in which entrepreneurial universities contribute to economic impact through teaching, research, and entrepreneurial activities, identified as human, knowledge, and entrepreneurship capital. Skills, capacities, and knowledge generation should be the result of the entrepreneurial university strategy with directed efforts on creating a sustainable, adaptable ecosystem. From a more systematic point of view, Gustomo and Ghina (2017) propose a framework which covers three important key stakeholders within a university: student, lecturer, and institution. This framework gives deeper analysis and detailed structure of roles and responsibilities of the key stakeholders. In the proposed framework, the stakeholders have their own roles and responsibility connected to the three core activities within university: teaching activities, research activities, and third-stream activities (Gustomo & Ghina, 2017). Furthermore, this framework proposes the concept which assures learning in all three core activities and guarantees the quality of core activities while providing the concrete outputs and measurement for each activity of the mission goals (Gustomo & Ghina, 2017). In this way, the framework enables universities to measure the impact through the output of the activities and can serve as one of the instruments for understanding and measuring the impact of entrepreneurial strategy of the institution.

The Babson Entrepreneurship Ecosystem consists of 12 elements (leadership, government, financial capital, success stories, societal norms, non-government institutions, support professions, infrastructure, educational institution, labor, networks, early customers) and 6 conductive elements (policy, markets, capital, human skills, culture, and supports). The author notes that "although they are idiosyncratic because they interact in very complex ways", each element is "always present if entrepreneurship is self-sustaining" (Isenberg, 2011, p. 6). Leading to the point where the combinations form unique link in order for there to be self-sustaining entrepreneurship, you need "conducive policy, markets, capital, human skills, culture, and supports", as it "impinges on the entrepreneur's perceptions, and this how it impacts the entrepreneur's decisions and success" (Isenberg, 2011 p. 7).

Observing the system from the entrepreneur perspective gives an important input for the development of this Evaluation Model, with the difference that the Evaluation Model will reflect the students' perceptions of the entrepreneurial ecosystem. The stakeholders and activities presented in all the mentioned frameworks were adapted for this Evaluation Model and some parts were used as questions in the survey for the awareness of entrepreneurial opportunities component.

We have identified the need to have a student aspect of entrepreneurship possibility because the fact is most universities have strategic goals and programs towards building a startup ecosystem. But to which point this is recognized by the students themselves is a very relevant factor because usually entrepreneurial possibilities on campus exist, but the question remains how they are perceived and used by the students themselves. The second reason is to identify the aspirations, motivations, and skills development towards the entrepreneurial mindset. The entrepreneurial mindset is not just needed for startup industry; it is a demanded concept of today's and future work posts, which include higher levels of knowledge and prospects for developing and applying soft and hard skills (Trilling & Fadel, 2009). Finally, with all the information and exchange knowledge possible on the internet today, our goal was to investigate the awareness of new technologies from the students' perspective and usage of online learning resources, besides the curriculum provided by the universities.

## 19.2 DESIGN OF THE EVALUATION MODEL

The proposed model of the Evaluation Entrepreneurial University Ecosystem from Talent perspective has four components: motivation towards entrepreneurship, awareness of entrepreneurial activities, skill set, and usage of external learning resource.

The entrepreneurial university represents an institution for talent acquisition in line with future trends (Guerrero et al., 2015), with the possibility for the broader university community to have entrepreneurial activities with the corresponding roles (Gustomo & Ghina, 2017) and have strategically designed activities to have more hands-on experience on entrepreneurship subjects (Isenberg, 2011). This suggests that the infrastructure of the university should have facilities, stakeholders, collaboration, and activities created to develop/enhance the entrepreneurial mindset and awareness of possibilities for founding a startup.

Motivation towards entrepreneurship was introduced in the Evaluation Model from the aspect of students' motivations for business startup, identified and adapted from (Shinnar, Pruett, & Toney, 2009) through self-assessment of entrepreneurial disposition and behavioral intentions towards startup founding. In this Evaluation Model, the options of working in the startup and connecting with other people to build a startup together were added to further investigate the students' aspirations. In addition, component of involvement level in university entrepreneurship activity was added to investigate level of students' interest and to point out the awareness of existing options on the campus.

To startup ecosystems, having entrepreneurial skills is one of the bases upon which other elements are built. In this context, we identify entrepreneurial skills and competencies as fundamental to have on current and future workplaces, and as such, skills are highly connected to startup development opportunities within higher educational institutions. Entrepreneurship skills can be defined as competence in opportunity identification, the ability to take advantage of recognized opportunities (BIS Entrepreneurship Skills, 2015), and being able to identify a current customer problem and capitalize on it.

Within this model, entrepreneurship skills consist of hard (technical), financial, and soft skills and competencies. The skills and competence framework was developed from the view of the necessities of today's workplace, the need for practical knowledge in startups and companies (soft, hard and financial skills), and the importance of 21st-century soft skills in this digital era of development. The purpose of presenting the skills set in this manner is to provide an overview of mostly needed skills in each category. The second reason is the sole nature of entrepreneurship skills as being an interconnected group linking quite diverse skill sets, from soft, 21st-century skills to competencies and practical technical and financial skills.

The Figure 19.1 shows the three skill sets and the new technologies integrated in the Evaluation Model, with the notion that new technologies are part of the usage of external learning resources component. Technology changes and creates different scenarios, with internet, mobile technologies, artificial intelligence, and other technological achievements constantly causing modifications in organizations and society. From this perspective, it is crucial to look at new technology development and connect new technologies with startup ecosystems. The fact is inevitable that future startups, founded either on university grounds or any other type of accelerator, will include some points of new technology in their business value proposition.

Community platforms, online courses, online peer-to-peer learning is setting its role as a new, continues way of learning. Different combinations of online learning and tools (from platforms to forums) used by students, researchers, and founders are today a consistent way of receiving knowledge. Different practices of e-learning have undergone a number of initiatives in the past, particularly through the possibilities and openness of the learning environment (Kikkas, Laanpere, & Põldoja, 2011). Within this evaluation model, the focus was on open-source learning available on different platforms. The intent was to identify the most frequent resources students

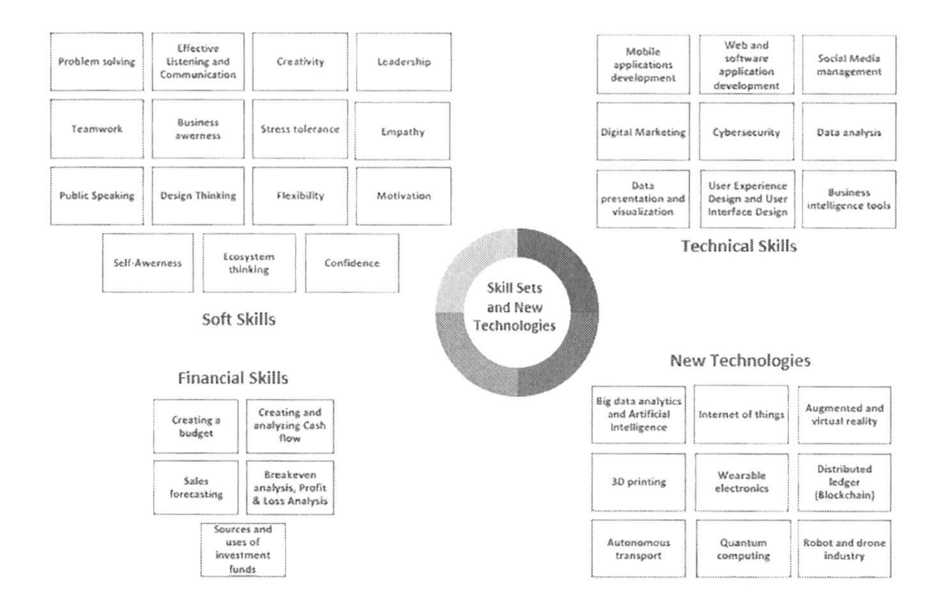


**FIGURE 19.1** Proposed skill set and new technology framework.

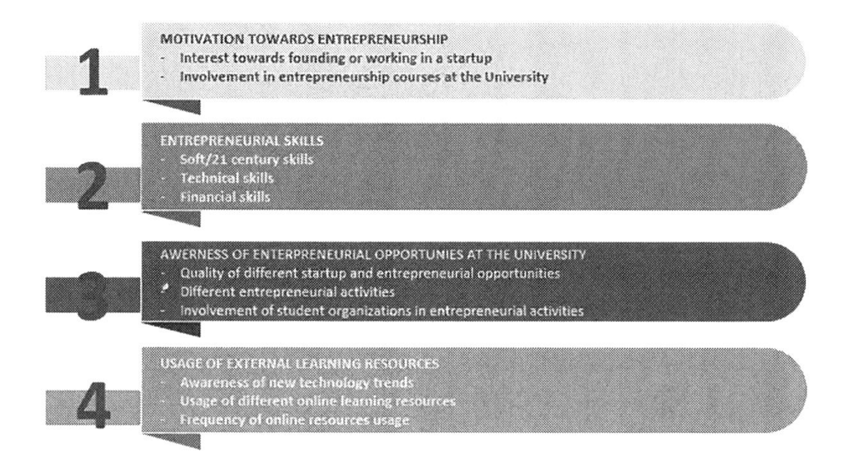


**FIGURE 19.2** Proposed model of Evaluating Entrepreneurial University Ecosystem Components.

use either to discover new or supplement the received knowledge on compulsory courses.

With the four components of the model presented in Figure 19.2 (along with sub-components), the Entrepreneurial Ecosystem Evaluation Model for Talent

Development is designed to provide inputs of entrepreneurial ecosystem through motivation towards entrepreneurship, skill set, awareness of entrepreneurial opportunities at the university, and usage of external learning resources from the perspective of the students.

## 19.3 DATA COLLECTION AND METHOD

Students of nine universities across Europe in Romania, Check Republic, Slovenia, Portugal, and Spain were surveyed, including students from one university in Israel. The surveys were distributed in online form to the students in faculties of various fields of study (humanities, social science, natural science, formal science, and professions and applied science), resulting in a total of 913 responses (in Table 19.1 the response rate per each university is included). The survey consisted of Likert-type scale questions, multiple-choice questions, and several demographic questions. In regard to skills section, the soft skills and competency questions framework were adapted from (Rainsbury et al., 2002) with added competencies connected to entrepreneurial areas. Financial, hard skills and competences added to the list were connected to the startup ecosystem necessities and skills demanded for future work. The data for technical skills were constructed out of the research on the most resourceful platforms for today's work recruiting and learning platforms, such as LinkedIn,[1] Udemy platform,[2] learntocodewith.me,[3] GitHub,[4] and Hired.com. The data for new technologies introduced in the survey were adapted from WEF Future of Jobs report (World Economic Forum, 2018). Also, some other parts of the survey were adapted from the earlier mentioned university entrepreneurial frameworks, followed by modifications for this Evaluation Model. The data were analyzed in order to get summed averages weighted on the scale from 1 to 5. The questions were created to clearly distinguish every answer (from 5 being the best option and 1 being the least favorable option), so the data could be analyzed by taking the weighted average sum of each question. Some questions were also given an option of "I cannot comment since I am not aware" for two reasons. The first reason was to reduce uninformed response, since it signals participants that they need not feel compelled to answer every questionnaire item, especially the ones that they are not familiar with. Secondly, if the mentioned activates exist and students are not aware of it, this signals an unawareness of the activities, which can serve as information for the university as well. For the preliminary analysis of the results, the mean score was used for formulating the results (considering the statistical limitation the mean score provides but for the purpose of the preliminary results the authors have identified this analytical tool as suitable). In order to present the variances of results on the 10-point scale, all components were multiplied by 2 and compared on the scale. Since main components of the model were constructed with 2, 3, or 4 sub-components, the mean score was used to produce final variance of each component, as presented in Table 19.1.

**TABLE 19.1**
**The Results with all the Components of the Evaluation Model within All Universities**

| Country | Romania | | | | Czech Republic | Slovenia | Spain | Israel | Portugal |
|---|---|---|---|---|---|---|---|---|---|
| University/Output | Alexandru Ioan Cuza University (UAIC) | Babes-Bolyai University (UBB) | West University of Timisoara (UVT) | Technic University (UTCN) | University of Economics | University of Maribor | University of Salamanca | University of Bar Ilan | University of Nova Lisboa |
| **Response rate:** | **240** | **132** | **104** | **64** | **170** | **53** | **49** | **35** | **66** |
| **1. Motivation towards entrepreneurship** | **4.46** | **5.08** | **4.54** | **5.14** | **5.12** | **5.21** | **4.80** | **5.49** | **4.83** |
| 1.a Please indicate your level of involvement in university entrepreneurship activity. | 2.49 | 3.23 | 2.63 | 3.63 | 3.38 | 3.74 | 3.55 | 3.94 | 4.33 |
| 1.b. How interested are you towards working or creating your own startup? | 6.43 | 6.94 | 6.46 | 6.66 | 6.87 | 6.68 | 6.04 | 7.03 | 5.33 |
| **2. Awareness of entrepreneurial opportunities on university** | **5.80** | **5.21** | **5.92** | **5.40** | **6.84** | **6.46** | **5.27** | **5.43** | **7.30** |
| 2.a. Please indicate how do you identify the quality of the activities in entrepreneurship courses at your university. | 5.68 | 4.93 | 5.73 | 4.51 | 7.42 | 6.55 | 5.26 | 6.80 | 7.84 |
| 2.b. Please indicate how do you identify the quality of the entrepreneurship activities of startup contests as hackathons, pitch events organized at your university. | 6.14 | 6.00 | 5.84 | 6.00 | 7.22 | 6.68 | 5.66 | 5.70 | 8.05 |

| | | | | | | | | | |
|---|---|---|---|---|---|---|---|---|---|
| 2.c. What kind of services are provided by your university, other than entrepreneurship courses and entrepreneurship/ startup events? | 4.84 | 4.08 | 5.78 | 4.15 | 6.22 | 5.83 | 4.51 | 5.00 | 5.84 |
| 2.d. How involved are your university student organizations in entrepreneurship activities? | 6.55 | 5.84 | 6.31 | 6.96 | 6.52 | 6.77 | 5.64 | 4.22 | 7.48 |
| **3. Skills Set** | **5.63** | **6.16** | **5.19** | **5.99** | **6.01** | **5.94** | **5.61** | **7.09** | **6.52** |
| 3.a. Please indicate your level of confidence in your soft skill and competences. | 6.59 | 7.30 | 6.25 | 6.95 | 7.03 | 6.55 | 7.07 | 8.26 | 8.71 |
| 3.b. Please identify your confidence in ability to execute the following financial skills. | 5.10 | 5.44 | 4.57 | 5.61 | 5.56 | 5.51 | 5.16 | 6.43 | 5.03 |
| 3.c. Please identify your awareness and ability to execute the following technical skills. | 5.20 | 5.75 | 4.76 | 5.40 | 5.45 | 5.75 | 4.61 | 6.58 | 5.81 |
| **4. Usage of external learning resources** | **5.84** | **6.21** | **5.31** | **6.12** | **5.67** | **5.75** | **5.89** | **5.74** | **6.11** |
| 4.a. Please identify your awareness and knowledge of new technologies in digital era. | 4.85 | 4.77 | 4.03 | 5.40 | 4.86 | 5.08 | 4.92 | 5.06 | 5.49 |
| 4.b. How actively do you use web resources to gain new knowledge? | 6.83 | 7.65 | 6.59 | 6.84 | 6.47 | 6.42 | 6.86 | 6.41 | 6.73 |
| **SUM:** | **21.73** | **22.67** | **20.96** | **22.66** | **23.65** | **23.35** | **21.57** | **23.74** | **24.76** |

## 19.4 RESULTS

This Evaluation Model was created in order to capture students' perceptions on entrepreneurial opportunities in the university ecosystem. For the purpose of the preliminary analysis, only the final scoring data are presented.

The first component, motivation towards entrepreneurship, can be identified as moderate, scoring from 4.46 to 5.49 (on the 10-point scale) for all universities. But what is interesting is the relation between the two sub-components, level of involvement in university entrepreneurship activity and interest towards working or creating a startup. Looking at Alexandru Ioan Cuza University scores, a high level of possible interest towards working or creating a startup can be identified (6.43). Another sub-component, involvement in entrepreneurial activities, has a lower score (2.49), signalizing the high motivation is not followed by involvement in entrepreneurship activities. Students' perceptions for awareness of entrepreneurial opportunities on university is higher, scaling from 5.21 to 7.30 among all the universities. The levels of perception for each of the four sub-components is even more important information for universities, since perception and awareness of different activities can indicate the students' views on activities regarding the strategic goals or programs universities are working on.

An interesting example of diversity in ecosystem can be identified in Technic University (Romania) where students have rated moderately high the quality of the entrepreneurship activities of startup contests as hackathons, pitch events (6.00) and student organizations in entrepreneurship activities (6.96), signalizing the importance of these activities for the ecosystem. Another example is University Nova Lisbon (Portugal), with the highest overall score of the Evaluation Model was for this component awareness of opportunities (7.30), in which the lowest score was given to the component of services provided by university other than entrepreneurship courses and entrepreneurship/startup event (5.84).

Skills set was designed in order to ask students what level of confidence and abilities they possess. In University of Bar-llan (Israel), students perceive themselves confident in soft, technical, and financial skills, with the highest score among all the universities. From the perspective of the ecosystem, this can be interpreted as an environment which has activities developing the identified skill sets, scoring higher in the area of soft skills (8.26) and lower in financial skills (6.43). Usage of external learning resources component varied from 5.31 to 6.12 scores between the universities.

As presented on Figure 19.3, we can identify the ecosystems based on all four components on a scale from 4.46 to 7.30. Meaning that all the universities can be identified as medium-level entrepreneurial ecosystems on the 10-point scale, from the students' perspective.

Interestingly, there is not much difference in motivation towards entrepreneurship across countries, ranging from 4.46 to 5.49, with the highest value in Bar-Ilan University (Israel).

Students' perception for the component awareness of entrepreneurial opportunities on university is higher, but still moderate, scaling from 5.21 to 7.30.

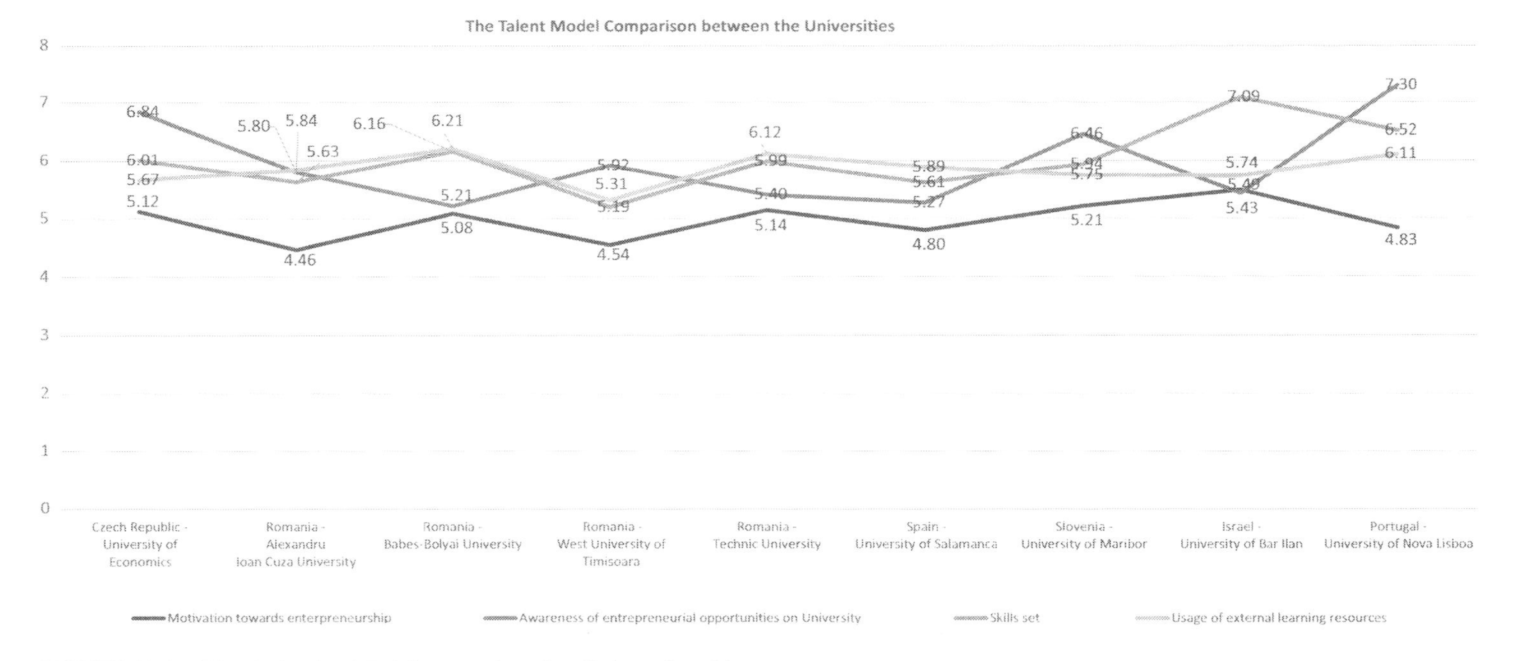


**FIGURE 19.3** The Evaluation Model comparison for all the universities.

Students in Nova University (Portugal), Bar-Ilan (Israel), and Babes-Bolyai University (Romania) reported the highest confidence in soft, financial, and technical skills among all the universities. What is also interesting is the fact that for Nova and Bar-Ilan Universities the skill set values are the highest values of the all components, identifying these environments as more developed towards the skills section. Focusing on the usage of external knowledge component, three out of four Romanian universities have the highest scores among all universities (from 5.84 to 6.21). With these values, Romanian students have identified their learning habits by moderately and actively using web resources to gain new knowledge. When it comes to the online learning platforms used for gaining new knowledge, YouTube channels and social media groups have been identified as the most frequently used sources among the students.

We can conclude that by the student's perception, all universities in this survey were identified as mostly moderate in their efforts of making the universities more entrepreneurial. This fact leaves space for further development of the components, especially the component of motivation towards entrepreneurship, which was the lowest-value component with all universities. Also, some components have higher values than others, giving universities information on areas in which the work on specific components has proven to be productive (as in Bar-Ilan University for skill sets, and Prague Economics and Nova University in awareness of entrepreneurial opportunities).

## 19.5 LIMITATIONS

For the purpose of preliminary analysis, the statistical method used in deliverance of data, mean score, poses a limitation. The average values are used to identify the main points of all components, but in future analyses detailed data will be explored and presented in accordance with more precise statistical analysis. Another limitation was the response rate per university and variation in response rates across universities.

## 19.6 FURTHER WORK

The research reported in this chapter represents preliminary analysis of the data gathering and analysis at the time of writing. Remaining related factors of the Evaluation Model will focus on the statistical analyses from the aspect of comparisons and impact of main components or sub-components in order to get the precise analytical information from the Evaluation Model. The aim of this research was to understand entrepreneurial ecosystems in different universities and countries from the students' perspective, because a successful university ecosystem is the result of collaboration and contribution from all stakeholders within the network university formed (Greene, Rice, & Fetters, 2010). Another step to consider would be to compare these components and have higher insights in each component. It is important to understand the perception and aspirations from students' point of view in order to point out the complexity, further development, and highlight successful factors of established ecosystems and specific areas.

## NOTES

1 https://blog.linkedin.com/2018/january/11/linkedin-data-reveals-the-most-promising-jobs-and-in-demand-skills-2018
2 https://business.udemy.com/blog/10-hot-it-skills-2018/
3 https://learntocodewith.me/posts/tech-skills-in-demand/
4 https://blog.github.com/2018-02-08-open-source-project-trends-for-2018/#new-skills

The authors would like to note the research was conducted as a part of MYGATEWAY project (780758), supported by EU Horizon 2020 fund. We would also like to show appreciations to all the partners of the project and the Universities for taking an active part in this research.

## REFERENCES

BIS Entrepreneurship Skills. (2015). *Entrepreneurship skills: Literature and policy review* (Vol. 236, pp. 1–49). Retrieved from https://assets.publishing.service.gov.uk/government/uploads/system/uploads/attachment_data/file/457533/BIS-15-456-entrepreneurship-skills-literature-and-policy-review.pdf

Greene, P. G., Rice, M., & Fetters, M. (2010). University-based entrepreneurship ecosystems: Framing the discussion. In *The development of university-based entrepreneurship ecosystems: Global practices* (pp. 1–11). Cheltenham: Edward Elgar.

Guerrero, M., Cunningham, J. A., & Urbano, D. (2015). Economic impact of entrepreneurial universities' activities: An exploratory study of the United Kingdom. *Research Policy, 44*(3), 748–764. doi:10.1016/j.respol.2014.10.008.

Gustomo, A., & Ghina, A. (2017). Building a systematic framework for an entrepreneurial university. *International Journal of Advanced and Applied Sciences, 4*(7), 116–123. doi:10.21833/ijaas.2017.07.017

Hannon, P. D. (2013). Why is the entrepreneurial university important? *Journal of Innovation Management, 1*(2), 10–17.

Isenberg, D. (2011). The entrepreneurship ecosystem strategy as a new paradigm for economic policy : Daniel Isenberg, Ph.D Professor of management practice, babson global executive director, the babson entrepreneurship ecosystem project. *Institute of International and European Affairs, 1*(781), 1–13. doi:10.1093/rfs/hhr098

Isenberg, D. J. (2010). The big idea: How to start an entrepreneurial revolution. *Harvard Business Review, 88*(6). doi:10.1353/abr.2012.0147

Kikkas, K., Laanpere, M., & Põldoja, H. (2011). *Open courses: The next big thing in eLearning. I.* Proceedings of the 10th European Conference on e- Learning, Reading: Academic Publishing Limited, pp. 370–376.

Kirby, D. A., Guerrero, M., & Urbano, D. (2011). Making universities more entrepreneurial: Development of a model. *Canadian Journal of Administrative Sciences, 28*(3), 302–316. doi:10.1002/CJAS.220.

OECD. (2012). *A guiding framework for entrepreneurial universities.* Retrieved from www.oecd.org/site/cfecpr/EC-OECD Entrepreneurial Universities Framework.pdf

Rainsbury, E., Hodges D., Burchell N., Lay, M. (2002). Ranking workplace competencies: Student and graduate perceptions. *Asia-Pacific Journal of Cooperative Education, 3*, 8–18.

Rombach, D., et al. (2008). Teaching disciplined software development. *Journal of Systems and Software, 81*(5), 747–763. doi:10.1016/j.jss.2007.06.004

Shinnar, R., Pruett, M., & Toney, B. (2009). Entrepreneurship education: Attitudes across campus. *Journal of Education for Business, 84*(3), 151–159. doi:10.3200/JOEB.84.3.151-159

Smith, M. O. (2003). Teaching basic business: An entrepreneurial perspective. *Business Education*, *58*, 23–25.
Trilling, B., & Fadel, C. (2009). *21St century skill*; Partnership for 21st Century Skills; Josey-Bass – A Willey Imprint.
World Economic Forum. (2018). The future of jobs report 2018. *Centre for the New Economy and Society*. doi:10.1177/1946756712473437

# 20 Research and Innovation in Teaching Pedagogy for Emerging Markets

*Soma Arora*

## CONTENTS



## AN EXPERIMENT IN BLENDED LEARNING

"We are, as a species, blended learners."

—Elliot Masie

## 20.1 INTRODUCTION

Achieving superior learning outcomes through higher student engagement has always been a challenge amongst academics. Several methods have been used to enhance teaching effectiveness in the field of International Business and Strategy. Some prominent methods involved using global virtual teams, such as the X-Culture project or collaborative consulting (as seen in globalview.org), in addition to the quintessential case studies and student-exchange programmes. Most of these methods have remained focused on harnessing the benefits of new age information and communication technologies that provided educators and learners with an innovative pedagogical environment.

The collaborators in these novel techniques attempted to address the following questions: (1) "Does the use of combined learning methods improve people's learning performance?"; (2) "What elements define the most effective combination?"; and (3) "Will any combination of modalities provide improved results or is there an 'optimum blend' in blended learning?"

Therefore, an optimal learning design can be created. The magic is in the mix—the meticulous blend of learning methods (Masie, 2002). An appropriate platform for blended learning can, therefore, enhance the experiential education of students learning International Marketing. If balance and harmony are the qualities sought in blended environments, it is crucial to first identify the elements that should be mixed together in a blended course (Osguthorpe & Graham, 2003). Studies have indicated that the creation of a new learning environment that blends the three aforementioned elements may fulfil the following goals: (1) pedagogical richness, (2) access to knowledge, (3) social interaction, (4) personal agency, (5) cost effectiveness, and (6) ease of simultaneous revision (Carman, 2002; Osguthorpe & Graham, 2003).

A new blended learning environment can therefore achieve a lot for its learners. Taking cue from the students' engagement process at crowdsourcing events and on-campus industry interactions, the author became interested in blending this into the classroom learning process. She observed a fall in students' span of attention due to the advent of smartphones in the classroom. However, if they were physically engaged in a task, it took their attention off the mobile lying in their pockets. The only way to engage them physically in a task was a workshop, like they did in engineering colleges. This would be truly "learning by doing". Studies proved that students' perception of importance of activity had a positive influence on their engagement (Florian & Black-Hawkins, 2011). Is it possible to engage the students in a task which they execute themselves and take pride in, to get optimal results? Can the workshop mode in the International Marketing Specialisation course be aimed to achieve higher goals of blended learning combining experiential education and classroom based teaching?

This chapter adds to the current literature on experiential education (Beard and Wilson 2002) by shifting the focus from traditional unilateral classroom-based teaching to blended learning (Bonk and Graham 2012; Garrison and Kanuka 2004). The workshop mode was used to improve the learning outcome through enhanced student engagement. The key areas of skill development within this pedagogy were (1) course content (i.e., knowledge regarding International Marketing); (2) prodigious use of online information; (3) cross-cultural communication; and (4) team cohesiveness leading to optimal output.

## 20.2 BACKGROUND

In recent years, research interest in experiential learning education is increasing because of the need to integrate business school curricula to "real world" capabilities. Similar developments were noted in emerging markets, such as India, through various e-learning portals and platforms (Goyal & Tambe, 2015; Dodani et al. 2009)

A lot of successful e-learning initiatives at the institution as well as policy level in India, (Bhattacharya and Sharma 2007), motivated the researcher towards the current study. For instance, an experiment with virtual biotechnology laboratories was conducted to improve students' performance (Radhamani et al., 2014). Cutrell et al. (2015) examined the blended learning approach (Graham 2006) in massively empowered classrooms for undergraduate technical education and obtained positive results because of the inclusion of videos as part of online learning that was blended with standard classroom teaching. However, research has also noted cautionary perspectives regarding blended learning, citing limitations, such as the need for a flexible framework (Redmond & Lock, 2006).

## 20.3 THE EXPERIMENT

The research methodology was unique because of the nature of the experiment. The workshop mode of instruction has been explained in several steps to elucidate the richness of the methodology.

The methodology explicitly focuses on the context because the successful implementation of generically described teaching methods in units of learning depends on the context. The organization of the workshop was divided into two phases—the concept stage and active stage. During the concept stage, the stage was set for a successful workshop.

The entire class of International Marketing students (the universal set of 75 students) was divided into two broad samples The number of participants in the intervention group, were limited to 35, according to the students' interest and the company's set restriction to deal with a small cohesive team. The second sample comprised 40 students who were engaged in an academic project that was aimed at delivering the same set of learning outcomes; this group was the control group. The control group worked in groups for their projects, but were not subject to physical presence of the corporate partner during the sessions like those in the workshop did. They were trained in the traditional method of classroom-based teaching. Both groups started work simultaneously and would receive 30% of their assessment on the basis of their project submission and presentation.

### 20.3.1 Active Phase

After the concept stage, the teaching activities commenced. Based on pedagogic frameworks developed using the experiential learning (Beard, 2008, 2010) and blended learning systems, (Bonk & Graham, 2012; Garrison & Vaughan, 2008), the author organized the workshop. The workshop was organized in three stages: Day 1 comprised of three sessions and marked the first stage; it consisted of the theme of building awareness amongst students in addition to the crucial task of team formation (Hassanien 2006; 2007). The Marketing Manager made a presentation regarding the company using suitable illustrations. She then explained the business challenges to the students through a phased method. The discussion progressed towards a multicultural involvement (Kelly 2009); the Country Head and MD then suggested a list of stakeholders across their target destinations that students would have to contact and negotiate with during the course (Melles, 2004). The students were excited at this opportunity to interact with overseas corporate partners as part of their workshop project. Because the interactions with overseas channel partners and marketing agents would be conducted online, the role of online information management became crucial (Muilenburg & Berge, 2005). The first day of the workshop was concluded with a task orientation, wherein students were asked to report their findings for all the related parameters on the second day of the workshop.

### 20.3.2 Second Stage

Day 2 of the workshop was held a week after the first stage. This provided the students with sufficient time to conduct their secondary research regarding several

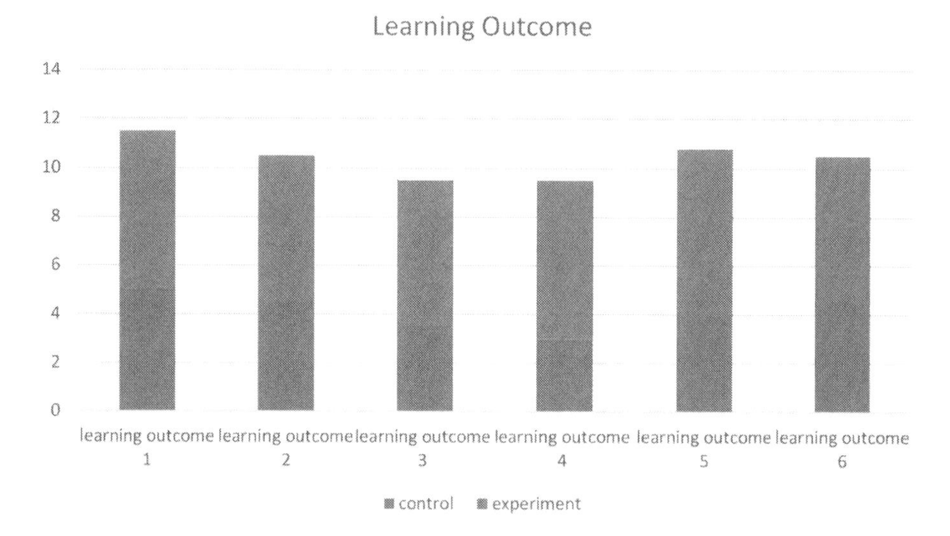


**FIGURE 20.1** Average scores obtained by students in the experiment and control group.

topics by using the internet. This was crucial for achieving all the learning outcomes as provided in the assessment rubrics

### 20.3.3 Third Stage

Day 3 of the workshop marked the assessment and evaluation stage for the participants. All students in the group were marked equally. For the assessment sheet provided by the corporate partner to the workshop, the author relied on the judgements of the Country Head, MD in consultation with his team of experts.

The projects submitted by students in the control group were evaluated by the faculty/instructor for the workshop. These reports were also subjected to the same assessment sheet presented in Table 20.1. Therefore, the corporate and peer perspectives of evaluation would not be included in this review. This is true for all the traditional classroom-based teaching methods wherein the faculty conducts the evaluations across all parameters in the assessment sheet.

To document the level of students' engagement in the workshop, an ethnographic study was conducted. An ethnographer accompanied the instructor to each of the sessions in both groups (control and experiment) to record the student engagement levels. The total number of sessions conducted for the course was 20, amongst which 10 sessions were converted into the workshop mode of training. Therefore, the ethnographer attended 10 sessions of the intervention group to observe and capture the attributes of student engagement, as described in Figure 20.1. Data were collected using the explicit field notes (classroom) of the ethnographer, along with pictures taken by her during the end of class. Similarly, she attended 10 sessions of the traditional classroom lectures conducted by the faculty member/instructor in the control group during the same period as the workshop. The control group classes were

**TABLE 20.1**
**Assessment Rubrics for the Workshop**

| Learning outcome | International Marketing course content and curriculum for teaching | Skills development (Rated on a scale of 1–7) of students | | | | Total scores obtained by workshop participants/students across each learning outcome based on skills developed | | | Total marks |
|---|---|---|---|---|---|---|---|---|---|
| | | Course content SK1 | Online information management SK2 | Team cohesiveness and output SK3 | Cross-cultural communication SK4 | Evaluation by corporate partner (10%) | Evaluation by faculty (10%) | Evaluation by peers (10%) | Max Marks (30%) |
| LO1 | Target market selection | | | | | | | | |
| LO2 | Product adaptation | | | | | | | | |
| LO3 | Pricing strategy | | | | | | | | |
| LO4 | Channel partner selection and management | | | | | | | | |
| LO5 | Promotion strategy and communication | | | | | | | | |
| LO6 | Country marketing plan/final report | | | | | | | | |

conducted in a different room as a separate section. These students were engaged in a live project assigned by the faculty, and prepared a presentation at the end of 10 sessions. The assessment of the report and presentation was their learning outcome for the International Marketing course that accounted for 30% of internal assessment in the subject. This was necessary for maintaining balance in the quantum of assessment and review in both groups.

To objectively report the level of student engagement (SE), the ethnographer ranked student learning as 1 for "Beginner", 2 for "Emerging", 3 for "Developed", and 4 for "Well—developed" across all the three parameters—active participation (AP), learning environment (LE) and formative tools and processes (FT). Likewise, she would rank them from 1 to 4 for instructional design across the three parameters—AP, LE, and FT.

The concept of student engagement has gained widespread acceptance in the West as it was found to be positively related to academic outcomes as well as personal development of the students (Carini, Kuh, & Klein, 2006; Lim, Morris and Kupritz, 2007). Studies done in the past linked the student engagement with "measurable outcomes" both inside and outside the classroom and with "high-quality learning outcomes" (Krause & Coates, 2008; Kuh, 2007). An educational institution has to create the opportunities for academic and developmental experience of students. Likewise, students have to show commitment to these opportunities (Manwaring, Larsen, Graham et al., 2017)

The researcher was however keen to discover student engagement within classrooms rather than at the institution level. So she integrated this student engagement scale developed in the Indian context (Singh & Srivastava, 2014) to the rubrics developed by International Center for Leadership in Education (see Figure 20.1) more suitable for classroom-based participation. The three parameters characterizing student engagement were: (1) active participation, (2) learning environment, and (3) formative tools and processes. Active participation in the task can be treated as display of sense of belonging towards the institution and its curricula.

## 20.4 DISCUSSION

The study explores how the workshop intervention in classroom teaching would improve students' learning outcome because of significantly higher student engagement. Learning outcome was measured based on the marks obtained by the students; this was the best method of evaluating their performance in the workshop and the International Marketing course. This became the independent variable recorded on the basis of mean scores achieved by the students. Student engagement was the dependent variable measured across four levels. The ethnographer provided the scores for student engagement in four distinct categories, wherein 1 is the lowest form of engagement (i.e. "Beginner") and 4 is the highest form of engagement (i.e. "Well-developed") (based on Table 20.2).

To eliminate any form of bias arising from familiarity between the object and subject, the authority for assessment and review (i.e. the faculty) was separated from the judgement of student engagement (i.e. the ethnographer). The performance scores were obtained using a two-step process. In the first step, evaluators were asked to

**TABLE 20.2**
**Rubrics for Student Engagement**

| **Active Participation** | **Beginning** | **Emerging** | **Developed** | **Well developed** |
|---|---|---|---|---|
| Student learning | Limited student engagement with the exception of raising hands. Lesson is teacher-led, with challenges to productivity. | Most students remain focused and on tasks during the session. Lesson is teacher-led and students indicate productive progress. | All students responding to frequent opportunities for active engagement throughout the session. Lesson is led both by teacher and student. | All students remain on task and proactively engaged throughout the lesson. Students take ownership of learning and actively seek to enhance performance. |
| Instructional design | Lesson relies solely on direct instruction with few opportunities for student engagement. | Lesson focused more on direct instruction instead of student engagement through application. | Lesson provides multiple strategies designed to maximise participation. | Lesson achieves a focus on student-centric engagement, wherein students can adjust and monitor their own participation. |
| **Learning environment** | **Beginning** | **Emerging** | **Developed** | **Well developed** |
| Student learning | Student rely on peers or teachers for answers to questions. Students demonstrate lack of respect for peers or teachers. | Students demonstrate some evidence of risk-taking and perseverance in learning rigorous content. | Students are encouraged to take risks and persevere through productive struggles. | Students demonstrate respect for peers, teachers, and learning environment. |
| Instructional design | Classroom learning procedures and routines are inconsistently communicated and/or implemented. | Classroom learning procedures are visible but inconsistently implemented. | Clear classroom routines and implementation. | Classroom learning procedures are clearly established and remain flexible to adaptation as per the task. |
| **Formative Tools and processes** | **Beginning** | **Emerging** | **Developed** | **Well developed** |
| Student learning | Lesson indicates few opportunities for formative assessment to evaluate mastery of content. Assessment results indicate student growth as minimal. | Students are partnered or grouped and receive some opportunities for differentiated learning. Assessment results indicate that student growth is progressing. | Students indicate mastery of content. Assessment results indicate students meeting expectations. | Assessment results indicate students exceeding expected outcomes. |

(*Continued*)

**TABLE 20.2**
**Rubrics for Student Engagement (Continued)**

| Formative Tools and processes | Beginning | Emerging | Developed | Well developed |
|---|---|---|---|---|
| Instructional design | Results from formative processes and tools are used to monitor progress. | Results from formative processes and tools are used to plan and implement aspects of differentiated instruction and monitor progress. | Results from formative processes and tools are used to strategically adjust instructional design, plan differentiated instruction, and monitor progress. | Results from formative processes and tools along with effective feedback are used to immediately adjust instructional pacing, plan differentiated instruction, and monitor progress. |

*Source*: International Center for Leadership in Education, Houghton Mifflin Harcourt.

rate the students on a scale of 1–7 (wherein 1 is the lowest score and 7 is the highest score) for their skill development under each learning outcome (Table 20.2). In the second step, the individual scores for SK1, SK2, SK3, and SK4 (see in assessment rubrics in Table 20.2) were added to derive the mean score for skills developed, which was a scale variable that was considered in the one-way ANOVA statistics in this study. Therefore, the study disregarded the role of absolute numbers or discrete marks provided to students for each learning outcome. The emphasis was instead placed on skill development for each learning outcome on a continuous basis. The learning outcome scores obtained by each student in both the groups were treated to the student engagement condition provided by the ethnographer. The F-statistic was very high at 6.9, thus indicating a significant difference between the means and role of the treatment variable leading to the difference. Therefore, the null hypothesis was rejected and an alternate hypothesis was accepted, thus indicating improvement in learning outcomes caused by higher levels of student engagement.

To report the difference between the mean scores of the two groups—control and experiment—across each learning outcome, a bar chart was plotted as presented in Figure 20.1. Across all the six learning outcomes, it was noted that the mean scores of the experiment group were significantly higher than that of the control group for all learning outcomes. The $x$-axis represented the learning outcomes for the International Marketing course and the $y$-axis represented the average mean score obtained by students in both groups.

## 20.5 STUDY LIMITATIONS

Although any of the factors associated with experiential learning may have contributed to the potential increase in mean scores, I believe that intervention through the workshop mode led to superior results in the experiment group students. The success of blended learning initiatives do not seem to stem from what the students consumed as learning content but instead the process of imparting that knowledge. Instructors have attempted the concept of flipped classrooms (Delozier & Rhodes, 2017) to improve learning outcome. The workshop method for training has been utilized in developing countries (Wilson et al., 2009). However, the blend of classroom teaching and a workshop for MBA students is a novel pedagogy for Indian students.

The findings revealed that students in the international marketing workshop indicated improved performance than their counterparts who worked on their own. The learning outcomes measured through their scores were invariably higher against the student engagement parameters. This study has contributed significantly to the field of higher education by institutionalizing the role of student engagement in classroom teaching. Thus far, all the studies were engaged in gathering student feedback and drawing results from their satisfaction after completing the process. However, the criterion for student engagement that was used in this experiment was within the course delivery process. Second, the assessing community tends to be obsessed with numbers (i.e., marks provided for performance) after the process is over. By shifting the focus to scaled criteria for skill set development, the assessor needed to partake in the process with the students. This methodology worked suitably in the workshop environment because the participants were working in tandem with the evaluators. The assessment rubrics used in the study is an essential contribution; the author has integrated all aspects of the international marketing curriculum as milestones to be achieved in the learning outcomes. This has also been evaluated from the perspective of skillset development, wherein the student uses more online information management skills in LO1 and less in LO6, whereas the team is presenting the country marketing plan.

The study findings can encourage other researchers to pursue this line of thought and repeat the experiment for other business schools in India. The inability to perform this experiment simultaneously across several other business schools in India is a significant limitation of this study. In the second phase of this study, the author has endeavoured to include business schools in and outside India to contrast the results in a developed country, such as the United States, and a developing country, such as India. The scope for future research is wide open because many business schools are inviting new ideas for pedagogical innovation. Once the experiment is successfully incorporated in a pan-India scenario, it will enable the author to recommend the workshop as a teaching pedagogy in the international marketing curriculum.

## REFERENCES

Arbaugh, J. B. (2000). Virtual classroom characteristics and student satisfaction with internet-based MBA courses. *Journal of Management Education*, *24*(1), 32–54.

Beard, C. (2008). *Experiential learning: The development of a pedagogic framework for effective practice* (Doctoral Dissertation), Sheffield Hallam University.

Beard, C. (2010). *The experiential learning toolkit: Blending practice with concepts*. London: Kogan Page Publishers.

Beard, C., & Wilson, J. P. (2002). *The power of experiential learning: A handbook for trainers and educators*. Herndon, VA: Stylus Publishing.

Bhattacharya, I., & Sharma, K. (2007). India in the knowledge economy – An electronic paradigm. *International Journal of Educational Management, 21*(6), 543–568.
Bonk, C. J., & Graham, C. R. (2012). *The handbook of blended learning: Global perspectives, local designs.* San Francisco, CA: John Wiley & Sons.
Carini, R. M., Kuh, G. D., & Klein, S. P. (2006). Student engagement and student learning: Testing the linkages. *Research in Higher Education, 7*(1), 1–32.
Carman, J. M. (2002). *Blended learning design: Five key ingredients.* Retrieved August 18, 2009.
Cutrell, E., O'Neill, J., Bala, S., et al. (2015). *Blended learning in Indian colleges with massively empowered classroom.* Proceedings of the Second ACM Conference on Learning Scale, Vancouver, BC, Canada, pp. 47–56.
Davies, W. M. (2009). Groupwork as a form of assessment: Common problems and recommended solutions. *Higher Education, 58*(4), 563–584.
DeLozier, S. J., & Rhodes, M. G. (2017). Flipped classrooms: A review of key ideas and recommendations for practice. *Educational Psychology Review, 29*(1), 141–151.
Dodani, S., Kazmi, K. A., Laporte, R. E., et al. (2009). Effectiveness of research training workshop taught by traditional and video-teleconference methods in a developing country. *Global Public Health, 4*(1), 82–95.
Flores, M. A., Veiga Simão, A. M., Barros, A., et al. (2015). Perceptions of effectiveness, fairness and feedback of assessment methods: A study in higher education. *Studies in Higher Education, 40*(9), 1523–1534.
Florian, L., & Black-Hawkins, K. (2011). Exploring inclusive pedagogy. *British Educational Research Journal, 37*(5), 813–828.
Garrison, D. R., & Vaughan, N. D. (2008). *Blended learning in higher education: Framework, principles, and guidelines.* John Wiley & Sons.
Goyal, E., & Tambe, S. (2015). Effectiveness of Moodle-enabled blended learning in private Indian business school teaching NICHE programs. *The Online Journal of New Horizons in Education, 5*(2), 14–22.
Kelly, P. (2009). Group work and multicultural management education. *Journal of Teaching in International Business, 20*(1), 80–102.
Krause, K., & Coates, H. (2008). Students' engagement in first-year University. *Assessment and Evaluation in Higher Education, 33*(5), 227–304.
Kuh, G. D. (2007). How to help students achieve. *Chronicle of Higher Education, 53*(41), B12–13.
Masie, E. (2002). Blended learning: The magic is in the mix. In *The ASTD e-learning handbook* (pp. 58–63). New York: McGrawHill.
Mazany, P., Francis, S., & Sumich, P. (1995). Evaluating the effectiveness of an outdoor workshop for team building in an MBA programme. *Journal of Management Development, 14*(3), 50–68.
McGraw, P., & Tidwell, A. (2001). Teaching group process skills to MBA students: A short workshop. *Education + Training, 43*(3), 162–171.
Melles, G. (2004). Understanding the role of language/culture in group work through qualitative interviewing. *The Qualitative Report, 9*(2), 216–240.
Muilenburg, L. Y., & Berge, Z. L. (2005). Student barriers to online learning: A factor analytic study. *Distance Education, 26*(1), 29–48.
Osguthorpe, R. T., & Graham, C. R. (2003). Blended learning environments: Definitions and directions. *Quarterly Review of Distance Education, 4*(3), 227–233.
Panda, S. (2005). Higher education at a distance and national development: Reflections on the Indian experience. *Distance Education, 26*(2), 205–225.
Phatak, D. B. (2015). *Adopting MOOCs for quality engineering education in India.* Proceedings of the International Conference on Transformations in Engineering Education, Springer, New Delhi, pp. 11–23.

Radhamani, R., Sasidharakurup, H., Sujatha, G., Nair, B., Achuthan, K., & Diwakar, S. (2014). *Virtual labs improve student's performance in a classroom.* In International Conference on E-Learning, E-Education, and Online Training, Springer, Cham, pp. 138–146.

Redmond, P., & Lock, J. V. (2006). A flexible framework for online collaborative learning. *The Internet and Higher Education, 9*(4), 267–276.

Singh, H. (2003). Building effective blended learning programs. *Educational Technology-Saddle Brook Then, Englewood Cliffs, NJ, 43*(6), 51–54.

Singh, H., & Reed, C. (2001). A white paper: Achieving success with blended learning. *Centra Software, 1*, 1–11.

Thorne, K. (2003). *Blended learning: How to integrate online & traditional learning.* London: Kogan Page Publishers.

Wilson, K. A., Bedwell, W. L., Lazzara, E. H., Salas, E., Burke, C. S., Estock, J. L., . . . Conkey, C. (2009). Relationships between game attributes and learning outcomes: Review and research proposals. *Simulation & Gaming, 40*(2), 217–266.

# Index

Note: Page numbers in **bold** indicate tables and those in *italics* indicate figures.